Future Propulsion Systems and Energy
Sources in Sustainable Aviation

Aerospace Series

Future Propulsion Systems and Energy Sources in Sustainable Aviation

Saeed Farokhi

The University of Kansas
Lawrence, Kansas
USA

Registered Offices
John Wiley & Sons, Inc., 111 River Street, Hoboken, NJ 07030, USA
John Wiley & Sons Ltd, The Atrium, Southern Gate, Chichester, West Sussex, PO19 8SQ, UK

Editorial Office
The Atrium, Southern Gate, Chichester, West Sussex, PO19 8SQ, UK

For details of our global editorial offices, customer services, and more information about Wiley products visit us at www.wiley.com.

Wiley also publishes its books in a variety of electronic formats and by print-on-demand. Some content that appears in standard print versions of this book may not be available in other formats.

Library of Congress Cataloging-in-Publication Data

Names: Farokhi, Saeed, author.
Title: Future propulsion systems and energy sources in sustainable aviation /
    Saeed Farokhi, PhD, FRAeS, Chancellor's Club Teaching Professor &
    Professor of Aerospace Engineering, The University of Kansas, Lawrence, Kansas, USA.
Description: First edition. | Hoboken, NJ : John Wiley & Son, Inc., 2020. |
    Series: Aerospace series | Includes bibliographical references and index.
Identifiers: LCCN 2019024989 (print) | LCCN 2019024990 (ebook) |
    ISBN 9781119414995 (hardback) | ISBN 9781119414988 (adobe pdf) |
    ISBN 9781119415053 (epub)
Subjects: LCSH: Airplanes–Motors. | Airplanes–Fuel. | Electric airplanes. |
    Aeronatics–Environmental aspects. | Sustainable engineering.
Classification: LCC TL701 .F34 2020 (print) | LCC TL701 (ebook) |
    DDC 629.134/35–dc23
LC record available at https://lccn.loc.gov/2019024989
LC ebook record available at https://lccn.loc.gov/2019024990

Cover Design: Wiley
Cover Image: © NASA

Set in 10/12pt Warnock by SPi Global, Pondicherry, India
Printed and bound in Singapore by Markono Print Media Pte Ltd

10 9 8 7 6 5 4 3 2 1

To my lovely grandchildren:
Sophia, Sasha, Sydney, Melody, and Shiley

Contents

Preface

Sustainable aviation in broad terms means environmentally friendly air travel. This term implies that our current ways are neither sustainable nor environmentally friendly. Fossil-fuel burning turbofan engines produce greenhouse gases, among other pollutants such as NOx that contribute to the rising temperature on the Earth that is called *global warming*. Hence, NASA (National Aeronautics and Space Administration) and the European Union's Advisory Council for Aeronautics Research in Europe (ACARE) work with the aerospace industry and academia to seek alternative solutions to the current state of the art (SOA) in commercial air travel. The solutions by necessity are system-driven, composed of propulsion and power, airframe, and system integration, air travel management (ATM) and operations. The solutions to the propulsion and power components of sustainable aviation are found in the following:

- Fuel burn reduction in advanced-core ultra-high bypass (UHB) turbofan engines
- Open rotor architecture
- Alternative jet fuels from renewable sources
- Hybrid-electric propulsion system
- Electric propulsion (EP) with superconducting motors/generators, electric power transmission, energy storage, and the cryogenic thermal management system
- Nuclear propulsion through advancements in the next-generation of compact fusion reactor (CFR)

For airframe and system integration components, sustainable aviation is achieved through:

- High lift-to-drag ratio (L/D) airplane configurations, e.g. hybrid wing–body (HWB) aircraft
- Fully integrated airframe and propulsion system through distributed propulsion (DP) concept
- Boundary layer ingesting (BLI) propulsion system and integration
- Other drag reduction concepts, e.g. hybrid laminar flow control (HLFC), folding high aspect ratio wings or fluidic actuators

When we consider flight navigation and operations, the roadmap to sustainable aviation is described in the Federal Aviation Administration's (FAA) vision, called *NextGen* program. The goal is to develop and implement clean, quiet, and energy efficient operational procedure through the following:

1) Advanced ATM capabilities
2) Gate-to-gate and surface operational procedures

As we expect from a highly integrated system, it takes all three areas of propulsion and power systems, airframe configuration, and operations to achieve the lofty environmental goals that are set for commercial aviation. The advanced concepts in propulsion and power and the airframe contribute roughly 35–45% to sustainable aviation goals, whereas ATM and operations contribute roughly 15–20%.

Another aviation area of concern is noise pollution. Airport's neighboring communities are subject to severe restrictions on noise emissions from aircraft takeoff and landing, ground operations, and noise pollution due to airport support vehicles and services. The noise mitigation strategies in sustainable aviation are focused on eight factors:

1) Airframe noise reduction through BLI propulsion system and control
2) Landing gear, flaps, and slat noise mitigation in the landing–takeoff (LTO) cycle
3) Fan noise reduction through lower pressure ratio design and higher bypass ratio engines
4) Jet noise mitigation through lower-speed jets, Chevrons, and other means to enhance jet mixing as well as shielding by the airframe
5) Power management and steep flight path angle in takeoff and landing
6) Engine noise mitigation through advanced design in swept-leaned stators in the fan exit duct, advanced acoustic liners, and flow path optimization
7) Gate and surface support operations
8) Airport traffic management

This book addresses sustainable aviation concepts and their subsequent promising technologies. It is intended as a resource for students and practicing engineers in aerospace industry. To be useful and self-contained, I start with a review chapter on aircraft propulsion (Chapter 1), followed by a review chapter on aircraft aerodynamics (Chapter 2). These chapters lay the foundation for the theory, terminology, and science of propulsion and fluid flow. The next four chapters address the essence of sustainable aviation, namely:

- Chapter 3: *Understanding Aviation's Impact on the Environment*
- Chapter 4: *Future Fuels and Energy Sources in Sustainable Aviation*
- Chapter 5: *Promising Technologies in Propulsion and Power*
- Chapter 6: *Pathways to Sustainable Aviation*

The presentation of the environmental impact of aviation is focused on the science of combustion, radiation physics with respect to greenhouse gases, NOx formation and the ozone layer, contrail formation, aviation-induced cloudiness (AIC), and radiative forcing (RF). The issues of air-quality standards and public health and safety are treated as universal concerns. Intentionally, the discussions are more focused on the scientific results and measurements and less on the debate of anthropogenic influences on climate or interference in global warming. We leave that debate, if there is one, to politicians. In this book, we identify the facts of aviation-related pollution and how we should, as citizen engineers and scientists, help responsibly reduce or eliminate this pollution, by design. The responsibility is especially acute in light of the growth in air travel, namely from 2.5 billion passengers in 2011 to the estimated 16 billion in 2050.

Since the title of the book uses the word *Future*, I could not limit the presentation in this book to the aircraft that currently fly the commercial aviation routes, i.e.

subsonic-transonic aircraft. At higher speeds, namely supersonic and hypersonic, we briefly address the low-boom supersonic flight technology as well as some promising propulsion systems for the *runway-to-orbit* or the *single-stage to orbit* (SSTO) commercial transport of the future.

Finally, the areas of technology that are presented here are rapidly evolving with new milestones and achievements, e.g. through laboratory and flight tests that appear almost on a monthly (if not weekly) basis in trade journals. The reader is thus encouraged to follow these technological advances and keep up with the current literature. This is certainly an exciting era in aviation.

Acknowledgments

I express my sincere appreciation to my professors at the University of Illinois at Urbana–Champaign and MIT Aeronautics and Astronautics Department. Their guidance and inspiration taught me the principles of thermal-fluid sciences and the curiosity to ask questions and push the boundaries. I owe my desire to learn and confidence to my teachers who are still my role models. Working in the gas turbine division of Brown, Boveri & Co. (in Switzerland) taught me the appreciation for the engineering design and manufacturing of complex systems. I learned hardware engineering in industry, for which I am grateful.

Since joining the Aerospace Engineering Department at the University of Kansas in 1984, I have received continuous support from my friends and colleagues in the department and have supervised the research of more than 50 of the most dedicated graduate students in aerospace engineering. In particular, I am grateful to my PhD students who stayed for many years and helped us reach a better understanding of our field. My research sponsors from the government and industry shared the same vision and curiosity and funded our work at KU for over 30 years. The research sponsors are truly the lifeblood of US graduate education. I express my gratitude to all of them.

Finally, my heartfelt appreciation and gratitude goes to my wife, Mariam, who has been a true supporter for over 40 years. Our lovely daughters and grandchildren have been the real inspiration for this work.

Abbreviations and Acronyms

A

ACARE	Advisory Council for Aeronautics Research in Europe
ACTE	adaptive compliant trailing edge
ADP	aerodynamic design point
AFRL	Air Force Research Laboratory
AIA	Aerospace Industry Association
AIAA	American Institute of Aeronautics and Astronautics
AIC	aviation-induced cloudiness
AHEAD	Advanced Hybrid Engines for Aircraft Development
AJF	alternative jet fuel
AoA	angle of attack
APU	auxiliary power unit
AR	aspect ratio
ASME	American Society of Mechanical Engineers
ASTM	American Society for Testing and Materials
ATM	air traffic management
ATP	advanced turboprop
ATRA	advanced technology research aircraft

B

BBSAN	broadband shock-associated noise
BF	block fuel
BFL	balanced field length
BLC	boundary layer control
BLI	boundary layer ingestion
BPF	blade passing frequency
BPR	bypass ratio
BTH	biomass to hydrogen
BTL	biomass to liquid
BWB	blended wing body

C

CAAFI (US)	Commercial Aviation Alternative Fuels Initiative
CAEP	Committee on Aviation Environmental Protection
CC	circulation control
CCD	climb-cruise-descent
CD	convergent-divergent
CFC	chlorofluorocarbons
CFD	computational fluid dynamics
CFR	compact fusion reactor
CJ	Chapman-Jouguet
CLEEN	continuous lower emissions, energy and noise
CNEL	community noise equivalent level
CNG	compressed natural gas
CO	carbon monoxide
CPS	cycles per second (Hz)
CROR	counter-rotating open rotor
CSP	concentrating solar power
CTL	coal to liquid

D

DD	drag divergence
DLR	Deutsche Luft und Raumfahrt
DME`	Dimethyl Ether
DNL	day-night level (24-hour average sound level)
DOC	direct operating cost
DOE	Department of Energy
DOF	degrees of freedom
DOT	Department of Transportation
DP	distributed propulsion
D-T	deuterium-tritium
DTPD	dry ton per day

E

EBF	externally blown flap
ECS	environmental control system
EEA	European Environment Agency
EI	emission index
EIA	Energy Information Administration
EIS	entry into service
EP	electric propulsion
EPA	Environmental Protection Agency
EPNL	effective perceived noise level
ERA	environmentally responsible aviation
ERA	environmentally responsible aviation

ESTOL	extreme short takeoff and landing
ETOPS	extended range twin operations
eVTOL	electric vertical takeoff and landing

F

FAA	Federal Aviation Administration
FAR	Federal Aviation Regulation
FPR	fan pressure ratio
FRL	fuel readiness level
FT	Fischer-Tropsch
FTL	Fischer-Tropsch liquid

G

GA	general aviation
GAO	Government Accounting Office
GED	gravimetric energy density
GHG	greenhouse gas
GIS	geographic information systems
GREET	greenhouse gases, regulated emissions, and energy use in transportation
GRC	Glenn Research Center
GT	gas turbine
GTF	geared turbofan
GTL	gas to liquid
GTOW	gross takeoff weight

H

HEFA	Hydro-processed esters and fatty acids
HEPS	hybrid electric propulsion system
HFC	hydrofluorocarbons
HHV	higher heating value
HLFC	hybrid laminar flow control
HPC	high-pressure compressor
HPT	high-pressure turbine
HRD	hydro-processed renewable diesel
HRJ	hydro-processed renewable jet
HSCT	high-speed civil transport
HTS	high-temperature superconductivity
HVO	hydro-treated vegetable oil
HWB	hybrid wing body

I

IATA	International Air Transport Association
IBF	internally blown flap

ICAO	International Civil Aviation Organization
IFR	instrument flight rules
IFSD	in-flight shutdown
IPC	intermediate-pressure compressor
IPCC	Intergovernmental Panel on Climate Change
IR	infrared
ISA	International Standard Atmosphere

K

KPP	key performance parameter
K-T	Kármán-Tsien
KTAS	knots true airspeed

L

LAA	Lighthill's acoustic analogy
LBL	laminar boundary layer
LCA	life cycle assessment
LCC	life cycle cost
LDAL	low-drag acoustic liner
LE	leading edge
LEAPTech	Leading-Edge Asynchronous Propellers Technology
LED	leading edge device
LES	large-eddy simulation
LFC	laminar flow control
LFG	landfill gas
LH2	liquid hydrogen
LHV	lower heating value
LM	Lockheed Martin
LNG	liquefied natural gas
LO_X	liquid oxygen
LPC	low-pressure compressor
LPT	low-pressure turbine
LRC	long-range cruise
LST	linear stability theory
LTO	landing–takeoff
LUC	land use change

M

MCFC	molten carbonate fuel cell
MCL	mean camber line
MDO	multidisciplinary design optimization
MFCC	multifunctional fuel cell
MIL-STD	military standard
MIT	Massachusetts Institute of Technology

MNE	mixed-nozzle ejector
MTF	mid-tandem fan
MTOGW	maximum takeoff gross weight

N

NAAQS	National Ambient Air Quality Standards
NACA	National Advisory Council for Aeronautics
NAS/NAE	National Academy of Science/National Academy of Engineering
NASA	National Aeronautics and Space Administration
NEF	noise exposure forecast
NLF	natural laminar flow
NOAA	National Oceanic and Atmospheric Administration
NOx	nitrogen oxides
NPR	nozzle pressure ratio
NPF	net propulsion force
NPSS	numerical propulsion system simulation
NREL	National Renewable Energy Laboratory

O

OEI	one engine inoperative
OEW	operating empty weight
OPR	overall pressure ratio
OSHA	Occupational Safety and Health Administration

P

PARTNER	Partnership for Air Transportation Noise and Emissions Reduction
PDE	pulse detonation engine
PDR	pulse detonation ramjet
PDRE	pulse detonation rocket engine
PEM	proton exchange membrane
PFC	perfluorocarbons
P-G	Prandtl-Glauert
PM	particulate matter
PPM	parts per million
PR	pressure ratio
PT	power turbine
PV	photovoltaic

R

RBAP	rocket-based airbreathing propulsion
RBCC	rocket-based combined cycle
RCEP	Royal Commission on Environmental Protection (UK)
RES	renewable energy sources

RF	radiative forcing
RTO	rolling takeoff

S

SABRE	Synergetic Air-Breathing Rocket Engine
SAE	Society of Automotive Engineers
SAF	sustainable aviation fuel
SBJ	supersonic business jet
SC	supercritical
SC	superconductivity
SCEPTOR	Scalable Convergent Electric Propulsion Technology and Operations Research
SESAR	Single European Sky ATM Research
SEL	sound exposure level
SFC	specific fuel consumption
SLFC	supersonic laminar flow control
SMR	steam methane reforming
SN	smoke number
SOA	state of the art
SOFC	solid-oxide fuel cell
SPL	sound pressure level
SSL	standard sea level
SSTO	single-stage to orbit
STJ	sugar-to-jet
STOL	short takeoff and landing
SUGAR	Subsonic Ultra-Green Aircraft Research
SWAFEA (Europe)	Sustainable Way for Alternative Fuel and Energy in Aviation

T

TBL	turbulent boundary layer
TE	trailing edge
TEA	techno-economic analysis
TET	turbine entry temperature
TeDP	turboelectric distributed propulsion
TF	turbofan
TJ	turbojet
TME	total mission energy
TOC	top of climb
TOGW	takeoff gross weight
TP	turboprop
TRJ	turbo-ramjet
TRL	technology readiness level
TS	Tollmien-Schlichting
TSFC	thrust-specific fuel consumption
TTR	total temperature ratio

TTW   tank-to-wake
TTW   tank-to-wheel (surface vehicle)

U

UAM   urban air mobility
UDF   unducted fan
UHB   ultra-high bypass
UHC   unburned hydrocarbon
USAF   US Air Force
USB   upper surface blowing
UTC   United Technologies Corporation
UV   ultraviolet

V

VB   vortex breakdown
VCE   variable-cycle engine
VED   volumetric energy density
VOC   volatile organic compounds
VTOL   vertical takeoff and landing

W

WTT   well-to-tank
WTW   well-to-wheel (surface vehicle)

Z

ZLL   zero-lift line

About the Companion Website

This book is accompanied by a companion website:

www.wiley.com/go/farokhi/power

The website includes:

- Figures
- Equations

Scan this QR code to visit the companion website.

1

Aircraft Engines – A Review

1.1 Introduction

A brief review of aircraft gas turbine propulsion is presented in Chapter 1. Most topics are fundamental to undergraduate propulsion education and thus serve as a refresher before we embark on more advanced propulsion concepts. *Aircraft Propulsion, 2nd edition,* by the author, provides the bulk of the material presented in this review chapter.

1.2 Aerothermodynamics of Working Fluid

In gas turbine (GT) engines the working fluid is treated as perfect gas, often air, which behaves as *continuum* rather than individual molecules. The density of the gas, ρ, is a fluid property that is defined on the basis of continuum. The perfect gas law that relates the pressure, density, and the absolute temperature of the gas may be derived rigorously from the kinetic theory of gases. It is stated here without proof:

$$p = \rho R T \tag{1.1}$$

Where R is known as the *gas constant*, which is inversely proportional to the molecular mass of the gas, i.e.

$$R \equiv \frac{\bar{R}}{MW} \tag{1.2a}$$

where $\bar{R}$ is the *universal gas constant* expressed in two systems of units:

$$\bar{R} = 8,314 \frac{J}{kmol.K} \quad \text{or} \quad \bar{R} = 4.97 \, x10^4 \, \frac{ft.lbf}{slug.mol.^\circ R} \tag{1.2b}$$

The thermodynamic relations for a perfect gas in terms of specific heats at constant pressure and volume are:

$$dh \equiv c_p dT \tag{1.3}$$

$$de \equiv c_v dT \tag{1.4}$$

Future Propulsion Systems and Energy Sources in Sustainable Aviation, First Edition. Saeed Farokhi.
© 2020 John Wiley & Sons Ltd. Published 2020 by John Wiley & Sons Ltd.
Companion website: www.wiley.com/go/farokhi/power

In aircraft engines, the specific heats at constant pressure and volume are often functions of gas temperature only:

$$c_p = c_p(T) \quad \text{and} \quad c_v = c_v(T) \tag{1.5}$$

The gas is then called *thermally perfect* gas. There is a simplifying assumption of constant specific heats, which is a valid approximation to gas behavior in a narrow temperature range. In this case,

$$c_p = \text{Constant} \quad \text{and} \quad c_v = \text{Constant} \tag{1.6}$$

The gas is referred to as *calorically perfect* gas.

The first law of thermodynamics is the statement of conservation of energy for a system of fixed mass, m, namely,

$$\delta q = de + \delta w \tag{1.7}$$

where the element of heat transferred to the system from the surrounding is considered positive and on a per-unit-mass basis is δq with a unit of energy/mass, e.g. $\text{J}\,\text{kg}^{-1}$. The element of work done by the gas on the surrounding is considered positive and per-unit-mass of the gas is depicted by δw. The net energy interaction with the system results in a change of energy of the system, which again on a per-unit-mass basis is referred to as de. The three terms of the first law of thermodynamics have dimensions of energy per mass, e.g. $\text{J}\,\text{kg}^{-1}$. The elemental heat and work exchange are shown by delta, δ, instead of an exact differential, d, as in de. This is in recognition of *path dependent* nature of heat and work exchange.

The application of the first law to a closed cycle is of importance to engineering and represents a balance between the heat and work exchange in a cyclic process, i.e.

$$\oint \delta q = \oint \delta w \tag{1.8}$$

Also, for an adiabatic process, i.e. $\delta q = 0$ with no mechanical exchange of work, i.e. $\delta w = 0$ the energy of a system remains constant, namely $e_1 = e_2 = \text{constant}$. We shall use this principle in conjunction with a control volume approach in the study of aircraft engine inlets and exhaust systems.

The second law of thermodynamics introduces the absolute temperature scale and a new thermodynamic property, s, the entropy. It is in fact a statement of impossibility of a heat engine exchanging heat with a single reservoir and producing mechanical work continuously. It calls for a second reservoir at a lower temperature where heat is rejected by the heat engine. In this sense, the second law of thermodynamics distinguishes between heat and work. It asserts that all mechanical work may be converted into system energy, whereas not all heat transfer to a system may be converted into system energy continuously. A corollary to the second law incorporates the new thermodynamic property, s, and the absolute temperature, T, into an inequality, known as the Clausius inequality,

$$Tds \geq \delta q \tag{1.9}$$

where the equal sign holds for a *reversible* process. The concept of irreversibility ties in closely with frictional losses, viscous dissipation and the appearance of shock waves in supersonic flow. The pressure forces within the fluid perform reversible work and the viscous stresses account for dissipated energy of the system (into heat). Hence, the reversible work done by a system per unit mass is the work done by pressure

$$\delta w_{rev.} = pdv \tag{1.10}$$

where v is the specific volume, which is the inverse of fluid density, ρ. A combined first and second law of thermodynamics is known as the *Gibbs equation*, which relates fluid property, entropy, to other thermodynamic properties, namely

$$Tds = de + pdv \tag{1.11}$$

Although it appears that we have substituted the reversible forms of heat and work into the first law to obtain the Gibbs equation, it is applicable to irreversible processes as well (see Farokhi 2014). We introduce a derived thermodynamic property known as *enthalpy*, h, as

$$h \equiv e + pv \tag{1.12}$$

This derived property, i.e. h, combines two forms of fluid energy, namely internal energy (or thermal energy) and what is known as the flow work, pv, or the pressure work. The other forms of energy such as kinetic energy and potential energy are still unaccounted by the enthalpy, h. We shall account for the other forms of energy by a new variable called the *total enthalpy* in Section 1.2.7.

We take the differential of Eq. (1.12) and substitute it in the Gibbs equation, to get:

$$Tds = dh - vdp \tag{1.13}$$

By expressing enthalpy in terms of specific heat at constant pressure, via Eq. (1.3), and dividing both sides of Eq. (1.13) by temperature, T, we get:

$$ds = c_p \frac{dT}{T} - \frac{v}{T} dp = c_p \frac{dT}{T} - R \frac{dp}{p} \tag{1.14}$$

We incorporated the perfect gas law in the last term of Eq. (1.14). We may now integrate this equation between states 1 and 2, in a perfect gas, to arrive at:

$$s_2 - s_1 \equiv \Delta s = \int_1^2 c_p \frac{dT}{T} - R \ln \frac{p_2}{p_1} \tag{1.15}$$

An assumption of calorically perfect gas (i.e. c_p = constant) will enable us to integrate the first term on the right-hand side of Eq. (1.15):

$$\Delta s = c_p \ln \frac{T_2}{T_1} - R \ln \frac{p_2}{p_1} \tag{1.16}$$

Otherwise, we need to use a tabulated thermodynamic function, ϕ, defined as:

$$\int_1^2 c_p \frac{dT}{T} \equiv \phi_2 - \phi_1 \tag{1.17}$$

From the definition of enthalpy, let us replace the flow work, pv, term by its equivalent from the perfect gas law, i.e. RT, and then take the differential of the equation as

$$dh \equiv c_p dT = de + RdT = c_v dT + RdT \tag{1.18}$$

Dividing through by the temperature differential, dT, we get

$$c_p = c_v + R \quad \text{Or} \quad \frac{c_p}{R} = \frac{c_v}{R} + 1 \tag{1.19}$$

This provides valuable relations among the gas constant and the specific heats at constant pressure and volume. The ratio of specific heats is given a special symbol, γ, due to its frequency of appearance in compressible flow analysis, i.e.

$$\gamma \equiv \frac{c_p}{c_v} = \frac{c_v + R}{c_v} = 1 + \frac{1}{c_v / R} \tag{1.20}$$

In terms of the ratio of specific heats, γ, and R, we express c_p and c_v as

$$c_p = \frac{\gamma}{\gamma - 1} R \quad \text{and} \quad c_v = \frac{1}{\gamma - 1} R \tag{1.21}$$

The ratio of specific heats is related to the degrees of freedom of the gas molecules, n, via

$$\gamma = \frac{n+2}{n} \tag{1.22}$$

The degrees of freedom of a molecule are represented by the sum of the energy states that a molecule possesses. For example, atoms or molecules possess kinetic energy in three spatial directions. If they rotate as well, they have kinetic energy associated with their rotation. In a molecule, the atoms may vibrate with respect to each other, which then create kinetic energy of vibration as well as the potential energy of intermolecular forces. Finally, the electrons in an atom or molecule are described by their own energy levels (both kinetic energy and potential) that depend on their position around the nucleus. As the temperature of the gas increases, the successively higher *energy states* are excited; thus, the degrees of freedom increases. A monatomic gas, which may be modeled as a sphere, has at least three degrees of freedom, which represent translational motion in three spatial directions. Hence, for a monatomic gas, under "normal" temperatures, the ratio of specific heats is

$$\gamma = \frac{5}{3} \cong 1.667 \quad \text{Monatomic gas at "normal" temperatures} \tag{1.23}$$

A monatomic gas has negligible rotational energy about the axes that pass through the atom due to its negligible moment of inertia. A monatomic gas will not experience a vibrational energy, as vibrational mode requires at least two atoms. At higher temperatures the electronic energy state of the gas is affected, which eventually leads to ionization of the gas.

For a diatomic gas, which may be modeled as a dumbbell, there are five degrees of freedom, under "normal" temperature conditions, three of which are in translational motion and two are in rotational direction. The third rotational motion along the intermolecular axis of the dumbbell is negligibly small. Hence, for a diatomic gas such as air (near room temperature), hydrogen, nitrogen etc., the ratio of specific heats is

$$\gamma = \frac{7}{5} = 1.4 \quad \text{Diatomic gas at "normal" temperatures} \tag{1.24a}$$

At high temperatures, molecular vibrational modes and the excitation of electrons add to the degrees of freedom and that lowers γ. For example, at ~600 K (i.e. typical of HPC, high-pressure compressor, environment in a GT engine) vibrational modes in air are excited; thus, the degrees of freedom of diatomic gases are initially increased by one, i.e. it becomes $5 + 1 = 6$, when the vibrational mode is excited. Therefore, the ratio of specific heats for diatomic gases at elevated temperatures becomes:

$$\gamma = \frac{8}{6} \approx 1.33 \quad \text{Diatomic gas at elevated temperatures} \tag{1.24b}$$

The vibrational mode represents two energy states corresponding to the kinetic energy of vibration and the potential energy associated with the intermolecular forces. When fully excited, the vibrational mode in a diatomic gas, such as air, adds two to the degrees of freedom; namely, it becomes 7. Therefore the ratio of specific heats becomes

$$\gamma = \frac{9}{7} \approx 1.29 \quad \text{Diatomic gas at higher temperatures} \tag{1.24c}$$

For example, air at 2000 K has its translational, rotational, and vibrational energy states fully excited. This temperature level describes the combustor, turbine, or after-burner environment. Gases with a more complex structure than a diatomic gas have higher degrees of freedom and thus their ratio of specific heats is less than 1.4. Figure 1.1 (from Anderson 2003) shows the behavior of a diatomic gas from 0 to 2000 K. The nearly constant specific heat ratio in 3–600 K range represents the calorically perfect gas behavior of a diatomic gas such as air with $\gamma = 1.4$. Note that near absolute zero (0 K), $c_v/R \rightarrow 3/2$, therefore, a diatomic gas ceases to rotate and thus behaves like a monatomic gas, i.e. it exhibits the same degrees of freedom as a monatomic gas, i.e. $n = 3$, $\gamma = 5/3$.

The composition of *dry air* at normal temperatures is often approximated as the mixture of two diatomic gasses, i.e. 21% O_2 and 79% N_2, by volume. However, this description applies to zero humidity. Water vapor, i.e. $H_2O_{(g)}$, a triatomic gas, is present in humid air, with the subsequent change in mixture properties, molecular weight, c_p, c_v,

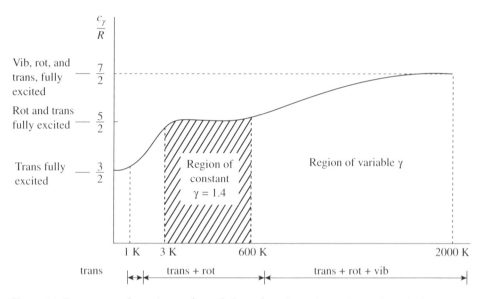

Figure 1.1 Temperature dependence of specific heats for a diatomic gas. *Source:* From Anderson 2003.

and γ. Also, air at elevated temperatures may dissociate to form oxygen and nitrogen atoms and other products such as O, OH, NO_2, N_2O, among others.

1.2.1 Isentropic Process and Isentropic Flow

For a reversible, adiabatic process, entropy remains constant, hence the designation: *isentropic process*. The Gibbs equation then may be used to relate the pressure and temperature ratios by an isentropic exponent, namely

$$\frac{p_2}{p_1} = \left(\frac{T_2}{T_1}\right)^{\frac{c_p}{R}} = \left(\frac{T_2}{T_1}\right)^{\frac{\gamma}{\gamma-1}} \left[\text{Calorically perfect gas}\right] \tag{1.25}$$

Now, using the perfect gas law by replacing the temperature ratio by pressure and density ratios in Eq. (1.25) and simplifying the exponents, we get:

$$\frac{p_2}{p_1} = \left(\frac{\rho_2}{\rho_1}\right)^{\gamma} \left[\text{Calorically perfect gas}\right] \tag{1.26}$$

1.2.2 Conservation of Mass

One of the pillars of the *great conservation laws* (the term coined by Richard Feynman) in Newtonian mechanics is the conservation of mass. The integral form is derived in elementary fluids/propulsion books (e.g. see Farokhi 2014) and is repeated here for reference:

$$\iiint_{C.V.} \frac{\partial \rho}{\partial t} dV + \iint_{C.S.} \rho \vec{V}.\hat{n} dS = 0 \tag{1.27}$$

$\vec{V}$ is the local fluid velocity vector, dV is the element of volume in the volume integral, dS is the element of surface in the surface integral and $\hat{n}$ is the unit vector normal to the control surface and pointing out. The integrals in Eq. (1.27) are thus performed over the control volume (C.V.) and the control surface (C.S.). The first integral represents the mass accumulation/depletion with time within the control volume. For steady flows (i.e. when $\frac{\partial}{\partial t} \rightarrow 0$), the first integral vanishes. The second integral represents the net mass flux through the control surface, surrounding the control volume. For steady flows, the second integral suggests that the mass flow rate into a control volume is balanced by the same mass flow rate out of the control volume. We can write the law of conservation of mass as the *continuity equation* as:

$$\dot{m}_{out} = \dot{m}_{in} \tag{1.28}$$

For steady uniform flows, the continuity equation may be integrated and written as:

$$\sum (\rho VA)_{out} = \sum (\rho VA)_{in} \tag{1.29}$$

where the summation is performed over all the "inlets" and "exits" of the control surface, e.g. in an aircraft engine, we have air and fuel inlets and one or more nozzle exits.

1.2.3 Conservation of Linear Momentum

Newton's secondnd law of motion, in vector form, written for a control volume is:

$$\iiint_{C.V.} \frac{\partial}{\partial t}\left(\rho \vec{V}\right)dV + \oiint_{C.S.} \rho \vec{V}\left(\vec{V}.\hat{n}\right)dS = \vec{F}_{net} \tag{1.30}$$

Here, $\vec{F}_{net}$ represents the net external forces acting on the fluid. These external forces are divided into *body forces* that acts over the fluid volume and the *surface forces* that act on the surface. Gravitational force or electromagnetic force are examples of the body force and the pressure and shear are the examples of the surface force. The first integral in Eq. (1.30) measures the unsteady momentum within the control volume and vanishes identically for a *steady flow*. The second integral is the net flux of momentum in and out of the control surface. Assuming uniform flow at the boundaries of the control surface inlets and outlets, we may simplify the momentum equation to a very useful engineering form, namely

$$\left(\dot{m}\vec{V}\right)_{out} - \left(\dot{m}\vec{V}\right)_{in} = \vec{F}_{net} \tag{1.31}$$

Since Eq. (1.34) is a vector equation, it is satisfied at the component level, namely:

$$\left(\dot{m}V_x\right)_{out} - \left(\dot{m}V_x\right)_{in} = F_{net,x} \tag{1.32a}$$

$$\left(\dot{m}V_y\right)_{out} - \left(\dot{m}V_y\right)_{in} = F_{net,y} \tag{1.32b}$$

$$\left(\dot{m}V_z\right)_{out} - \left(\dot{m}V_z\right)_{in} = F_{net,z} \tag{1.32c}$$

In cylindrical coordinates (r, θ, z), we may write the momentum equation as:

$$\left(\dot{m}V_r\right)_{out} - \left(\dot{m}V_r\right)_{in} = F_{net,r} \tag{1.33a}$$

$$\left(\dot{m}V_\theta\right)_{out} - \left(\dot{m}V_\theta\right)_{in} = F_{net,\theta} \tag{1.33b}$$

$$\left(\dot{m}V_z\right)_{out} - \left(\dot{m}V_z\right)_{in} = F_{net,z} \tag{1.33c}$$

Note that the cylindrical coordinate system best suits the turbomachinery environment and is thus of particular interest in gas turbine propulsion.

1.2.4 Conservation of Angular Momentum

By taking the moment of Newton's second law of motion (Eq. (1.30)) about an arbitrary point O in the fluid, we arrive at the law of conservation of angular momentum about the point O (in vector form), according to

$$\iiint_{C.V.} \frac{\partial}{\partial t}\left(\rho \vec{V}x\vec{r}\right)dV + \oiint_{C.S.} \rho \vec{V}x\vec{r}\left(\vec{V}.\hat{n}\right)dS = \vec{F}_{net}x\vec{r} \tag{1.34}$$

In turbomachinery, the fluid angular momentum undergoes a change interacting with the rotating (rotor) and stationary (stator) blade rows. The arbitrary point O is on the axis of turbomachinery. In cylindrical coordinates [r, θ, z], the angular momentum equation for *uniform steady flows* is simplified to:

$$\left(\dot{m}rV_\theta\right)_{out} - \left(\dot{m}rV_\theta\right)_{in} = rF_{net,\theta} \tag{1.35}$$

More discussions are included in the compressor section.

1.2.5 Conservation of Energy

The law of conservation of energy for a control volume starts with the first law of thermodynamics written for a system. On a per unit time basis, we get the *rate* of energy transfer to the fluid, namely:

$$\dot{Q} = \frac{dE}{dt} + \dot{W} \tag{1.36}$$

where the Eq. (1.36) is written for the entire mass of the system. The energy E is now represented by the internal energy, e, times mass, as well as the kinetic energy of the gas in the system and the potential energy of the system. Internal energy, e, is a fluid property that is revealed through the first law of thermodynamics. The contribution of potential energy change in gas turbine engines or most other aerodynamic applications is negligibly small and often ignored. We write the energy term, dE/dt, in Eq. (1.36) as the mass integral of specific energy, e, over the volume of the system, and apply *Leibnitz's rule of integration* to get the control volume version, namely,

$$\frac{dE}{dt} = \frac{d}{dt} \iiint\limits_{V(t)} \rho\left(e + \frac{V^2}{2}\right) dV = \iiint\limits_{C.V.} \frac{\partial}{\partial t}\left[\rho\left(e + \frac{V^2}{2}\right)\right] dV + \oiint\limits_{C.S.} \rho\left(e + \frac{V^2}{2}\right) \overline{V}.\hat{n} dS \tag{1.37}$$

The first term on the right-hand side (RHS) of Eq. (1.37) is the time rate of change of energy within the control volume, which identically vanishes for a steady flow. The second integral represents the net flux of fluid power (i.e. the *rate* of energy) crossing the boundaries of the control surface. The rates of heat transfer to the control volume and the rate of mechanical energy transfer by the gases inside the control volume on the surrounding are represented by $\dot{Q}$ and $\dot{W}$ terms in the energy Eq. (1.37). Now, let us examine the forces at the boundary that contribute to the rate of energy transfer. These surface forces are due to pressure and shear acting on the boundary, as noted earlier. The pressure forces act normal to the boundary and point inward, i.e. opposite to the convention on $\hat{n}$,

$$-p\hat{n} dS \tag{1.38}$$

To calculate the rate of work done by a force, we take the scalar product of the force and the velocity vector, namely

$$-p\overline{V}.\hat{n} dS \tag{1.39}$$

Now, we need to sum this elemental rate of energy transfer by pressure forces over the surface, via a closed surface integral, i.e.

$$\oiint\limits_{C.S.} -p\overline{V}.\hat{n} dS \tag{1.40}$$

Since the convention on the rate of work done in the first law is positive when it is performed "on" the surroundings, and Eq. (1.40) represents the rate of work done "by" the surroundings on the control volume, we need to incorporate an additional negative factor for this term in the energy equation. The rate of energy transfer by the shear forces is divided into a shaft power, $\wp_s$ that crosses the control surface in the form of shaft torque (due to moment of shear stresses in a rotating shaft) multiplied by the shaft angular speed and the viscous shear stresses on the boundary of the control volume. Hence,

$$\dot{W} = \oiint_{C.S.} p\overline{V}.\hat{n}dS + \wp_s + \dot{W}_{\text{viscous-shear}} \tag{1.41}$$

We arrive at a useful form of the energy equation for a control volume when we combine Eq. (1.41) with Eqs. (1.37) and (1.36), i.e.

$$\oiiint_{C.V.} \frac{\partial}{\partial t}\left[\rho\left(e + \frac{V^2}{2}\right)\right]dV + \oiint_{C.S.} \rho\left(e + \frac{V^2}{2}\right)\overline{V}.\hat{n}dS + \oiint_{C.S.} p\overline{V}.\hat{n}dS = \dot{Q} - \wp_s - \dot{W}_{\text{visc}} \tag{1.42}$$

The closed surface integrals on the LHS may be combined and simplified to:

$$\oiiint_{C.V.} \frac{\partial}{\partial t}\left[\rho\left(e + \frac{V^2}{2}\right)\right]dV + \oiint_{C.S.} \rho\left(e + \frac{V^2}{2} + \frac{p}{\rho}\right)\overline{V}.\hat{n}dS = \dot{Q} - \wp_s - \dot{W}_{\text{visc}} \tag{1.43}$$

We may replace the internal energy and the flow work terms in Eq. (1.43) by enthalpy, h, and define the sum of the enthalpy and the kinetic energy as the total or stagnation enthalpy, to get:

$$\oiiint_{C.V.} \frac{\partial}{\partial t}\left[\rho\left(e + \frac{V^2}{2}\right)\right]dV + \oiint_{C.S.} \rho h_t \overline{V}.\hat{n}dS = \dot{Q} - \wp_s - \dot{W}_{\text{visc.}} \tag{1.44}$$

Where the *total or stagnation enthalpy*, h_t, is defined as:

$$h_t \equiv h + \frac{V^2}{2} \tag{1.45}$$

In *steady flows* the volume integral that involves a time derivative vanishes. The second integral represents the net flux of fluid power across the control volume. The terms on the RHS are the external energy interaction terms, which serve as the *drivers* of the energy flow through the control volume. In adiabatic flows, the rate of heat transfer through the walls of the control volume vanishes. In the absence of shaft work, as in inlets and nozzles of a jet engine, the second term on the RHS of Eq. (1.44) vanishes. The rate of energy transfer via the viscous shear stresses is zero on solid boundaries (since velocity on solid walls obeys the no slip boundary condition) and nonzero at the inlet and exit planes. The contribution of this term over the inlet and exit planes is, however, small compared to the net energy flow in the fluid, hence neglected.

The integrated form of the energy equation for a control volume, assuming uniform flow over the inlets and outlets, yields a practical solution for quick engineering calculations,

$$\sum\left(\dot{m}h_t\right)_{\text{out}} - \sum\left(\dot{m}h_t\right)_{\text{in}} = \dot{Q} - \wp_s \tag{1.46}$$

The summations in Eq. (1.46) account for multiple inlets and outlets of a general control volume. In flows that are adiabatic and involve no shaft work, the energy equation simplifies to:

$$\sum(\dot{m}h_t)_{out} = \sum(\dot{m}h_t)_{in} \tag{1.47}$$

For a single inlet and a single outlet the energy equation is even further simplified, as the mass flow rate also cancels out in Eq. (1.47), to yield:

$$h_{t\text{-exit}} = h_{t\text{-inlet}} \tag{1.48}$$

Total or stagnation enthalpy then remains constant for adiabatic flows that involve no shaft power, such as inlets and nozzles or across shock waves.

1.2.6 Speed of Sound and Mach Number

Sound waves are infinitesimal pressure waves propagating in a medium. The propagation of sound waves, or acoustic waves, is *reversible and adiabatic*, hence *isentropic*. Since sound propagates through collision of fluid molecules, the speed of sound is higher in liquids than gas. Written for a gas, speed of sound, *a*, is related to pressure, density via

$$a^2 = \frac{dp}{d\rho} = \left(\frac{\partial p}{\partial \rho}\right)_s \tag{1.49}$$

In a perfect gas, Eq. (1.49) may be written as:

$$a = \sqrt{\frac{\gamma p}{\rho}} = \sqrt{\gamma RT} = \sqrt{(\gamma-1)c_p T} \tag{1.50}$$

The speed of sound is a *local* parameter, which depends on *local* absolute temperature of the gas. Its value changes with gas temperature, hence it drops when fluid accelerates (or *expands*) and increases when the gas decelerates (or *compresses*). The speed of sound in air at standard sea level conditions is $\sim 340\,\mathrm{m\,s^{-1}}$ or $\sim 1100\,\mathrm{ft\,s^{-1}}$. The molecular structure of gas also affects the speed of propagation of sound through its molecular mass. We may observe this behavior by the following substitution for gas constant, *R*,

$$a = \sqrt{\gamma RT} = \sqrt{\gamma\left(\frac{\bar{R}}{MW}\right)T} \tag{1.51}$$

A light gas, like hydrogen (H_2) with a molecular mass of 2 ($\mathrm{kg\,kmol^{-1}}$, or $\mathrm{g\,mol^{-1}}$), causes an acoustic wave to propagate *faster* than a heavier gas, such as air with (a mean) molecular weight of 29 ($\mathrm{kg\,kmol^{-1}}$, or $\mathrm{g\,mol^{-1}}$). If we substitute these molecular weights in Eq. (1.51) we note that sound propagates in gaseous hydrogen *nearly four times faster* than air. Since both hydrogen, H_2, and air (mixture of N_2 and O_2) are diatomic gases, the ratio of specific heats, *γ*, remains (nearly) the same for both gases at the same temperature.

The ratio of local gas speed to the speed of sound is called *Mach number*, *M*:

$$M \equiv \frac{V}{a} \tag{1.52}$$

Mach number is often used as a measure of the compressibility in a gas.

1.2.7 Stagnation State

We define the stagnation state of a gas as the state reached by decelerating a flow to rest reversibly and adiabatically and without any external work. Thus, the stagnation state is reached *isentropically*. This state is also referred to as the *total* state of the gas. The symbols for stagnation state in this book use a subscript t for total. The total pressure is p_t, the total temperature is T_t and the total density is ρ_t. Since the stagnation state is reached isentropically, the static and total entropy of the gas remain the same, i.e. $s_t = s$. Based on the definition of stagnation state, the total energy of the gas does not change in the deceleration or acceleration process; hence, the stagnation enthalpy, h_t, takes on the form:

$$h_t \equiv h + \frac{V^2}{2} \tag{1.53}$$

which we defined earlier in this chapter. Assuming a *calorically perfect gas*, we may simplify the total enthalpy relation (1.53) by dividing through by c_p to get an expression for total temperature according to:

$$T_t = T + \frac{V^2}{2c_p} \quad \left[\text{Calorically perfect gas} \right] \tag{1.54}$$

This equation is very useful in converting the local static temperature and gas speed into the local stagnation temperature. To nondimensionalize Eq. (1.54), we divide both sides by the static temperature, i.e.

$$\frac{T_t}{T} = 1 + \frac{V^2}{2c_p T} \tag{1.55}$$

The denominator of the kinetic energy term on the RHS is proportional to the square of the local speed of sound, a^2, according to Eq. (1.50), which simplifies to:

$$\frac{T_t}{T} = 1 + \left(\frac{\gamma - 1}{2} \right) \frac{V^2}{a^2} = 1 + \left(\frac{\gamma - 1}{2} \right) M^2 \tag{1.56}$$

The ratio of stagnation to static temperature of a (calorically perfect) gas is a unique function of local Mach number, according to Eq. (1.56). From isentropic relations between the pressure and temperature ratio that we derived earlier based on Gibbs equation of thermodynamics, we relate the ratio of stagnation to static pressure to local Mach number:

$$\frac{p_t}{p} = \left(\frac{T_t}{T} \right)^{\frac{\gamma}{\gamma - 1}} = \left[1 + \left(\frac{\gamma - 1}{2} \right) M^2 \right]^{\frac{\gamma}{\gamma - 1}} \quad \left[\text{Calorically perfect gas} \right] \tag{1.57}$$

Also the stagnation density is higher than the static density according to:

$$\frac{\rho_t}{\rho} = \left(\frac{p_t}{P} \right)^{\frac{1}{\gamma}} = \left[1 + \left(\frac{\gamma - 1}{2} \right) M^2 \right]^{\frac{1}{\gamma - 1}} \quad \left[\text{Calorically perfect gas} \right] \tag{1.58}$$

1.3 Thrust and Specific Fuel Consumption

An aircraft engine is designed to produce thrust, F (or sometimes lift in VTOL/STOL aircraft, e.g. the lift fan in the Joint Strike Fighter, F-35). In air-breathing engines, mass flow rate of air, $\dot{m}_0$, and fuel, $\dot{m}_f$, are responsible for creating that thrust. In a liquid rocket engine, the air is replaced with an on-board oxidizer, $\dot{m}_{ox}$, which then reacts with an on-board fuel, $\dot{m}_f$, to produce thrust. Figure 1.2 is a schematic drawing of a two-spool afterburning turbojet (TJ-AB) engine. The air is brought in through the air intake, or inlet, system, where station 0 designates the unperturbed flight condition, station 1 is at the inlet (or cowl) lip, and station 2 is at the exit of the air intake system, which corresponds to the inlet of the compressor (or fan). The compression process from stations 2 to 3 is divided into a low-pressure compressor (LPC) spool and a HPC spool. The exit of LPC is designated by station 2.5 and the exit of the HPC is station 3. The HPC is designed to operate at a higher shaft rotational speed than the LPC spool. The compressed gas enters the main or primary burner at 3 and is combusted with the fuel to produce hot, high-pressure gas at 4 to enter the high-pressure turbine (HPT). Flow expansion through the HPT and the low-pressure turbine (LPT) produces the shaft power for the HPC and LPC, respectively. An afterburner is designated between stations 5 and 7, where an additional fuel is combusted with the turbine discharge flow before it expands in the exhaust nozzle. Station 8 is at the throat of the nozzle and station 9 designates the nozzle exit.

To derive an expression for the engine thrust, it is most convenient to describe a control volume surrounding the engine and apply momentum principles to the fluid flow crossing the boundaries of the control volume. From a variety of choices that we have in describing the control volume, we may choose one that shares the same exit plane as the engine nozzle and its inlet is far removed from the engine inlet so as not to be disturbed by the nacelle lip. These choices are made for convenience.

As for the sides of the control volume, we may either choose stream surfaces, with the advantage of no flow crossing the sides, or a constant-area box, which has a simple geometry but fluid flow crosses the sides. Figure 1.3 depicts a control volume in the shape of a box, with cross-sectional area, A.

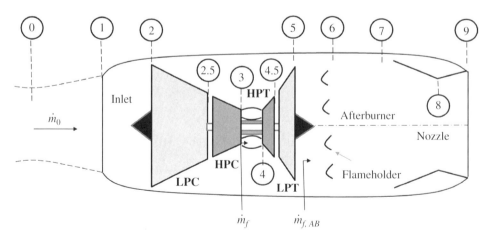

Figure 1.2 Station numbers for a turbojet with afterburner (TJ-AB) engine.

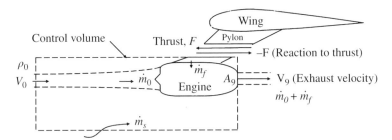

Figure 1.3 Schematic drawing of an air-breathing engine with a box-like control volume positioned around the engine (Note the flow of air to the engine, $\dot{m}_0$, through the sides, $\dot{m}_s$ and the fuel flow rate, $\dot{m}_f$).

The pylon that transmits the engine force to the aircraft, by necessity is cut by the control volume, which is enclosing the engine. The thrust force, F, and its reaction are shown in Figure 1.3. We may assume the sides of the control volume are not affected by the flowfield around the nacelle, i.e. the static pressure distribution on the sides is nearly the same as the ambient static pressure, p_0. In the same spirit, we may assume that the exit plane also sees an ambient pressure of p_0 with an exception of the plane of the jet exhaust, where a different static pressure, namely p_9, may prevail.

Application of continuity to the control volume for steady uniform flow, gives:

$$\rho_0 V_0 A + \dot{m}_s + \dot{m}_f = (\dot{m}_0 + \dot{m}_f) + \rho_0 V_0 (A - A_9) \tag{1.59}$$

which simplifies to:

$$\dot{m}_s = \dot{m}_0 - \rho_0 V_0 A_9 \tag{1.60}$$

The x-momentum balance on the control volume, assuming steady uniform flow, gives

$$\sum (\dot{m} V_x)_{out} - \sum (\dot{m} V_x)_{in} = \sum (F_x)_{fluid} \tag{1.61}$$

Equation (1.61) states that the difference between the fluid (time rate of change of) momentum out of the box and into the box is equal to the net forces acting on the fluid in the x-direction on the boundaries and within the box. These are:

$$\sum (\dot{m} V_x)_{out} = (\dot{m}_0 + \dot{m}_f) V_9 + \left[\rho_0 V_0 (A - A_9) \right] V_0 \tag{1.62}$$

$$\sum (\dot{m} V_x)_{in} = (\rho_0 V_0 A) V_0 + \dot{m}_s V_0 \tag{1.63}$$

$$\sum (F_x)_{fluid} = (-F_x)_{fluid} - (p_9 - p_0) A_9 \tag{1.64}$$

Substituting Eqs. (1.62), (1.63), and (1.64) into (1.61) and for $\dot{m}_s$ from Eq. (1.60), we get:

$$(\dot{m}_0 + \dot{m}_f) V_9 + \rho_0 V_0 (A - A_9) V_0 - \rho_0 V_0 A V_0 - (\dot{m}_0 - \rho_0 V_0 A_9) V_0 =$$
$$= (-F_x)_{fluid} - (p_9 - p_0) A_9 \tag{1.65}$$

which simplifies to:

$$(\dot{m}_0 + \dot{m}_f) V_9 - \dot{m}_0 V_0 = (-F_x)_{fluid} - (p_9 - p_0) A_9 \tag{1.66}$$

Now, through the action-reaction principle of Newtonian mechanics, we know that an equal and opposite axial force is exerted on the engine, by the fluid, i.e.

$$\left(-F_x\right)_{\text{fluid}} = \left(F_x\right)_{\text{pylon}} = \left(F_x\right)_{\text{engine}} \tag{1.67}$$

Therefore, calling the axial force of the engine "thrust," or simply *F*, we get the following expression for the engine thrust:

$$F = \left(\dot{m}_0 + \dot{m}_f\right)V_9 - \dot{m}_0 V_0 + \left(p_9 - p_0\right)A_9 \tag{1.68}$$

This expression for the thrust is referred to as the *net uninstalled thrust* and sometimes a subscript *n* is placed on *F* to signify the "net" thrust. Therefore, the thrust expression of Eq. (1.68) is often written as:

$$F_n)_{\text{uninstalled}} = \left(\dot{m}_0 + \dot{m}_f\right)V_9 - \dot{m}_0 V_0 + \left(p_9 - p_0\right)A_9 \tag{1.69}$$

We first note that the RHS of Eq. (1.69) is composed of two momentum terms and one pressure-area term. The first momentum term is the exhaust momentum through the nozzle, contributing positively to the engine thrust. The second momentum term is the inlet momentum, which contributes negatively to the engine thrust in effect it represents a drag term. This drag term is called *ram drag*. It is often given the symbol of D_{ram} or D_r and expressed as:

$$D_{\text{ram}} = \dot{m}_0 V_0 \tag{1.70}$$

The last term in Eq. (1.69), is a pressure-area term, which acts over the nozzle exit plane, i.e. area A_9, and will only contribute to the engine thrust if there is an imbalance of static pressure between the ambient and the exhaust jet. As we remember from our aerodynamic studies, a nozzle with a subsonic jet will always expand the gases to the same static pressure as the ambient condition and a sonic or supersonic exhaust jet may or may not have the same static pressure in their exit plane as the ambient static pressure. Depending on the *mismatch* of the static pressures, we categorized the nozzle flow as follows:

If $p_9 < p_0$ the nozzle is *overexpanded,*
which can only happen in supersonic jets
(i.e. in convergent-divergent nozzles with area ratio *larger* than needed for perfect expansion).

If $p_9 = p_0$ the nozzle is *perfectly expanded,*
which is the case for *all subsonic jets* and sometimes in sonic or supersonic jets (i.e. with the "right" nozzle area ratio).

If $p_9 > p_0$ the nozzle is *underexpanded,*
which can only happen in sonic or supersonic jets (i.e. with inadequate nozzle area ratio).

Examining the various contributions from Eq. (1.69) to the net engine (uninstalled) thrust, we note that the thrust is the difference between the nozzle contributions (both momentum and pressure-area terms) and the inlet contribution (the momentum term). The nozzle contribution to thrust is called *gross thrust* and is given a symbol F_g, i.e.

$$F_g \equiv \left(\dot{m}_0 + \dot{m}_f\right)V_9 + \left(p_9 - p_0\right)A_9 \tag{1.71}$$

and, as explained earlier, the inlet contribution was negative and was called *ram drag, D_{ram}*:

$$F_n)_{uninstalled} = F_g - D_{ram} \tag{1.72}$$

We may generalize the result expressed in Eq. (1.72) to aircraft engines with more than a single stream, as for example the turbofan engine with separate exhausts. This task is very simple as we account for all gross thrusts produced by all the exhaust nozzles and subtract all the ram drag produced by all the air inlets to arrive at the engine uninstalled thrust, i.e.

$$F_n)_{uninstalled} = \sum (F_g)_{nozzles} - \sum (D_{ram})_{inlets} \tag{1.73}$$

Equation (1.73) is the balance the momentum of the exhaust stream and the inlet momentum with the pressure thrust at the nozzle exit planes and the net uninstalled thrust of the engine. In a turbofan engine, the captured airflow is typically divided into a *core* flow, where the combustion takes place, and a fan flow, where the so-called *bypass* stream of air is compressed through a fan and later expelled through a fan exhaust nozzle. This type of arrangement (i.e. bypass configuration) leads to a higher overall efficiency of the engine and lower fuel consumption. A schematic drawing of the engine is shown in Figure 1.4.

In this example, the inlet consists of a single stream and the exhaust streams are split into a primary and a fan nozzle. We may readily write the uninstalled thrust produced by the engine following the momentum principle, namely;

$$F_n)_{uninstalled} = \dot{m}_9 V_9 + \dot{m}_{19} V_{19} + (p_9 - p_0) A_9 + (p_{19} - p_0) A_{19} - \dot{m}_0 V_0 \tag{1.74}$$

The first four terms account for the momentum and pressure thrusts of the two nozzles (what is known as gross thrust) and the last term represents the inlet ram drag.

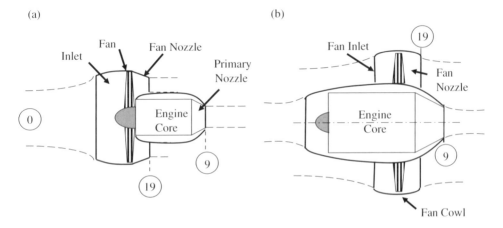

Figure 1.4 Schematic drawing of a turbofan engine with separate exhausts: (a) front-fan configuration; (b) aft-fan configuration.

1.3.1   Takeoff Thrust

At takeoff, the air speed V_0 ("flight" speed) is often ignored in the thrust calculation, therefore the ram drag contribution to engine thrust is neglected:

$$F_{\text{takeoff}} \approx F_g = \left(\dot{m}_0 + \dot{m}_f\right)V_9 + \left(p_9 - p_0\right)A_9 \tag{1.75}$$

For a perfectly expanded nozzle, the pressure thrust term vanishes to give:

$$F_{\text{takeoff}} \approx \left(\dot{m}_0 + \dot{m}_f\right)V_9 \approx \dot{m}_0 V_9 \tag{1.76}$$

Therefore, the takeoff thrust is proportional to the captured airflow.

1.3.2   Installed Thrust – Some Bookkeeping Issues on Thrust and Drag

As indicated by the description of the terms *installed thrust* and *uninstalled thrust*, they refer to the *actual propulsive force* transmitted to the aircraft by the engine and the thrust produced by the engine if it had zero external losses, respectively. Therefore, for the installed thrust, we need to account for the installation losses to the thrust such as the nacelle skin friction and pressure drags, which are to be included in the propulsion-side of the drag bookkeeping. On the other hand, the pylon and the engine installation, which affects the wing aerodynamics (in podded nacelle, wing-mounted configurations), namely by altering its "clean" drag polar characteristics, causes an "interference" drag that is accounted for in the aircraft drag polar. In the study of propulsion, we often concentrate on the engine's "internal" performance, i.e. the uninstalled characteristics, rather than the installed performance because the external drag of the engine installation depends not only on the engine nacelle geometry but also on the engine-airframe integration. Therefore, accurate installation drag accounting will require CFD analysis and wind tunnel testing at various flight Mach numbers and engine *throttle settings*. In its simplest form, we can relate the installed and uninstalled thrust according to:

$$F_{\text{installed}} = F_{\text{uninstalled}} - D_{\text{nacelle}} \tag{1.77}$$

In our choice of the control volume, as depicted in Figure 1.3, we made certain assumptions about the exit boundary condition imposed on the aft surface of the control volume. We made assumptions about the pressure boundary condition as well as the velocity boundary condition. About the pressure boundary condition, we stipulated the static pressure of flight, p_0, imposed on the exit plane, except at the nozzle exit area of A_9; therefore, we allowed for an under- or overexpanded nozzle. With regard to the velocity boundary condition at the exit plane, we stipulated that the flight velocity, V_0, is prevailed, except at the nozzle exit, where the jet velocity of V_9 prevails. In reality, the aft surface of the control volume is by necessity downstream of the nacelle and pylon, and therefore it is in the middle of the wake generated by the nacelle and the pylon. This implies that there would be a momentum deficit in the wake and the static pressure nearly equals the free stream pressure. On the other hand, the force transmitted through the pylon to the aircraft is not the *uninstalled* thrust, as we called it in our earlier derivation, rather the *installed* thrust and pylon drag. But the integral of momentum deficit and the pressure imbalance in the wake is exactly equal to the nacelle and pylon drag contributions, i.e. we have

$$D_{\text{nacelle}} + D_{\text{pylon}} = \iint \rho V\left(V_0 - V\right)dA + \iint \left(p_0 - p\right)dA \tag{1.78}$$

where the surface integral is taken over the exit plane downstream of the nacelle and pylon.

Further breakdown of Eq. (1.78) separates the installation effects as:

$$F_{\text{uninstalled}} - D_{\text{nacelle}} - D_{\text{pylon}} =$$
$$(\dot{m}_0 + \dot{m}_f)V_9 - \dot{m}_0 V_0 + (p_9 - p_0)A_9 - \iint \rho V(V_0 - V)dA - \iint (p_0 - p)dA \qquad (1.79)$$

After canceling the integrals on the RHS with the drag terms of the LHS in Eq. (1.79), we recover our Eq. (1.69), which we derived earlier for the *uninstalled thrust*, i.e.

$$(F_n)_{\text{uninistalled}} = (\dot{m}_0 + \dot{m}_f)V_9 - \dot{m}_0 V_0 + (p_9 - p_0)A_9$$

The pressure integral on the captured streamtube is called the inlet *additive drag:*

$$D_{\text{add}} \equiv \oiint_0^1 (p - p_0)dA_n \qquad (1.80)$$

where dA_n is the element of flow cross-sectional area normal to the flow direction. For subsonic inlets, additive drag is, for the most part, balanced by the cowl lip thrust and the difference is called the *spillage drag*. Figure 1.5 shows a schematic drawing of the flow near a blunt cowl lip and the resultant force. In general, for well-rounded cowl lips of subsonic inlets, the spillage drag is rather small and it becomes significant only for supersonic, sharp-lipped inlets. The nacelle frictional drag also contributes to the external force as well as the aft-end pressure drag of the boattail, which may be written as the sum of six contributions:

$$\sum (F_{x\text{-external}})_{\text{control-surface}} = D_{\text{spillage}} + \iint_{M-9} (p - p_0)dA_n + \iint_{1-9} \tau_w dA_x$$
$$(I) \qquad\qquad (II) \qquad\qquad (III)$$
$$+ D_{\text{pylon}} - F_{x,\text{fluid}} - (p_9 - p_0)A_9 \qquad (1.81)$$
$$(IV) \qquad (V) \qquad (VI)$$

The first term on the RHS, i.e. (I), is the spillage drag that we discussed earlier, the second term (II) is the pressure drag on the nacelle aft end or boat tail pressure drag, the third term (III) is the nacelle viscous drag, the fourth term (IV) is the pylon drag, the

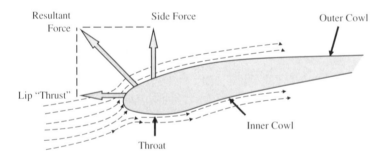

Figure 1.5 Flow detail near a blunt cowl lip showing lip thrust component and side force.

fifth term (V) is the reaction to the installed thrust force and pylon drag acting on the fluid, and the last term (VI) is the pressure thrust due to imperfect nozzle expansion. To summarize:

- Terms I + II + III are propulsion system installation drag losses.
- Term IV is accounted for in the aircraft drag polar.
- Terms V and VI combine with I, II, III, and IV to produce the uninstalled thrust.

Figure 1.6 (adapted from Lotter, 1977) shows the various contributions to net installed thrust in an air-breathing jet engine.

1.3.3 Air-Breathing Engine Performance Parameters

The engine thrust, mass flow rates of air and fuel, the rate of kinetic energy production across the engine or the mechanical power/shaft output, and engine dry weight, among other parameters, are combined to form a series of important performance parameters, known as the propulsion system figures of merit.

1.3.3.1 Specific Thrust

The size of the air intake system is a design parameter that establishes the flow rate of air, $\dot{m}_0$ Accordingly, the fuel pump is responsible for setting the fuel flow rate in the engine, $\dot{m}_f$. Therefore, in producing thrust in a "macro-engine," the engine size seems to be a "scalable" parameter. The only exception in scaling the jet engines is the "micro-engines" where the component losses do not scale. In general, the magnitude of the thrust

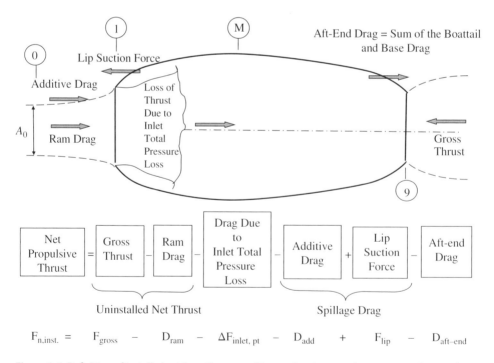

Figure 1.6 Definition of *installed net thrust* in terms of internal and external parameters of an engine nacelle. *Source:* adapted from Lotter, K. 1977.

produced is directly proportional to the mass flow rates of the fluid flow through the engine. Then, it is logical to study thrust per unit mass flow rate as a *figure of merit* of a candidate propulsion system. In case of an air-breathing engine, the ratio of thrust to air mass flow rate is called *specific thrust* and is considered to be an engine performance parameter:

Air-breathing engine performance parameter #1: $\dfrac{F}{\dot{m}_0}$ "specific thrust"

With a metric unit of: [N.s/kg] and a British unit of: [lbf.s/lbm]

The target for this parameter, i.e. specific thrust, in a cycle analysis is usually to be maximized, i.e. to produce thrust with the least quantity of airflow rate, or equivalently to produce thrust with a minimum of engine frontal area. However, with subsonic cruise Mach numbers, the drag penalty for engine frontal area is far less severe than their counterparts in supersonic flight. Consequently, the specific thrust as a figure of merit in a commercial transport aircraft (e.g. Boeing 777 or Airbus A-340) takes a back seat to the lower fuel consumption achieved in a very large bypass ratio turbofan engine at subsonic speeds. As noted, specific thrust is a dimensional quantity with the unit of force per unit mass flow rate. A nondimensional form of the specific thrust, which is useful for graphing purposes and engine comparisons, is (following Kerrebrock 1992)

$$\text{Nondimensional specific thrust} \equiv \frac{F}{\dot{m}_0 a_0} \qquad (1.82)$$

where a_0 is the ambient speed of sound taken as the reference velocity.

1.3.3.2 Specific Fuel Consumption and Specific Impulse

The ability to produce thrust with a minimum of fuel expenditure is another parameter, which is considered to be a performance parameter in an engine. In the commercial world, e.g. the airline business, specific fuel consumption represents perhaps the most important parameter of the engine. After all, the money spent on fuel is a major expenditure in operating an airline, for example. However, the reader is quickly reminded of the unspoken parameters of *reliability and maintainability* that have a direct impact on the cost of operating commercial engines and, therefore, they are at least as important, if not more important, than the engine specific fuel consumption. In the military world, the engine fuel consumption parameter takes a decidedly second role to other aircraft performance parameters, such as stealth, agility, maneuverability, and survivability. For an air-breathing engine, the ratio of fuel flow rate per unit thrust force produced is called thrust-specific fuel consumption, TSFC, or the specific fuel consumption, i.e. sfc, and is defined as:

Air-breathing engine performance parameter #2: $\dfrac{\dot{m}_f}{F}$ "thrust-specific fuel consumption,"

TSFC, with a metric unit of: [mg/s/N] and a British unit of: [lbm/h/lbf]

The target for this parameter, i.e. TSFC, in a cycle analysis is to be minimized, i.e. to produce thrust with a minimum of fuel expenditure. This parameter, too, is dimensional. For a rocket, on the other hand, the oxidizer as well as the fuel both contribute to the "expenditure" in the engine to produce thrust, and as such the oxidizer flow rate, $\dot{m}_{ox}$, needs to be accounted for as well.

The word *propellant* is used to reflect the combination of oxidizer and fuel in a liquid propellant rocket engine or a solid propellant rocket motor. It is customary to define a corresponding performance parameter in a rocket as thrust per unit propellant *weight flow rate*. This parameter is called *specific impulse, I_s*, i.e.

$$I_s \equiv \frac{F}{\dot{m}_p g_0} \tag{1.83}$$

where $\quad \dot{m}_p \equiv \dot{m}_f + \dot{m}_{ox}$ $\hspace{4cm}$ (1.84)

And g_0 is the gravitational acceleration on Earth's surface, i.e. $9.8\,\mathrm{m\,s}^{-2}$ or $32.2\,\mathrm{ft\,s}^{-2}$. The dimension of $\dot{m}_p g_0$ in denominator of Eq. (1.83) is then the *weight flow rate* of the propellant based on Earth's gravity, or force per unit time. Consequently, the dimension of specific impulse is "Force/Force/second," which simplifies to just the "second." All propulsors, rockets, and air breathers, then could be compared using a unifying figure of merit, namely their *specific impulse* in seconds. An added benefit of the specific impulse is that regardless of the units of measurement used in the analysis, i.e. either metric or the British on both sides of the Atlantic, specific impulse comes out as seconds in both systems. The use of specific impulse as a unifying figure of merit is further justified in the twenty-first century as we attempt to commercialize the space with potentially reusable rocket-based combined cycle (RBCC) powerplants in propelling a variety of single-stage-to-orbit (SSTO) vehicles. To study advanced propulsion concepts where the mission calls for multimode propulsion units, as in air-ducted rockets, ramjets, and scramjets, all combined into a single "package," the use of specific impulse becomes even more obvious.

In summary,

Specific impulse: $I_s \equiv \dfrac{F}{\dot{m}_f g_0}$ [sec] for an air-breathing engine

Specific impulse: $I_s \equiv \dfrac{F}{\dot{m}_p g_0}$ [sec] for a rocket

---

An important goal in an engine cycle design is to maximize specific impulse, i.e. the ability to produce thrust with the least amount of fuel or propellant consumption in the engine.

1.4 Thermal and Propulsive Efficiency

1.4.1 Thermal Efficiency

The ability of an engine to convert the thermal energy inherent in the fuel to net kinetic energy gain of the working medium, is called the engine *thermal efficiency, η_{th}*

$$\eta_{th} \equiv \frac{\Delta K\dot{E}}{\wp_{thermal}} = \frac{\dot{m}_9 \dfrac{V_9^2}{2} - \dot{m}_0 \dfrac{V_0^2}{2}}{\dot{m}_f Q_R} = \frac{\left(\dot{m}_0 + \dot{m}_f\right)V_9^2 - \dot{m}_0 V_0^2}{2\dot{m}_f Q_R} \tag{1.85}$$

where $\dot{m}$ signifies mass flow rate corresponding to stations 0 and 9 and subscript f signifies fuel, and Q_R is the fuel heating value. The unit of Q_R is energy per unit mass of the fuel (e.g. kJ kg^{-1} or BTU lbm^{-1}) and is tabulated as a fuel property. Eq. (1.85) compares the *mechanical power* production in the engine to the *thermal power* investment in the engine. Figure 1.7 graphically depicts the energy sources in an air-breathing engine. The thermal energy production in an engine is not actually "lost," as it shows up in the hot jet exhaust stream, rather, this energy is "wasted" and we were unable to convert it to a *useful* power. It is important to know, i.e. to quantify, this inefficiency in our engine.

Equation (1.85), which defines thermal efficiency, is simply the ratio of "net mechanical output" to the "thermal input," as we learned in thermodynamics.

To lower the exhaust gas temperature, we can place an additional turbine wheel in the high-pressure, hot gas stream and produce shaft power. This shaft power can then be used to power a propeller, a fan, or a helicopter rotor, for example. The concept of additional turbine stages to extract thermal energy from the combustion gases and powering a fan in a jet engine has led to turbofan engines. Therefore, the mechanical output of these engines is enhanced by the additional shaft power. The thermal efficiency of a cycle that produces a shaft power can therefore be written as:

$$\eta_{th} = \frac{\wp_{shaft} + \Delta K\dot{E}}{\dot{m}_f Q_R} \tag{1.86}$$

A schematic drawing of an aircraft gas turbine engine, which is configured to produce shaft power, is shown in Figure 1.8. In part (a), the *power turbine* provides shaft power to a propeller, whereas in part (b) the power turbine provides the shaft power to a helicopter main rotor.

The gas generator in Figure 1.8 refers to the compressor, burner, and the turbine combination, i.e. the basic building block of a GT engine. In turboprops and turboshaft engines the mechanical output of the engine is dominated by the shaft power that in the definition of the thermal efficiency of such cycles the rate of kinetic energy increase is neglected, i.e.

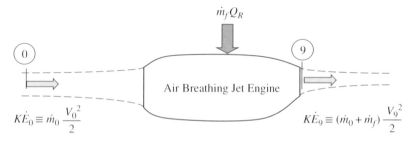

Figure 1.7 Thermal power input (by fuel) and the mechanical power production (output) by the engine.

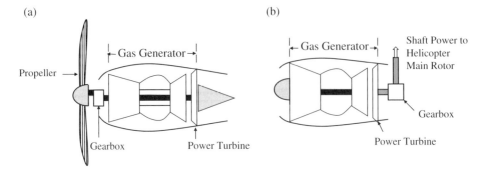

Figure 1.8 Schematic drawing of a power turbine placed in the exhaust of a gas turbine engine: (a) gas turbine engine with a power turbine providing shaft power to a propeller (turboprop); (b) power turbine providing shaft power to a helicopter rotor.

$$\eta_{th} \equiv \frac{\wp_s}{\dot{m}_f Q_R} \quad \text{In turboprop and turboshaft engines} \qquad (1.87)$$

In addition to a shaft-power turbine concept, we can lower the exhaust gas temperatures by placing a heat exchanger in the exhaust stream to preheat the compressor air prior to combustion. The exhaust gas stream is cooled as it heats the cooler compressor gas and less fuel is needed to achieve a desired turbine entry temperature (TET). This scheme is referred to as *regenerative cycle* and is shown in Figure 1.9.

All of the cycles shown in Figures 1.8 and 1.9 produce less waste heat in the exhaust nozzle; consequently, they achieve a higher thermal efficiency than their counterparts without the extra shaft power or the heat exchanger.

We remember that the highest thermal efficiency attainable in a heat engine operating between two temperature limits was that of a Carnot cycle operating between those temperatures. Figure 1.10 shows the Carnot cycle on a T-s diagram.

As noted in Figure 1.10, both heat rejection at absolute zero ($T_1 = 0$) and heating to infinite temperatures ($T_2 = \infty$) are impossibilities; therefore, we are thermodynamically bound by the ideal Carnot thermal efficiency as the maximum, which for peak temperature of 2500 K corresponds to stoichiometric combustion of Jet-A, and the low temperature of 288 K, which corresponds to standard ambient condition, is 88%. Brayton cycle, which represents gas turbine engines, experiences a much lower thermal efficiency than the corresponding Carnot cycle., typically in the 30s.

1.4.2 Propulsive Efficiency

The fraction of the net mechanical output of the engine that is converted into thrust power is called the *propulsive efficiency*. The net mechanical output of the engine is $\Delta \dot{KE}$ for perfectly expanded nozzle and the thrust power is $F \cdot V_0$, therefore, the propulsive efficiency is defined as their ratio:

$$\eta_p \equiv \frac{F \cdot V_0}{\Delta \dot{KE}} \qquad (1.88)$$

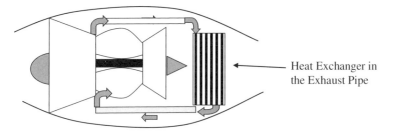

Figure 1.9 Schematic drawing of a gas turbine engine with a regenerative scheme.

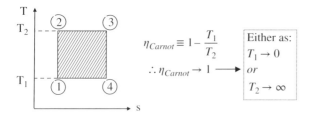

$$\eta_{Carnot} \equiv 1 - \frac{T_1}{T_2}$$

$$\therefore \eta_{Carnot} \rightarrow 1$$

Either as:

$$T_1 \rightarrow 0$$

or

$$T_2 \rightarrow \infty$$

Figure 1.10 T-s diagram of a Carnot cycle.

A graphical depiction is shown in Figure 1.11 to help the reader to remember the definition of propulsive efficiency.

Although the thrust power represented by $F \cdot V_0$ in Eq. (1.88) is based on the *installed* thrust, for simplicity, it is often taken as the uninstalled thrust power to highlight a very important, and at first astonishing, result about the propulsive efficiency. Now, let us substitute the uninstalled thrust of a perfectly expanded jet in the above definition to get:

$$\eta_p \approx \frac{\left[\left(\dot{m}_0 + \dot{m}_f \right) V_9 - \dot{m}_0 V_0 \right] V_0}{\left(\dot{m}_0 + \dot{m}_f \right) \dfrac{V_9^2}{2} - \dot{m}_0 \dfrac{V_0^2}{2}} \tag{1.89}$$

We recognize that the fuel-flow rate is but a small fraction (~2–3%) of the airflow rate and thus may be ignored relative to the airflow rate so Eq. (1.89) may be simplified to

$$\eta_p \approx \frac{\left(V_9 - V_0 \right) V_0}{\dfrac{1}{2} \left(V_9^2 - V_0^2 \right)} = \frac{2 V_0}{V_9 + V_0} = \frac{2}{1 + \dfrac{V_9}{V_0}} \tag{1.90}$$

Equation (1.90) as an approximate expression for the propulsive efficiency of a jet engine is cast in terms of a single parameter, namely the jet-to-flight velocity ratio, V_9/V_0. We further note that 100% propulsive efficiency (within the context of approximation presented in the derivation) is mathematically possible and will be achieved by engines whose exhaust velocity is as fast as the flight velocity, i.e. V_9/V_0. Thrust production demands $V_9 > V_0$. To achieve a small velocity increment across a gas turbine engine for a given fuel-flow rate, we need to drain the thermal energy in the combustion gas further and convert it to additional shaft power. In turn, shaft power supplied to a fan

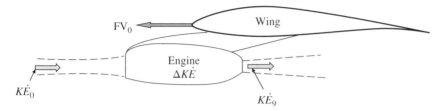

Figure 1.11 Schematic drawing of an aircraft engine installation showing the mechanical power produced by the engine and thrust power transmitted to aircraft.

(or propeller) impacts a larger mass flow rate of air, in a secondary or a bypass stream, which produce the desired level of thrust.

Propulsive efficiency of a turboprop engine is defined as the fraction of mechanical power, which is converted to the total thrust (i.e. the sum of propeller and engine nozzle thrust) power, namely;

$$\eta_p \equiv \frac{F \cdot V_0}{\wp_s + \Delta \dot{KE}} \approx \frac{F \cdot V_0}{\wp_s} \quad [\text{Turboprop}] \tag{1.91}$$

Again, this definition compares the propulsive *output* ($F \cdot V_0$) to the mechanical power *input* (shaft and change in jet kinetic power) in an aircraft engine. The fraction of shaft power delivered to the propeller, converted to the propeller thrust, is called *propeller efficiency*, η_{pr},

$$\eta_{pr} \equiv \frac{F_{\text{prop}} \cdot V_0}{\wp_{s,\text{prop}}} \tag{1.92}$$

Due to large diameters of propellers as compared to power turbine, it is necessary to reduce their rotational speed to avoid severe tip shock-induced losses. Consequently, the power turbine rotational speed is mechanically reduced in a reduction gearbox and a small fraction of shaft power is lost in the gearbox, which is referred to as gearbox efficiency, i.e.

$$\eta_{gb} \equiv \frac{\wp_{s,\text{prop}}}{\wp_{s,\text{turbine}}} \tag{1.93}$$

1.4.3 Engine Overall Efficiency and Its Impact on Aircraft Range and Endurance

The product of the engine thermal and propulsive efficiency is called overall efficiency, η_0

$$\eta_0 \equiv \eta_{th} \cdot \eta_p = \frac{\Delta \dot{KE}}{\dot{m}_f Q_R} \frac{F \cdot V_0}{\Delta \dot{KE}} = \frac{F \cdot V_0}{\dot{m}_f Q_R} \tag{1.94}$$

The overall efficiency of an aircraft engine is therefore the fraction of the fuel thermal power, which is converted into the thrust power of the aircraft. Again, a useful output is compared to the input investment in this efficiency definition. In an aircraft performance course that typically precedes the aircraft propulsion class, the engine overall

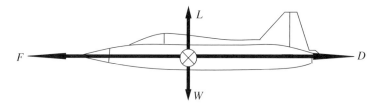

Figure 1.12 Aircraft in unaccelerated level flight showing a balance between thrust and drag forces and lift and weight forces.

efficiency is tied with the aircraft range, through the Breguet range equation. The derivation of the Breguet range equation is both fundamental to our studies and is surprisingly simple enough for us to repeat it here for review purposes. An aircraft in level flight cruising at the speed V_0 (see Figure 1.12) experiences a drag force that is entirely balanced by the engine installed thrust, i.e.

$$F_{engine} = D_{aircraft} \tag{1.95}$$

The aircraft lift, L, is also balanced by the aircraft weight to maintain level flight, i.e.

$$L = W \tag{1.96}$$

We can multiply Eq. (1.95) by the flight speed V_0 and then replace the resulting thrust power, $F \cdot V_0$, by $\eta_0 \, \dot{m}_f \, Q_R$, via the definition of the engine overall efficiency, to get

$$F.V_0 = \eta_0 \, \dot{m}_f \, Q_R = D.V_0 \tag{1.97}$$

Now, let us divide the RHS term in Eq. (1.97) by lift, L, and the middle term of Eq. (1.97) by aircraft weight (which is the same as lift), to get

$$\frac{\eta_0 \dot{m}_f Q_R}{W} = \frac{D}{L}.V_0 \tag{1.98}$$

Noting that the fuel-flow rate, $\dot{m}_f = -\dfrac{1}{g_0}\dfrac{dW}{dt}$, i.e. the rate at which the aircraft is losing mass (thus the negative sign), we can substitute this expression in Eq. (1.98) and rearrange to get

$$-\frac{\eta_0 Q_R}{g_0}\frac{dW}{W} = \frac{D}{L}.V_0 dt = \frac{dR}{L/D} \tag{1.99}$$

where g_0 is Earth's gravitational acceleration and $V_0 dt$ is interpreted as the aircraft elemental range, dR, which is the distance traveled in time dt by the aircraft flying at speed V_0. Now, we can proceed to integrate Eq. (1.99) by making the assumptions of constant lift-to-drag ratio and constant engine overall efficiency, over the cruise period, to derive the Breguet range equation as:

$$R = \eta_0.\frac{Q_R}{g_0}\frac{L}{D}.\ell n \frac{W_i}{W_f} \tag{1.100}$$

where W_i is the aircraft initial weight and W_f is the aircraft final weight (note that the initial weight is larger than the final weight by the weight of the fuel burned in flight). As noted, Eq. (1.100) is known as the *Breguet range equation* that shows both aerodynamic efficiency (L/D) and overall propulsion efficiency, η_o, impact aircraft range. This equation owes its simplicity and elegance to our assumptions of (i) unaccelerated level flight and (ii) constant lift-to-drag ratio and engine overall efficiency. Also note that the range segment contributed by the takeoff and landing distances is not accounted for in the Breguet range equation. The direct proportionality of the engine overall efficiency and range is demonstrated by the Breguet range equation, i.e.

Aircraft range, $R \propto \eta_0$

By replacing the overall efficiency of the engine in the range equation by the ratio of thrust power to the thermal power in the fuel, we get:

$$R = \frac{F_n V_0}{\dot{m}_f Q_R} \frac{Q_R}{g_0} \frac{L}{D} \ell n \frac{W_i}{W_f} \tag{1.101}$$

We may express the flight speed in terms of a product of flight Mach number and the speed of sound; in addition, we may substitute the TSFC for the ratio of fuel flow rate to the engine net thrust, to get

$$R = \left(M_0 \frac{L}{D} \right) \frac{a_0 / g_0}{TSFC} \ell n \frac{W_i}{W_f} \tag{1.102}$$

The result of this representation of the aircraft range is the emergence of (ML/D) as the aerodynamic figure of merit for aircraft range optimization, known as the *range factor*, or *cruise efficiency*:

$$\text{Aircraft range}, R \propto M_0 \frac{L}{D} \tag{1.103}$$

$$\text{Aircraft range}, R \propto \frac{1}{TSFC} \tag{1.104}$$

By using a more energetic fuel than the current jet aviation fuel (e.g. hydrogen), we will be able to reduce the thrust-specific fuel consumption, or we can see the effect of fuel energy content on the range following Eq. (1.105), which shows

$$\text{Aircraft range}, R \propto Q_R \tag{1.105}$$

Equivalently, we may seek out the effect of engine overall efficiency, or the specific fuel consumption on aircraft endurance, which for our purposes is the ratio of aircraft range to the flight speed:

$$\text{Aircraft endurance} = \frac{R}{V_0} = \frac{\eta_0}{V_0} \frac{Q_R}{g_0} \frac{L}{D} \ell n \frac{W_i}{W_f} \tag{1.106}$$

This again points out the importance of engine overall efficiency on aircraft performance parameters such as endurance. In terms of TSFC, we get:

$$\text{Aircraft endurance} = \left(\frac{L}{D} \right) \frac{1 / g_0}{TSFC} \ell n \frac{W_i}{W_f} \tag{1.107}$$

The engine thrust-specific-fuel consumption appears in the denominator as in the engine impact on the range equation, and this time the aerodynamic figure of merit is aerodynamic efficiency, L/D, instead of ML/D, as expected for the aircraft endurance.

$$\text{Aircraft endurance} \propto \frac{1}{TSFC} \tag{1.108}$$

$$\text{Aircraft endurance} \propto \frac{L}{D} \tag{1.109}$$

1.5 Gas Generator

At the heart of an aircraft gas turbine engine *is a gas generator*. It is composed of three major components, a compressor, a burner (sometimes referred to as combustor or combustion chamber) followed by a turbine. The schematics of a gas generator and station numbers are shown in Figure 1.13. The parameters that define the physical characteristics of a gas generator are noted in Table 1.1.

Compressor total pressure ratio, π_c, is a design parameter. An aircraft engine designer has the design choice of the compressor staging, i.e. the number and type of compressor stages. That choice is a strong function of flight Mach number, or what we will refer to as *ram pressure ratio*. As a rule of thumb, the higher the flight Mach number, the lower the compressor pressure ratio the cycle requires to operate efficiently. In fact, at the upper supersonic Mach numbers, i.e. $M_0 \geq 3$, an air-breathing engine will not even require *any* mechanical compression, i.e. the compressor is totally unneeded. Such engines work on the principle of ram compression and are called *ramjets*. We will refer to this later on in our analysis.

The compressor air mass flow rate is the sizing parameter, which basically scales the engine face diameter. The takeoff gross weight of the aircraft is the primary parameter that most often *sizes* the engine. Other parameters, which contribute to engine sizing, are the engine-out rate of climb requirements, the transonic acceleration and the allowable use of afterburner, among other mission specification parameters. The burner fuel flow rate is the fuel energy release rate parameter, which may be replaced by the TET, T_{t4}. Both of these parameters establish a *thermal limit* identity for the

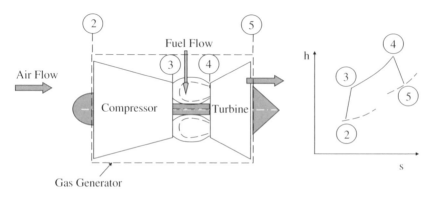

Figure 1.13 Schematic drawing of a gas generator.

Table 1.1 The parameters in a gas generator.

1.	Compressor pressure ratio, $\pi_c \equiv \dfrac{p_{t3}}{p_{t2}}$	["design" parameter]
2.	Compressor air mass flow rate, $\dot{m}_0$	["size" parameter]
3.	Combustor fuel flow rate, $\dot{m}_f$, or turbine entry temperature, T_{t4}	["temperature limit" parameter]
4.	Fuel lower heating value, Q_R	["ideal fuel energy" parameter]
5.	Component efficiency	["irreversibility" or loss parameter]

engine, which dictates the material and the cooling technologies to be employed in the engine *hot section* (i.e. the turbine and nozzle) at the design stage. Fuel heating value, or heat of reaction, Q_R, represents the (ideal) fuel energy density, i.e. the fuel thermal energy per unit mass of fuel. Finally, the component efficiencies are needed to describe the extent of losses or stated in thermodynamic language "irreversibility" in each component.

1.6 Engine Components

1.6.1 The Inlet

The *basic* function of the inlet is to deliver the air to the fan/compressor at the *right* Mach number, M_2, and the *right* quality, i.e. low distortion. The subsonic compressors are designed for an axial Mach number of $M_2 = 0.5 - 0.6$. Therefore, if the flight Mach number is higher than 0.5 or 0.6, which includes all commercial (fixed-wing) transports and military (fixed-wing) aircraft, then the inlet is required to *decelerate* the air efficiently. Therefore the main function of an inlet is to *diffuse or decelerate* the flow, and hence it is also called a diffuser. Flow deceleration is accompanied by the static pressure rise, or what is known as the *adverse pressure gradient* in fluid dynamics.

As one of the first principles of fluid mechanics, we learned that the boundary layers, being of a low-energy and momentum deficit zone, facing an adverse pressure gradient environment tend to separate. Therefore, one of the challenges facing an inlet designer is to prevent inlet boundary layer separation. One can achieve this by tailoring the geometry of the inlet to avoid rapid diffusion or possibly through variable geometry inlet design. Now, it becomes obvious why an aircraft inlet designer faces a bigger challenge if the inlet has to decelerate a Mach 2 or 3 stream to the compressor face Mach number of 0.5 than an aircraft that flies at Mach 0.8 or 0.9.

Figure 1.14 shows the thermodynamic states of air in an inlet, where state 0 corresponds to flight or freestream condition, and station 2 is at the exit of the intake system, which corresponds to fan/compressor inlet in a GT engine and combustor inlet in a ramjet.

Irreversibility in an inlet stems from viscous effects and presence of shocks in supersonic flows. The real flow is assumed to be adiabatic. Consequently, total enthalpy remains constant in an inlet and total pressure suffers a loss.

$$\frac{p_{t2}}{p_{t0}} = e^{-\frac{s_2 - s_0}{R}} = e^{-\frac{\Delta s}{R}} \tag{1.110}$$

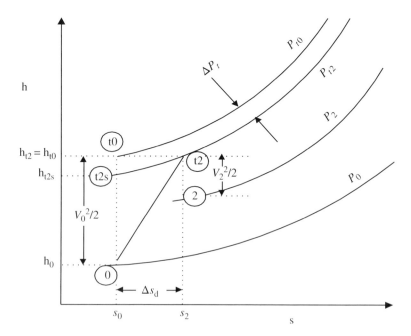

Figure 1.14 Enthalpy-entropy (h-s) diagram of an aircraft engine inlet flow under real and ideal conditions.

The ratio of p_{t2}/p_{t0} is the inlet total pressure ratio (often referred to as inlet recovery), π_d. We may define inlet adiabatic efficiency, η_d as

$$\eta_d \equiv \frac{h_{t2s} - h_0}{h_{t2} - h_0} = \frac{\left(V^2/2\right)_{ideal}}{V_0^2/2} \tag{1.111}$$

The practical form of the above definition is derived when we divide the numerator and denominator by h_0 to get:

$$\eta_d = \frac{\dfrac{h_{t2s}}{h_0} - 1}{\dfrac{h_{t2}}{h_0} - 1} = \frac{\dfrac{T_{t2s}}{T_0} - 1}{\dfrac{h_{t0}}{h_0} - 1} = \frac{\left(\dfrac{p_{t2}}{p_0}\right)^{\frac{\gamma-1}{\gamma}} - 1}{\dfrac{\gamma-1}{2} M_0^2} \tag{1.112}$$

where we have used the isentropic relation between the states $(t2s)$ and (0). Note that the only unknown in Eq. (1.112) is p_{t2}, for a given flight altitude, p_0, flight Mach number, M_0 and an inlet adiabatic efficiency, η_d. We can separate the unknown term, p_{t2}, and write the following expression:

$$\frac{p_{t2}}{p_0} = \left\{ 1 + \eta_d \frac{\gamma-1}{2} M_0^2 \right\}^{\frac{\gamma}{\gamma-1}} \tag{1.113}$$

It is interesting to note that Eq. (1.113) recovers the isentropic relation for a 100% efficient inlet, or $\eta_d = 1.0$. Another parameter, or a *figure of merit*, which describes the inlet performance is the total pressure ratio between the compressor face and the (total) flight condition. This is called π_d, and is often referred to as the *inlet total pressure recovery*:

$$\pi_d \equiv \frac{p_{t2}}{p_{t0}} \tag{1.114}$$

As expected, the two figures of merit for an inlet, i.e. η_d or π_d are not independent from each other and we can derive a relationship between η_d and π_d working the left-hand side of Eq. (1.113), as follows:

$$\frac{p_{t2}}{p_0} = \frac{p_{t2}}{p_{t0}} \frac{p_{t0}}{p_0} = \left\{ 1 + \eta_d \frac{\gamma-1}{2} M_0^2 \right\}^{\frac{\gamma}{\gamma-1}} \tag{1.115a}$$

$$\pi_d = \frac{\left\{ 1 + \eta_d \dfrac{\gamma-1}{2} M_0^2 \right\}^{\frac{\gamma}{\gamma-1}}}{\dfrac{p_{t0}}{p_0}} = \left\{ \frac{1 + \eta_d \dfrac{\gamma-1}{2} M_0^2}{1 + \dfrac{\gamma-1}{2} M_0^2} \right\}^{\frac{\gamma}{\gamma-1}} \tag{1.115b}$$

Therefore, Eq. (1.115b) relates the inlet total pressure recovery, π_d, to the inlet adiabatic efficiency, η_d, at any flight Mach number, M_0. We note that in Eq. (1.115b) as $\eta_d \to 1$, then $\pi_d \to 1$ as well, as expected. Figure 1.14 also shows the static state 2, which shares the same entropy as the total state, $t2$, and lies below it by the kinetic energy at 2, namely, $V_2^2/2$.

1.6.2 The Nozzle

The primary function of an aircraft engine exhaust system is to accelerate the gas efficiently. The nozzle parameter that is of utmost importance in *propulsion* is the gross thrust, F_g. The expression we had earlier derived for the gross thrust was

$$F_g = \dot{m}_9 V_9 + \left(p_9 - 9_0 \right) A_9 \tag{1.116}$$

In this equation, the first term on the RHS is called the momentum thrust and the second term is called the pressure thrust. It is interesting to note that the nozzles produce a "signature," typically composed of infrared radiation, thermal plume, smoke, and acoustic signatures, which are key design features of a stealth aircraft exhaust system *in addition* to the main propulsion requirement of the gross thrust.

As the fluid accelerates in a nozzle, the static pressure drops and hence a *favorable pressure gradient* environment is produced in the nozzle. This is in contrast to diffuser flows where an *adverse pressure gradient* environment prevails. Therefore, boundary layers are, by and large, well behaved in the nozzle and less cumbersome to treat than the inlet. For a subsonic exit Mach number, i.e. $M_9 < 1$, the nozzle expansion process will continue all the way to the ambient pressure, p_0. This important result means that in subsonic streams, the static pressure inside and outside of the jet are the same. In fact, there is no mechanism for a pressure jump in a subsonic flow, which is in contrast to the supersonic flows where shock waves and expansion fans allow for static pressure discontinuity. We have depicted a convergent nozzle in Figure 1.15 with its exhaust stream (i.e. a jet) emerging in the ambient gas of static state (0). The outer shape of the nozzle is called a boattail, which affects the *installed performance* of the exhaust system. The external aerodynamics of the nozzle installation belongs to the propulsion system integration studies and does not usually enter the discussions of the *internal performance,* i.e. the cycle analysis. However, we need to be aware that our decisions for the internal flow path optimization, e.g. the nozzle exit-to-throat area ratio, could have adverse effects on the installed performance, which may offset any gains that may have been accrued as a result of the internal optimization.

We have learned in aerodynamics that a convergent duct, as shown above, causes flow acceleration in a subsonic stream to a maximum Mach number of 1, which can only be reached at the minimum area of the duct, namely at its exit. So we stipulate that for all subsonic jets, i.e. $M_9 < 1$, there is a static pressure equilibrium in the exhaust stream and the ambient fluid, i.e. $p_9 = p_0$. Let us think of this as the *Rule 1* in nozzle flow.

> $Rule\,1: If\ M_{jet} < 1, then\ p_{jet} = p_{ambient.}$

The expression "jet" in the above rule should not be confusing to the reader, as it relates to the flow that emerges from the nozzle. In that context, M_{jet} is the same as M_9 and $p_{ambient}$ is the same as p_0. We recognize that not only it is entirely possible but also desirable for the sonic and supersonic jets to expand to the ambient static pressure as well. Such nozzle flows are called *perfectly expanded.* Actually the nozzle gross thrust can be maximized if the nozzle flow is perfectly expanded. We state this principle here without proof, but we will address it again, and actually prove it, in the next chapters. Let us think of this as the *Rule 2* in nozzle flow.

> $Rule\,2: If\ p_{jet} = p_{ambient}$, then we have a perfectly expanded nozzle which results in $F_{g,max}$.

Here, the stipulation is only on the *static pressure match* between the jet exit and the ambient static pressure and not *perfect flow* inside the nozzle. A real nozzle flow experiences total pressure loss due to viscous dissipation in the boundary layer as well as shock

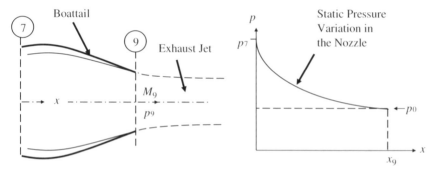

Figure 1.15 Schematic drawing of a subsonic nozzle with its static pressure distribution.

waves, and yet it is possible for it to *perfectly expand* the gas to the ambient condition. Remembering compressible duct flows in aerodynamics, we are reminded that the exit pressure, p_9, is a direct function of the nozzle area ratio, A_{exit}/A_{throat}, which in our notation, it becomes A_9/A_8, and the nozzle pressure ratio (NPR). Recalling the definition of NPR,

$$NPR \equiv \frac{p_{t7}}{p_0} \tag{1.117}$$

we will demonstrate a critical value of NPR that will result in the choking condition at the nozzle throat, i.e. $M_8 = 1.0$, when $NPR \geq (NPR)_{\text{critical}}$. Let us think of this as Rule 3 in convergent or convergent-divergent nozzle aerodynamics. These rules do not apply if the divergent section of a C-D nozzle becomes a subsonic a diffuser.

> $Rule\,3:$ *If* $NPR \geq (NPR)_{\text{critical}}$, then the nozzle throat velocity is sonic
> $\left(\text{i.e. choked}\right), M_8 = 1.0.$

A schematic drawing of a convergent-divergent supersonic nozzle is shown in Figure 1.16.

The nozzle throat becomes choked at a static pressure of about 50% of the nozzle total pressure, i.e. $p_{\text{throat}} \sim \frac{1}{2}\,p_{t7}$. This fact provides for an important *rule of thumb* in nozzle flows, which is definitely worth remembering. Let us consider this, as Rule 4.

> $Rule\,4:$ *If* $\dfrac{p_{t7}}{p_0} \geq \sim 2$, then the nozzle throat can be choked, i.e. $M_8 = 1.0.$

The two possible solutions in static pressure distribution along the nozzle axis are shown in Figure 1.16. One is a subsonic solution downstream of the nozzle throat and the second is a supersonic flow solution downstream of the throat. The subsonic solution is clearly a result of high backpressure, i.e. p_0, and causes a flow deceleration in the divergent duct downstream of the throat. Therefore, the divergent portion of the duct is actually a *diffuser* and not a nozzle. The supersonic solution is a result of low backpressure and therefore the flow continues to expand (i.e. accelerates) beyond the throat to supersonic speeds at the exit. Also note that only a perfectly expanded nozzle flow is shown, as the supersonic branch of the nozzle flow in Figure 1.16. For ambient pressures in between the two pressures shown in Figure 1.16, there are a host of shock solutions, which occur as an oblique shock at the lip or normal shock inside the nozzle.

To examine the efficiency of a nozzle in expanding the gas to an exit (static) pressure, p_9, we create an enthalpy–entropy diagram, very similar to an inlet. In the inlet studies, we took the ambient static condition, 0, and compressed it to the total intake exit condition, $t2$. But since a nozzle may be treated as a reverse-flow diffuser (and vice versa), we take the nozzle inlet gas, at the total state $t7$, and expand it to the exit static condition, 9. This process is shown in Figure 1.17. A solid line connecting the gas total state $t7$ to the exit static state, 9, shows the actual nozzle expansion process. As the real nozzle flows may still be treated as adiabatic, the total enthalpy, h_t, remains constant in a nozzle, i.e

$$h_{t7} = h_{t9} \tag{1.118}$$

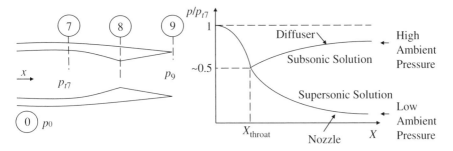

Figure 1.16 A choked C-D nozzle and the static pressure distribution along the nozzle axis.

An ideal exit state, 9s, is reached isentropically from the total state, t7, to the same exit pressure p_9, in Figure 1.17. The vertical gap between a total and static enthalpy states, in an h-s diagram, represents the kinetic energy of the gas, $V^2/2$, which is also shown in Figure 1.17. Due to frictional and shock losses in a nozzle, the flow will suffer a total pressure loss, i.e.

$$p_{t9} \prec p_{t7} \tag{1.119}$$

as depicted in Figure 1.17. We define the total pressure ratio across a nozzle as

$$\pi_n \equiv \frac{p_{t9}}{p_{t7}} \tag{1.120}$$

The entropy rise in an adiabatic nozzle is:

$$\frac{\Delta s_n}{R} = -\ln \pi_n \tag{1.121}$$

We define a nozzle adiabatic efficiency, η_n, very similar to the inlet adiabatic efficiency, as

$$\eta_n \equiv \frac{h_{t7} - h_9}{h_{t7} - h_{9s}} = \frac{V_9^2/2}{V_{9s}^2/2} \tag{1.122}$$

Let us interpret the above definition in physical terms. The fraction of an ideal nozzle exit kinetic energy, $V_{9s}^2/2$, which is realized in a real nozzle, $V_9^2/2$, is called the nozzle adiabatic efficiency. The loss in total pressure in a nozzle manifests itself as a loss of kinetic energy. Consequently, the adiabatic efficiency, η_n, which deals with the loss of kinetic energy in a nozzle, is related to the nozzle total pressure ratio, π_n, following:

$$\eta_n = \frac{1 - \dfrac{h_9}{h_{t7}}}{1 - \dfrac{h_{9s}}{h_{t7}}} = \frac{1 - \dfrac{h_9}{h_{t7}}}{1 - \left(\dfrac{p_9}{p_{t7}}\right)^{\frac{\gamma-1}{\gamma}}} = \frac{1 - \left(\dfrac{p_9}{p_{t9}}\right)^{\frac{\gamma-1}{\gamma}}}{1 - \left(\dfrac{p_9}{p_{t7}}\right)^{\frac{\gamma-1}{\gamma}}} \tag{1.123}$$

It is interesting to note that as the nozzle exit total pressure, p_{t9}, approaches the value of the nozzle inlet total pressure, p_{t7}, the nozzle adiabatic efficiency will approach 1. We

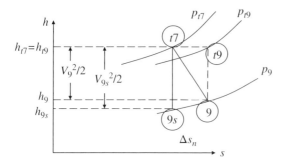

Figure 1.17 Enthalpy-entropy (h-s) diagram of nozzle flow expansion.

may treat Eq. (1.123) as an equation with one unknown, p_{t9}. Therefore, for a given nozzle exit static pressure, p_9, and the nozzle inlet total pressure, p_{t7}, the nozzle adiabatic efficiency, η_n, will result in a knowledge of the total pressure at the exit of the nozzle, p_{t9}. Now, if we multiply the numerator and denominator of the right-hand side of the Eq. (1.123) by $\left(p_{t7} / p_9 \right)^{\frac{\gamma-1}{\gamma}}$, we will reach our goal of relating the two figures of merit in a nozzle, i.e.

$$\eta_n = \frac{\left(\dfrac{p_{t7}}{p_9} \right)^{\frac{\gamma-1}{\gamma}} - \pi_n^{\frac{\gamma-1}{\gamma}}}{\left(\dfrac{p_{t7}}{p_9} \right)^{\frac{\gamma-1}{\gamma}} - 1} \tag{1.124}$$

There are three parameters in Eq. (1.124). The two figures of merit, η_n and π_n, and the ratio of nozzle inlet total pressure to the nozzle exit static pressure, p_{t7}/p_9. The last parameter is a known quantity, as we know the nozzle inlet total pressure, p_{t7}, from the upstream component analysis (in a turbojet, it is the turbine, $p_{t7} = p_{t5}$) and the nozzle exit pressure, p_9, is a direct function of the nozzle area ratio, A_9/A_8. To express the Eq. (1.124) in terms of the NPR, we may split p_{t7}/p_9 into

$$\frac{p_{t7}}{p_9} = \frac{p_{t7}}{p_0} \frac{p_0}{p_9} = NPR.\left(\frac{p_0}{p_9} \right) \tag{1.125}$$

Substituting the above expression in Eq. (1.124), yields:

$$\eta_n = \frac{\left\{ NPR\left(\dfrac{p_0}{p_9} \right) \right\}^{\frac{\gamma-1}{\gamma}} - \pi_n^{\frac{\gamma-1}{\gamma}}}{\left\{ NPR\left(\dfrac{p_0}{p_9} \right) \right\}^{\frac{\gamma-1}{\gamma}} - 1} \tag{1.126}$$

A plot of Eq. (1.126) is shown in Figure 1.18 for a perfectly expanded nozzle, i.e. $p_9 = p_0$.

In Eq. (1.126), the nozzle adiabatic efficiency will approach 1, as the nozzle total pressure ratio approaches 1. The parameter p_0/p_9 represents a measure of mismatch between the nozzle exit static pressure and the ambient static pressure. For $p_9 > p_0$ the flow is considered to be *underexpanded*. For $p_9 < p_0$, the flow is defined as *overexpanded*. In the underexpanded scenario, the nozzle area ratio is not adequate, i.e. not large enough, to expand the gas to the desired ambient static pressure. In the overexpanded nozzle flow case, the nozzle area ratio is too large for the perfect expansion. As noted earlier, a perfectly expanded nozzle will have $p_0 = p_9$; therefore, Eq. (1.126) is further simplified to:

$$\eta_n = \frac{\{NPR\}^{\frac{\gamma-1}{\gamma}} - \pi_n^{-\frac{\gamma-1}{\gamma}}}{\{NPR\}^{\frac{\gamma-1}{\gamma}} - 1} \tag{1.127}$$

For a perfectly expanded nozzle. It is instructive to show the three cases of nozzle expansion, i.e. under-, over-, and perfectly expanded cases, on an h-s diagram (see Figure 1.19).

We note in Figure 1.19, that all nozzle expansions are depicted as irreversible processes with an associated entropy rise; therefore *perfect expansion* is not to be mistaken as *perfect (isentropic) flow*.

In summary, we learned that:

- The primary function of a nozzle is to accelerate the gas efficiently.
- The gross thrust parameter, F_g, signifies nozzle's contribution to the thrust production.
- The gross thrust reaches a maximum when the nozzle is perfectly expanded; i.e. $p_9 = p_0$.
- Real nozzle flows may still be considered as adiabatic.
- A NPR that causes a Mach-1 flow at the throat (i.e. choking condition) is called the *critical nozzle pressure ratio*, and as a rule of thumb, we may remember an $(NPR)_{crit}$ of ~2.
- There are two efficiency parameters that quantify losses or the degree of irreversibility in a nozzle, and they are related.
- Nozzle losses manifest themselves as the total pressure loss.
- All subsonic exhaust streams have $p_{jet} = p_{ambient}$.

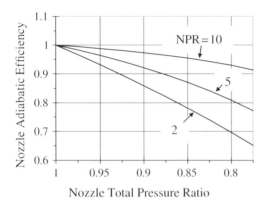

Figure 1.18 Two figures of merit in a nozzle plotted as a function of NPR ($\gamma = 1.33$).

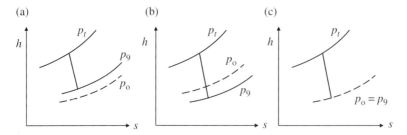

Figure 1.19 Enthalpy-entropy (h-s) diagram for the three possible nozzle expansions: (a) underexpanded; (b) overexpanded; (c) perfectly expanded.

- A perfect nozzle expansion means that the nozzle exit (static) pressure and the ambient pressure are equal.
- An imperfect nozzle expansion is caused by a mismatch between the nozzle area ratio and the altitude of operation.
- Underexpansion is caused by smaller-than-necessary nozzle area ratio, leading to $p_9 > p_0$.
- Overexpansion is caused by larger-than-necessary nozzle area ratio, leading to $p_9 < p_0$.

1.6.3 The Compressor

The thermodynamic process in a gas generator begins with the mechanical compression of air in the compressor. As the compressor discharge contains higher energy gas, i.e. the compressed air, it requires *external power* to operate. The power is delivered to compressor from the turbine via a shaft, as shown in Figure 1.2, in an operating gas turbine engine. Other sources of external power may be used to *start* the engine, which are in the form of electric motor, air-turbine, and hydraulic starters. The flow of air in a compressor is considered to be an essentially adiabatic process, which suggests that only a *negligible* amount of heat transfer takes place between the air inside and the ambient air outside the engine. Therefore, even in a real compressor analysis, we will still treat the flow as adiabatic.

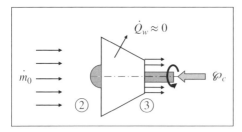

Perhaps a more physical argument in favor of neglecting the heat transfer in a compressor can be made by examining the order of magnitude of the energy transfer sources in a compressor. The power delivered to the medium in a compressor is achieved by one or more rows of rotating blades (called rotors) attached to one or more spinning shafts (typically referred to as *spools*). Each rotor blade, which changes the *spin* (or swirl) of a medium, will experience a countertorque as a reaction to its own action on the fluid.

If we denote the rotor torque as τ, then the power delivered to the medium by the rotor spinning at the angular speed, ω, follows Newtonian mechanics, namely

$$\wp = \tau \cdot \omega \tag{1.128}$$

A typical axial-flow compressor contains hundreds of rotor blades (assembled in several stages), which interact with the medium according to Eq. (1.128). Therefore, the rate of mechanical energy transfer in a typical modern compressor is usually measured in megaWatt (MW) and is several orders of magnitude larger than heat transfer through the compressor wall. Symbolically, we may present this as:

$$\wp_c >>> \dot{Q}_w \tag{1.129}$$

As a real process, however, the presence of wall friction acting on the medium through the boundary layer and shock waves caused by the relative supersonic flow through compressor blades will render the process inside the compressor irreversible. The measure of irreversibility in a compressor may be thermodynamically defined through some form of compressor efficiency. There are two methods of compressor efficiency definitions:

- compressor adiabatic efficiency, η_c
- compressor polytropic efficiency, e_c

To define the compressor adiabatic efficiency, η_c, we depict a "real" compression process on an h-s diagram, as shown in Figure 1.20 and compare it to an ideal, i.e. isentropic, process. The state $t2$ represents the *total* (or stagnation) state of the gas entering the compressor, typically designated by p_{t2} and T_{t2}. An actual flow in a compressor will follow the solid line from $t2$ to $t3$ thereby experiencing an entropy rise in the process, Δs_c. The actual total state of the gas is designated by $t3$ in Figure 1.20. The ratio p_{t3}/p_{t2} is known as the compressor pressure ratio, with a shorthand notation, π_c. The compressor total temperature ratio is depicted by the shorthand notation, $\tau_c \equiv T_{t3}/T_{t2}$. Since the state $t3$ is the actual state of the gas at the exit of the compressor and is not achieved via an isentropic process, we cannot expect the isentropic relation between τ_c and π_c, to hold, i.e.

$$\tau_c \neq \pi_c^{\frac{\gamma-1}{\gamma}} \tag{1.130}$$

Figure 1.20 Enthalpy-entropy (h-s) diagram of an actual and ideal compression process.

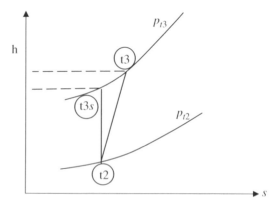

It can be seen from Figure 1.20 that the actual τ_c is higher than the ideal, i.e. isentropic, τ_c, which is denoted by the end state, T_{t3s}. This fact actually helps with the exponent memorization in a real compression process. The compressor adiabatic efficiency is the ratio of the ideal power required to the power consumed by the compressor, i.e.

$$\eta_c \equiv \frac{h_{t3s} - h_{t2}}{h_{t3} - h_{t2}} = \frac{\Delta h_{t,\text{isentropic}}}{\Delta h_{t,\text{actual}}} \tag{1.131}$$

The numerator in Eq. (1.131) is the power-per-unit mass flow rate in an *ideal compressor* and the denominator is the power-per-unit mass flow rate in the actual compressor. If we divide the numerator and denominator of Eq. (1.131) by h_{t2}, we get:

$$\eta_c = \frac{T_{t3s}/T_{t2} - 1}{T_{t3}/T_{t2} - 1} \tag{1.132}$$

Since the thermodynamic states $t3s$ and $t2$ are on the same isentrope, the temperature and pressure ratios are then related via the isentropic formula:

$$\frac{T_{t3s}}{T_{t2}} = \left(\frac{p_{t3s}}{p_{t2}}\right)^{\frac{\gamma-1}{\gamma}} = \left(\frac{p_{t3}}{p_{t2}}\right)^{\frac{\gamma-1}{\gamma}} = \pi_c^{\frac{\gamma-1}{\gamma}} \tag{1.133}$$

Therefore, compressor adiabatic efficiency may be expressed in terms of compressor pressure and temperature ratios as:

$$\eta_c = \frac{\pi_c^{\frac{\gamma-1}{\gamma}} - 1}{\tau_c - 1} \tag{1.134}$$

Equation (1.134) involves three parameters, η_c, π_c, and τ_c. It can be used to calculate τ_c for a given compressor pressure ratio and adiabatic efficiency. As compressor pressure ratio and efficiency are typically known (π_c is a design parameter and η_c is assumed efficiency), the only unknown in Eq. (1.134) is τ_c.

A second efficiency parameter in a compressor is the *polytropic efficiency, e_c*. As might be expected, compressor adiabatic and polytropic efficiencies are related. The definition of compressor polytropic efficiency is:

$$e_c \equiv \frac{dh_{ts}}{dh_t} \tag{1.135}$$

It is interesting to compare the definition of compressor adiabatic efficiency, involving finite jumps (Δh_t), and the polytropic efficiency, which takes infinitesimal steps (dh_t). The conclusion can be reached that the polytropic efficiency is actually the adiabatic efficiency of a compressor with *small* pressure ratio. Consequently, compressor polytropic efficiency is also called *small-stage efficiency*. From the combined first and second law of thermodynamics, we have

$$T_t ds = dh_t - \frac{dp_t}{\rho_t} \tag{1.136}$$

We deduce that for an isentropic process, i.e. $ds = 0$, $dh_t = dh_{ts}$ and therefore

$$dh_{ts} = \frac{dp_t}{\rho_t}$$

(1.137)

If we substitute Eq. (1.137) in (1.135) and replace density with pressure and temperature from the perfect gas law, we get

$$e_c = \frac{\frac{dp_t}{p_t}}{\frac{dh_t}{RT_t}} = \frac{\frac{dp_t}{p_t}}{\frac{C_p dT_t}{RT_t}} = \frac{\frac{dp_t}{p_t}}{\frac{\gamma}{\gamma-1}\frac{dT_t}{T_t}}$$

(1.138)

$$\frac{dp_t}{p_t} = \frac{\gamma e_c}{\gamma-1}\frac{dT_t}{T_t}$$

(1.139)

which can now be integrated between the inlet and exit of the compressor to yield:

$$\frac{p_{t3}}{p_{t2}} = \pi_c = \left(\frac{T_{t3}}{T_{t2}}\right)^{\frac{\gamma e_c}{\gamma-1}} = (\tau_c)^{\frac{\gamma e_c}{\gamma-1}}$$

(1.140)

To express the compressor total temperature ratio in terms of compressor pressure ratio and polytropic efficiency, Eq. (1.140) can be rewritten as:

$$\tau_c = \pi_c^{\frac{\gamma-1}{\gamma e_c}}$$

(1.141)

The presence of e_c in the denominator of the above exponent (Eq. (1.141)) causes the exponent of π_c to be larger than its isentropic exponent (which is $\frac{\gamma-1}{\gamma}$), therefore

$$\tau_{c,real} \succ \tau_{c,isentropic} \quad \text{or} \quad T_{t3} > T_{t3s}$$

(1.142)

The physical argument for higher actual T_t than the isentropic T_t (to achieve the same compressor pressure ratio) can be made on the ground that *lost work* to overcome the irreversibility in the real process (friction, shock) is converted into heat therefore, a higher exit T_t is reached in a real machine due to dissipation. On the other hand, for a given compressor pressure ratio, π_c, an ideal compressor consumes less power than an actual compressor (the factor being η_c, as defined earlier). Again, the absence of dissipative mechanisms, leading to *lost work*, is cited as the reason for a reversible flow machine to require less power to run.

We relate the two types of compressor efficiency description, e_c and η_c, as

$$\eta_c = \frac{\pi_c^{\frac{\gamma-1}{\gamma}} - 1}{\tau_c - 1} = \frac{\pi_c^{\frac{\gamma-1}{\gamma}} - 1}{\pi_c^{\frac{\gamma-1}{\gamma \cdot e_c}} - 1}$$

(1.143)

Equation (1.143) is plotted in Figure 1.21.

The compressor adiabatic efficiency, η_c, is a function of compressor pressure ratio, while the polytropic efficiency is independent of it. Consequently in a cycle analysis, we

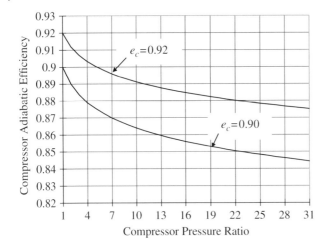

Figure 1.21 Variation of compressor adiabatic efficiency with the pressure ratio and polytropic efficiency.

usually assume the polytropic efficiency, e_c, as the figure of merit for a compressor (and turbine) and then we may keep e_c as constant in our engine off-design analysis. Typical values for the polytropic efficiency in modern compressors range in the 90–92%.

In summary, we have learned that:

- A real compressor flow may be considered adiabatic, i.e. $\dot{Q}_{compressor} \approx 0$.
- The energy transfer to the fluid due to shaft, in a compressor, is several orders of magnitude higher than any heat transfer that takes place through the casing; thus, heat transfer is neglected.
- Viscous dissipation in the wall boundary layer and shocks account for the sources of irreversibility in a compressor.
- Two figures of merit describe the compressor efficiency, the adiabatic compressor efficiency, η_c, and the polytropic or small-stage efficiency, e_c.
- The two compressor efficiencies are interrelated, i.e. $\eta_c = \eta_c(\pi_c, e_c)$.
- The compressor polytropic efficiency is independent of compressor pressure ratio, π_c.
- The compressor adiabatic efficiency is a function of π_c and decreases with increasing pressure ratio.
- To achieve a high pressure ratio in a compressor, multistaging and multispool configurations are needed.
- In a gas turbine engine, the compressor power is derived from a shaft that is connected to a turbine.

1.6.4 The Combustor

In the combustor, the air is mixed with the fuel and an *exothermic* chemical reaction ensues, resulting in a heat release. The ideal burner is considered to behave like a reversible heater, which means very slow burning, $M_b \approx 0$, and with no friction acting on its walls. Under such circumstances, the combustor total pressure remains constant.

In a real combustor, however, due to wall friction, turbulent mixing and chemical reaction at finite Mach number, the total pressure drops, i.e.

$$\pi_b = \frac{p_{t4}}{p_{t3}} \prec 1 \quad \text{"\textbf{Real} combustion chamber"} \tag{1.144}$$

$$\pi_b = 1 \quad \text{"\textbf{Ideal} combustion chamber"} \tag{1.145}$$

Kerrebrock (1992) gives an approximate expression for π_b in terms of the average Mach number of the gas in the burner, M_b, as

$$\pi_b \approx 1 - \varepsilon \frac{\gamma}{2} M_b^2 \quad \text{where} \quad 1 \prec \varepsilon \prec 2 \tag{1.146}$$

The total pressure loss in a burner is then proportional to the average dynamic pressure of the gases inside the burner, i.e. $\propto \frac{\gamma}{2} M_b^2$, where the proportionality coefficient is ε. Assuming an average Mach number of gases of 0.2 and $\varepsilon = 2$, we get $\pi_b \approx 0.95$ (for a $\gamma \approx 1.33$). This formula is obviously not valid for supersonic combustion throughflow; rather, it points out the merits of *slow* combustion in a conventional burner. A schematic diagram of a combustion chamber, and its essential components, is shown in Figure 1.22.

A preliminary discussion of the components of the combustion chamber is useful at this time. The inlet diffuser decelerates the compressor discharge flow to a Mach number of about 0.2–0.3. The low-speed flow will provide an efficient burning environment in the combustor. Mixing improvement with the fuel in the combustor primary zone is achieved via the air swirler. A recirculation zone is created that provides the necessary stability in the primary combustion zone (i.e. by increasing residence time). To create a fuel-rich environment in the primary zone to sustain combustion, a large percentage of air is diverted around a dome-like structure. The airflow that has bypassed the burner primary zone will enter the combustor mainly as the cooling flow through a series of cooling and dilution holes (as shown). The fuel–air mixture is ignited in the combustor primary zone via an igniter properly positioned in the dome area.

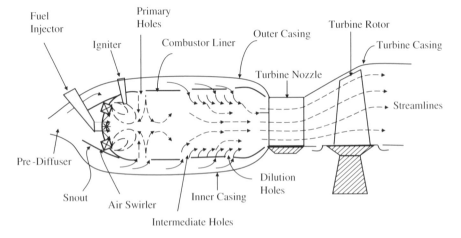

Figure 1.22 Components of a conventional combustion chamber and the first turbine stage.

For the purposes of *cycle analysis*, a combustor flow is only analyzed at its inlet and outlet. Thus, we will not consider the details of combustion processes such as atomization, vaporization, mixing, chemical reaction, and dilution in the cycle analysis phase. Nor do we consider pollutant formation and the means of reducing them in this chapter. A block diagram representation of a combustor, useful in cycle analysis, is shown in Figure 1.23.

Figures 1.23a,b are the steady-state mass and the energy balance applied to the combustion chamber, i.e.

$$\dot{m}_4 = \dot{m}_0 + \dot{m}_f = \dot{m}_0 (1 + f) \tag{1.147}$$

where f is the fuel-to-air ratio, $f \equiv \dfrac{\dot{m}_f}{\dot{m}_0}$, and energy balance in Figure 1.23b gives:

$$\dot{m}_0 h_{t3} + \dot{m}_f Q_R \eta_b = (\dot{m}_0 + \dot{m}_f) h_{t4} = \dot{m}_0 (1 + f) h_{t4} \tag{1.148}$$

The fuel is characterized by its *energy content per unit mass*, i.e. the amount of thermal energy inherent in the fuel, capable of being released in a chemical reaction. This parameter is heat of reaction and is given the symbol, Q_R. The unit for this parameter is energy/mass, which in the metric system is kJ kg^{-1} and in the British system of units is BTU lbm^{-1}. In an actual combustion chamber, primarily due to volume limitations, the entirety of the Q_R cannot be realized. The fraction that can be realized is called *burner efficiency* η_b. Therefore,

$$\eta_b \equiv \frac{Q_{R,\text{Actual}}}{Q_{R,\text{Ideal}}} \tag{1.149}$$

The ideal heat of reaction or lower heating value of typical hydrocarbon fuels (e.g. Jet-A), to be used in our cycle analysis, are:

$$Q_R \cong 43000 \text{kJ kg}^{-1} \quad \text{or} \quad Q_R \cong 18600 \text{BTU lbm}^{-1} \tag{1.150a}$$

However, the most energetic fuel (on mass basis) is the hydrogen, which is capable of releasing roughly three times the energy of typical hydrocarbon fuels per unit mass, i.e.

$$Q_R \cong 127500 \text{kJ kg}^{-1} \quad \text{or} \quad Q_R \cong 55400 \text{BTU lbm}^{-1} \tag{1.150b}$$

However, there are major drawbacks in fuels such as hydrogen. The low molecular weight of hydrogen makes it the lightest fuel, even in liquid form (LH$_2$), with a density

(a)   (b)

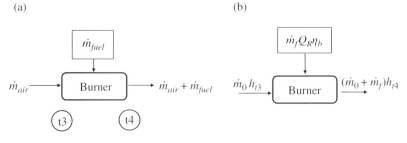

Figure 1.23 Block diagram of a burner with mass and energy balance: (a) mass balance; (b) energy balance.

ratio of about 1/10 of typical hydrocarbon fuels, such as Jet-A. This implies a comparatively very large volume requirement for LH_2. Consequently, *volumetric efficiency* (energy/volume) of the hydrogen is the lowest of all fuels. Second, hydrogen in liquid form is cryogenic, which means a very low boiling point temperature, i.e. $-423°F$ or $20K$ at ambient pressure. The cryogenic aspect of hydrogen requires thermally insulated fuel tanks, fuel lines, valves, and the associated weight penalty. Therefore, due to space/volume limitations and fuel system requirement, current commercial aviation uses fossil fuel (e.g. Jet-A) instead of hydrogen. In hypersonic air-breathing propulsion as well as in rockets, LH_2 has been a fuel of choice since its cryogenic property can be used in regenerative cooling of the vehicle.

Typically in modern gas turbine engines, the burner efficiency can be as high as 98–99%. In a cycle analysis, we need to make assumptions about the loss parameters in every component, which in a combustion chamber are: $\pi_b < 1$ and $\eta_b < 1$. The real and ideal combustion process can be depicted on a T-s diagram, which later will be used to perform cycle analysis. Figure 1.24 shows the burner thermodynamic process on a T-s diagram.

The isobars, p_{t3} and p_{t4}, drawn at the entrance and exit of the combustor in Figure 1.24, clearly show a total pressure drop in the burner, $\Delta p_{t, burner}$. The maximum temperature limit, T_{t4} is governed by the level of cooling technology, material selection and the thermal protective coating used in the turbine. Typical current values for the maximum T_{t4} is about 3200–3600°R or 1775–2000 K.

Another burner parameter is the temperature rise, ΔT_t, across the combustion chamber, as shown in Figure 1.24. The thermal power invested in the engine (by the fuel) is proportional to the temperature rise across the combustor, i.e. it is nearly equal to: $\dot{m}_0 c_p \left(\Delta T_t \right)_{burner}$.

The application of energy balance across the burner will yield the fuel-to-air ratio, f, as the only unknown parameter. To derive an expression for the fuel-to-air ratio, we will divide Eq. (1.148) by $\dot{m}_0$, the air mass flow rate, to get:

$$h_{t3} + f Q_R \eta_b = \left(1 + f \right) h_{t4} \tag{1.151}$$

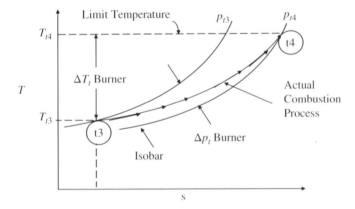

Figure 1.24 Actual flow process in a burner (note total pressure loss, Δp_t, across the burner).

The unknown parameter, f, can be isolated and expressed as:

$$f = \frac{h_{t4} - h_{t3}}{Q_R \eta_b - h_{t4}} \tag{1.152}$$

Knowing the fuel property, Q_R, assuming burner efficiency, η_b, and having specified a turbine inlet temperature, T_{t4}, the denominator of Eq. (1.152) is fully known. The compressor discharge temperature, T_{t3}, is established via compressor pressure ratio, efficiency, and inlet condition, as described in the compressor section, which renders the numerator of Eq. (1.152) fully known as well. Therefore, application of energy balance to a burner, usually results in the establishment of fuel-to-air ratio parameter, f. It is customary to express Eq. (1.152) in terms of nondimensional parameters, by dividing each term in the numerator and denominator by the flight static enthalpy, h_0, to get:

$$f = \frac{\dfrac{h_{t4}}{h_0} - \dfrac{h_{t3}}{h_0}}{\dfrac{Q_R \eta_b}{h_0} - \dfrac{h_{t4}}{h_0}} = \frac{\tau_\lambda - \tau_r \tau_c}{\dfrac{Q_R \eta_b}{h_0} - \tau_\lambda} \tag{1.153}$$

where we recognize the product $\tau_r \tau_c$ as h_{t3}/h_0 and τ_λ as the cycle thermal limit parameter, h_{t4}/h_0.

In summary, we learned that:

- The fuel is characterized by its lower heating value, Q_R (maximum releasable thermal energy per unit mass).
- The burner is characterized by its efficiency, η_b, and its total pressure ratio, π_b.
- Burning at finite Mach number, frictional losses on the walls, and turbulent mixing are identified as the sources of irreversibility, i.e. losses, in a burner.
- The fuel-to-air ratio, f, and the burner exit temperature, T_{t4}, are the thrust control/ engine design parameters.
- The application of the energy balance across the burner yields either f or T_{t4}.

1.6.5 The Turbine

The high pressure and temperature gas that leaves the combustor is directed into a turbine. The turbine may be thought of as a *valve*, because on one side, it has a high-pressure gas and on the other side, it has a low-pressure gas of the exhaust nozzle or the tailpipe. Therefore the first *valve*, i.e. the throttle station, in a gas turbine engine is at the turbine. The throat of an exhaust nozzle in a supersonic aircraft is the second and final throttle station in an engine. Thus, the flow process in a turbine (and exhaust nozzle) involves significant (static) pressure drop, and in harmony with it, the (static) temperature drop, which is called *flow expansion*. The flow expansion produces the necessary power for the compressor and the propulsive power for the aircraft. The turbine is connected to the compressor via a common shaft (see Figure 1.25), which provides the shaft power to the compressor. In drawing an analogy, we can think of the expansion process in a gas turbine engine as the counterpart of *power stroke* in an intermittent combustion engine. However in a turbine, the power transmittal is continuous.

Due to high temperatures of the combustor exit flow, e.g. 1600–2000 K, the first few stages of the turbine, i.e. the HPT, need to be cooled. The coolant is the air bleed from

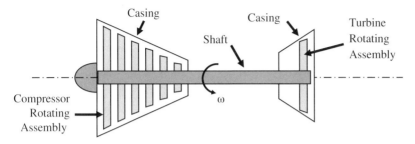

Figure 1.25 Common shaft in a gas generator connects the compressor and turbine (ω is the shaft angular speed in rad s^{-1} or rpm).

the compressor, which may be bled from different compression stages, e.g. between the LPC and HPC and/or at the compressor exit. It has been customary however to analyze an *uncooled* turbine in the preliminary cycle analysis, then followed by an analysis of the cycle with cooling effects in the turbine and the exhaust nozzle. Other cooling media, such as water, have been used in stationary gas turbine power plants. However, in a design-to-weight environment of aircraft, carrying extra water to cool the turbine blades is not feasible. A cooling solution, which uses the engine cryogenic fuel, such as hydrogen or methane, as the coolant to cool the engine and aircraft components is called *regenerative cooling* and has proven its effectiveness in liquid propellant chemical rocket engines for decades.

A real flow process in an uncooled turbine involves irreversibility such as frictional losses in the boundary layer, tip clearance flows, and shock losses in transonic turbine stages. The viscous dominated losses, i.e. boundary layer separation, reattachment, and tip vortex flows, are concentrated near the end walls; thus, a special attention in turbine flow optimization is made on the *end-wall* regions. Another source of irreversibility in a real turbine flow is related to the cooling losses. Coolant is typically injected from the blade attachment (to the hub or casing) into the blade, which provides internal convective cooling and usually external film cooling on the blades. The turbulent mixing associated with the coolant stream and the hot gases is the primary mechanism for (the turbine stage) cooling losses.

The thermodynamic process for an uncooled turbine flow may be shown in an h-s diagram (see Figure 1.26). The actual expansion process in the turbine is depicted by the solid line connecting the total (or stagnation) states *t4* and *t5* (see Figure 1.26). The isentropically reached exit state, *t5s*, represents the ideal, loss-free flow expansion in the turbine to the same backpressure, p_{t5}. The relative height, on the enthalpy scale in Figure 1.26, between the inlet total and outlet total condition of the turbine represents the power production potential (i.e. ideal) and the actual power produced in a turbine. The ratio of these two heights is called the turbine adiabatic efficiency, η_t, i.e.

$$\wp_{t,actual} = \dot{m}_t \left(h_{t4} - h_{t5} \right) = \dot{m}_t \Delta h_{t,actual} \tag{1.154}$$

$$\wp_{t,ideal} = \dot{m}_t \left(h_{t4} - h_{t5s} \right) = \dot{m}_t \Delta h_{t,isentropic} \tag{1.155}$$

$$\eta_t \equiv \frac{h_{t4} - h_{t5}}{h_{t4} - h_{t5s}} = \frac{\Delta h_{t,actual}}{\Delta h_{t,isentropic}} \tag{1.156}$$

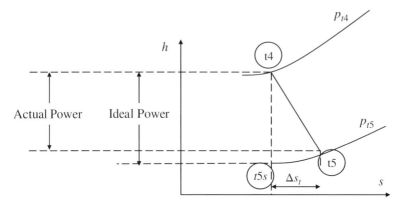

Figure 1.26 Expansion process in an uncooled turbine (note the entropy rise across the turbine, Δs_t).

In Eqs. (1.154) and (1.155), the turbine mass flow rate is identified as $\dot{m}_t$, which accounts for the air and fuel mass flow rate that emerge from the combustor and expand through the turbine, i.e.

$$\dot{m}_t = \dot{m}_0 + \dot{m}_f = (1+f)\dot{m}_0 \tag{1.157}$$

The numerator in Eq. (1.156) is the actual power produced in a *real uncooled turbine* and the denominator is the *ideal* power that in a reversible and adiabatic turbine could be produced. If we divide the numerator and denominator of Eq. (1.156) by h_{t4}, we get:

$$\eta_t = \frac{1 - T_{t5}/T_{t4}}{1 - T_{t5s}/T_{t4}} \tag{1.158}$$

Since the thermodynamic states $t5s$ and $t4$ are on the same isentrope, the temperature and pressure ratios are then related via the isentropic formula, i.e.

$$\frac{T_{t5s}}{T_{t4}} = \left(\frac{p_{t5s}}{p_{t4}}\right)^{\frac{\gamma-1}{\gamma}} = \left(\frac{p_{t5}}{p_{t4}}\right)^{\frac{\gamma-1}{\gamma}} = \pi_t^{\frac{\gamma-1}{\gamma}} \tag{1.159}$$

Therefore, turbine adiabatic efficiency may be expressed in terms of the turbine total pressure and temperature ratios as:

$$\eta_t = \frac{1 - \tau_t}{1 - \pi_t^{\frac{\gamma-1}{\gamma}}} \tag{1.160}$$

We may also define a *small-stage* efficiency for a turbine, as we did in a compressor, and call it the turbine polytropic efficiency, e_t. For a small expansion, representing a small stage, we can replace the finite jumps, i.e. Δs, with incremental step, d, in Eq. (1.156) and write:

$$e_t \equiv \frac{dh_t}{dh_{ts}} = \frac{dh_t}{\frac{dp_t}{\rho_t}} \tag{1.161}$$

Expressing enthalpy in terms of temperature and specific heat at constant pressure as $dh_t = C_p dT_t$, which is suitable for a perfect gas, and simplify Eq. (1.161) similar to the compressor section, we get:

$$\tau_t = \pi_t^{\frac{(\gamma-1)e_t}{\gamma}} \quad \text{or} \quad \pi_t = \tau_t^{\frac{\gamma}{(\gamma-1)e_t}} \tag{1.162}$$

$$\tau_t \equiv T_{t5}/T_{t4} \tag{1.163}$$

$$\pi_t \equiv p_{t5}/p_{t4} \tag{1.164}$$

We note that in Eq. (1.162), in the limit of e_t approaching 1, i.e. isentropic expansion, we will recover the isentropic relationship between the temperature and pressure ratio, as expected. Replacing π_t in Eq. (1.159) by its equivalent expression (from Eq. (1.162)), we derive a relation between the two types of turbine efficiencies, η_t and e_t, as

$$\eta_t = \frac{1-\tau_t}{1-\tau_t^{1/e_t}} \tag{1.165}$$

Equation (1.165) is plotted in Figure 1.27. For a small-stage efficiency of 90%, i.e. $e_t = 0.90$, the turbine adiabatic efficiency, η_t, grows with the inverse of turbine expansion parameter, $1/\tau_t$.

The temperature ratio parameter, τ_t, across the turbine is established via a power balance between the turbine, compressor, and other shaft power extraction (e.g. electric generator) on the gas generator. Let us first consider the power balance between the turbine and compressor in its simplest form and then try to build on the added parameters. Ideally, the compressor absorbs all the turbine shaft power, i.e.

$$\wp_t = \wp_c \tag{1.166a}$$

$$\dot{m}_0(1+f)(h_{t4}-h_{t5}) = \dot{m}_0(h_{t3}-h_{t2}) \tag{1.166b}$$

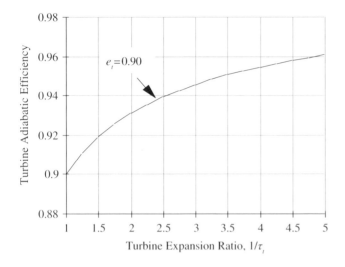

Figure 1.27 Variation of turbine adiabatic efficiency, η_t, with the inverse of turbine expansion ratio, $1/\tau_t$.

which simplifies to the following nondimensional form:

$$(1+f)\tau_\lambda(1-\tau_t)=\tau_r(\tau_c-1) \tag{1.166c}$$

The only unknown in the above equation is τ_t, as all other parameters either flow from upstream components, e.g. combustor will provide f, the compressor produces τ_c, etc. or are design parameters such as τ_λ or τ_r. Hence, the turbine expansion parameter, τ_t, is:

$$\tau_t=1-\frac{\tau_r(\tau_c-1)}{(1+f)\tau_\lambda} \tag{1.166d}$$

Next, we consider the practical issue of hydrodynamic (frictional) losses in bearings holding the shaft in place and provide dynamic stability under operating conditions to the rotating assemblies of turbine and compressor. Therefore, a small fraction of the turbine power output is dissipated through viscous losses in the bearings, i.e.

$$\wp_t=\wp_c+\Delta\wp_{\text{bearings}} \tag{1.167}$$

Where, $\Delta\wp_{\text{bearings}}$ is the power loss due to bearings. In addition, an aircraft has electrical power needs for its flight control system and other aircraft subsystems, which requires tapping into the turbine shaft power. Consequently, the power balance between the compressor and turbine should account for the electrical power extraction, which usually accompanies the gas generator, i.e.

$$\wp_t=\wp_c+\Delta\wp_{\text{bearing}}+\Delta\wp_{\text{electric generator}} \tag{1.168a}$$

$$\wp_c=\wp_t-\Delta\wp_{\text{bearing}}-\Delta\wp_{\text{electric generator}} \tag{1.168b}$$

In a simple cycle analysis, it is customary to lump all power dissipation and power extraction terms into a single *mechanical efficiency* parameter, η_m, that is multiplied by the turbine shaft power, to derive the compressor shaft power, i.e.

$$\wp_c=\eta_m\wp_t \tag{1.169}$$

where η_m is the mechanical efficiency parameter, which is assumed, e.g. $\eta_m=0.995$.

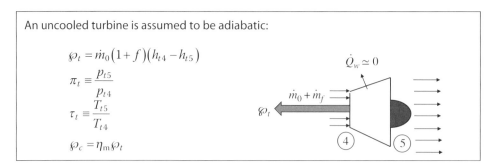

An uncooled turbine is assumed to be adiabatic:

$$\wp_t=\dot{m}_0(1+f)(h_{t4}-h_{t5})$$

$$\pi_t\equiv\frac{p_{t5}}{p_{t4}}$$

$$\tau_t\equiv\frac{T_{t5}}{T_{t4}}$$

$$\wp_c=\eta_m\wp_t$$

$$\dot{Q}_w\simeq0$$

$$\dot{m}_0+\dot{m}_f$$

To cool the high-pressure turbine stages, a small fraction of compressor air can be diverted from various stages of the compression. Engine cooling is essentially a *pressure-driven* process, which calls for a pressure scheduling of the coolant to achieve the highest cooling efficiency. For example, to cool the first nozzle and the first rotor in a turbine, we need to tap the compressor exit air, as it has the *right* pressure. Earlier stages of the compressor have not yet developed the necessary pressure to overcome downstream pressure in the high-pressure end of a turbine. To cool the medium-pressure turbine stages, we need to divert compressor air from the medium-pressure compressor. Therefore, the problem of *pressure matching* between the coolant stream and the hot gas in a turbine is real and significant. The consequence of a mismatch is either inadequate cooling flow in a turbine blade with a possibility of even a *reverse flow* of hot gases into the cooling channels or the need for excessive throttling of the coolant stream and associated cycle losses. Figure 1.28 shows the schematic drawing of turbine blade-casing cooling.

We can simplify the investigation of the thermodynamics of cooled turbine stages significantly by assuming that each blade row coolant is discharged at the blade row exit, i.e. trailing edge ejection. This simplifies the thermodynamics as we consider *dumping* the entire coolant used in a given blade row at the trailing edge of the blades and then apply conservation laws to the coolant-hot gas mixing process. This approach avoids the discrete coolant-hot gas mixing that may be distributed over the suction and pressure surfaces of the blade row, as in film-cooled blades. Application of the energy equation to the coolant-hot gas mixing process will result in a lower mixed-out temperature for the hot gas. Consequently, the hot-gas stream will experience a reduced entropy state after mixing with the coolant. In return, the coolant stream, which is heated by the hot gas, will experience an entropy rise. The expansion process in a turbine with two cooled stages followed by an uncooled low-pressure turbine is depicted in an h-s diagram in Figure 1.29. The mixing process at the exit of each blade row is identified with negative entropy production and the cooling path is roughly drawn in as a nearly constant pressure process. In contrast, the coolant stream will undergo a heating process from the blade and then through mixing with the hot gases in the turbine. The thermodynamic state of the coolant is shown in Figure 1.29.

The h-s diagram for the hot gas (shown in Figure 1.29) shows a constant h_t process for the first and second turbine nozzle. Constant total enthalpy indicates an adiabatic process with no mechanical energy exchange. These two criteria are met in an uncooled turbine nozzle, which is stationary and consequently exchanges no mechanical energy with the fluid. Across the nozzle, due to frictional and shock losses, a total pressure

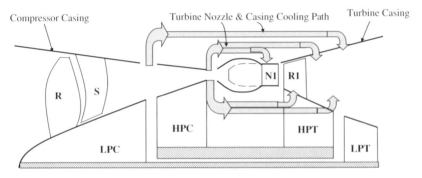

Figure 1.28 Two-spool GT engine with two stations of compressor bleed for cooling purposes (LPC: low-pressure compressor, HPC: high-pressure compressor, HPT: high-pressure turbine, LPT: low-pressure turbine, R: rotor, S: stator, N: nozzle).

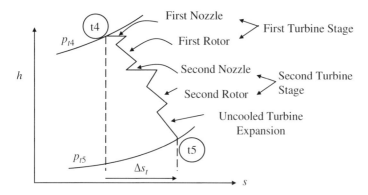

Figure 1.29 Enthalpy-entropy (h-s) diagram for a turbine with two cooled stages followed by an uncooled LPT (A negative Δs and a temperature drop indicate blade row exit mixing with the coolant).

drop is indicated in Figure 1.29. The cooling process results in a lower total temperature of the mixture of hot and cold gas, as well as a total pressure drop due to turbulent mixing of the two streams. This process can be seen in Figure 1.30. The drop of total enthalpy across the rotor is proportional to the shaft power production by the rotor blade row. This shaft power may be written as:

$$\wp_{\text{rotor}} = \tau_{\text{rotor}} \omega = \dot{m}_{\text{rotor}} \Delta h_t \tag{1.170}$$

The product of rotor torque τ, and the angular speed, ω, of the rotor is the shaft power. The irreversibility in the flow through the rotor is shown via a process involving entropy rise across the rotor in Figure 1.30. Again, the flow expansion after the rotor is cooled by the blade row coolant. To establish the mixed-out state of the coolant and the hot gas, we will apply the conservation principles, as noted earlier. The law of conservation of energy applied to a coolant stream of $\dot{m}_c$ mass flow rate carrying a total enthalpy of $h_{t,c}$ mixing with a hot gas stream of $\dot{m}_g$ with a total enthalpy of $h_{t,g}$ is written as:

$$\dot{m}_c h_{t,c} + \dot{m}_g h_{t,g} = \left(\dot{m}_c + \dot{m}_g \right) h_{t,\text{mixed-out}} \tag{1.171}$$

where the only unknown is the mixed-out enthalpy state of the coolant-hot gas mixture, $h_{t,\text{mixed-out}}$. In nondimensional form, Eq. (1.171) is:

$$\frac{h_{t,\text{mixed out}}}{h_{t,g}} = \frac{1 + \dfrac{\dot{m}_c}{\dot{m}_g} \dfrac{h_{t,c}}{h_{t,g}}}{1 + \dfrac{\dot{m}_c}{\dot{m}_g}} \approx \left(1 - \frac{\dot{m}_c}{\dot{m}_g} \right) \left(1 + \frac{\dot{m}_c}{\dot{m}_g} \frac{h_{t,c}}{h_{t,g}} \right) \tag{1.172}$$

In Eq. (1.172), two nondimensional parameters emerge that basically govern the energetics of a hot and a cold mixture:

- $\dfrac{\dot{m}_c}{\dot{m}_g}$ The coolant mass fraction (typically ~10%), and

- $\dfrac{h_{t,c}}{h_{t,g}}$ Cold-to-hot total enthalpy ratio (typically, ~0.5)

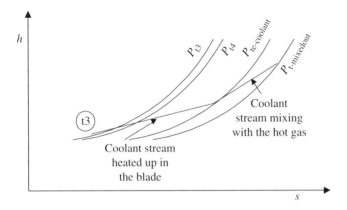

Figure 1.30 h-s diagram of the coolant stream, first inside the blade, then mixing with the hot turbine gas.

We have also used a mathematical approximation in Eq. (1.172), in the form of binomial expansion:

$$\frac{1}{1+\varepsilon} \approx 1 - \varepsilon \tag{1.173}$$

where $\varepsilon \ll 1$ and the largest neglected term in the series is $O(\varepsilon^2)$.

The coolant stream experiences a total pressure drop in the cooling channels of the blade as a result of frictional losses and a total pressure drop as a result of turbulent mixing with the hot gas. The enthalpy-entropy diagram for the coolant stream is depicted in Figure 1.30.

In summary we learned that:

- The flow expansion in the turbine produces the needed shaft power for the compressor and other propulsion system needs, e.g. an electric generator.
- There are two figures of merit in a turbine, which measure the extent of irreversibility in a turbine, η_t and e_t, and they are related.
- The gas expansion in an uncooled turbine is treated as adiabatic.
- The frictional losses on the blades and the casing as well as any shock losses, in relative supersonic passages, are the sources of irreversibility in an uncooled turbine.
- Turbulent mixing losses between the coolant and the hot gas is an added source of irreversibility in a cooled turbine.
- The TET, T_{t4}, is a design parameter that sets the stage for the turbine material and cooling requirements.
- The power balance between the turbine, the compressor, and other known power drainage, establishes the turbine expansion ratio, τ_t.
- Cooling of the high-pressure turbine is achieved through compressor air bleed that is injected through the blade root in the rotor and the casing for the turbine nozzle.
- The turbine nozzle is choked (i.e. the throat Mach number is one), over a wide operating range of the engine, and as such is the first *throttle station* of the engine.

1.7 Performance Evaluation of a Turbojet Engine

The TJ performance parameters are:

- Specific thrust, $F/\dot{m}$
- Thrust-specific fuel consumption, TSFC, or specific impulse, I_s
- Thermal, propulsive, and overall efficiencies

Our approach to calculating the performance parameters of an aircraft engine is to *march through* an engine, component by component, until we calculate the target parameters, which are:

- Fuel-to-air ratio, f
- Exhaust velocity, V_9

Once we establish these unknowns, i.e. V_0, f, and V_9, the specific thrust, as a figure of merit for an aircraft engine, may be written as:

$$\frac{F_n}{\dot{m}_0} = (1+f)V_9 - V_0 + \frac{(p_9 - p_0)A_9}{\dot{m}_0} \tag{1.174}$$

This can be expressed in terms of the calculated parameters by recognizing that the air mass flow rate is only a factor $(1+f)$ away from the exhaust mass flow rate, i.e.

$$\dot{m}_0 = \frac{\dot{m}_9}{1+f} = \frac{\rho_9 A_9 V_9}{1+f} = \frac{p_9 A_9 V_9}{RT_9(1+f)} \tag{1.175}$$

Now, substituting Eq. (1.175) in (1.174), we get:

$$\frac{F_n}{\dot{m}_0} = (1+f)V_9 - V_0 + \frac{(p_9 - p_0)A_9}{\dfrac{p_9 A_9 V_9}{RT_9(1+f)}} = (1+f)V_9 - V_0 + \frac{RT_9(1+f)}{V_9}\left(1 - \frac{p_0}{p_9}\right) \tag{1.176a}$$

Note that we have calculated all the terms on the RHS of the Eq. (1.176a), namely, f, V_9, V_0, and T_9. Therefore,

$$\frac{F_n}{\dot{m}_0} = (1+f)V_9 - V_0 + \frac{RT_9(1+f)}{V_9}\left(1 - \frac{p_0}{p_9}\right) \tag{1.176b}$$

The primary contribution to the specific thrust in an air-breathing engine comes from the first two terms in Eq. (1.176b), i.e. the momentum contribution. The last term would identically vanish if the nozzle is perfectly expanded, i.e. $p_9 = p_0$. Otherwise, its contribution is small compared to the momentum thrust.

It is very tempting to insert γ in the numerator and denominator of the last term in Eq. (1.176b), and identify the γRT_9 as the a_9^2 to recast the above equation in a more elegant form:

$$\frac{F_n}{\dot{m}_0} = (1+f)V_9\left(1 + \frac{1}{\gamma M_9^2}\left(1 - \frac{p_0}{p_9}\right)\right) - V_0 \tag{1.176c}$$

Thrust-specific fuel consumption was defined in a turbojet engine as:

$$TSFC \equiv \frac{\dot{m}_f}{F_n} = \frac{f}{F_n / \dot{m}_0} \tag{1.177}$$

and we have calculated the fuel-to-air ratio and the specific thrust, which are combined as in Eq. (1.177) to create the fuel efficiency parameter, or figure of merit, of the engine. The engine overall efficiency can be recast in terms of the TSFC:

$$\eta_o = \frac{(F_n / \dot{m}_0) V_0}{f Q_R} = \frac{V_0 / Q_R}{TSFC} \tag{1.178a}$$

The inverse proportionality between the engine overall efficiency and the thrust-specific fuel consumption, i.e. the lower the specific fuel consumption, the higher the engine overall efficiency, is noted in Eq. (1.178a), i.e.

$$\eta_o \propto \frac{1}{TSFC} \tag{1.178b}$$

In concluding the section on turbojet engines, it is appropriate to show a *real cycle* depicted on an T-s diagram. This is shown in Figure 1.31a,b.

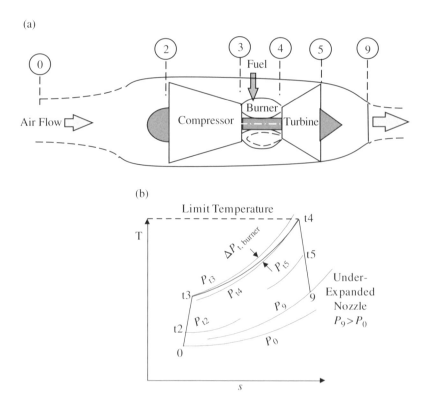

Figure 1.31 T-s diagram of a turbojet engine with component losses and Imperfect expansion in the nozzle: (a) a turbojet engine; (b) thermodynamic cycle of a turbojet engine.

1.8 Turbojet Engine with an Afterburner

1.8.1 Introduction

To augment the thrust of an aircraft gas turbine engine, an afterburner can be used. This solution is often sought in military aircraft, primarily due to its simplicity and effectiveness. It is interesting to note that an afterburner has the potential of nearly doubling the thrust produced in an aircraft gas turbine engine while in the process more than quadruple the engine fuel consumption rate. Figure 1.32 shows the schematics of an afterburning turbojet (AB-TJ) engine. The new station numbers 7 and 8 refer to the exit of the afterburner and the nozzle throat, respectively. Suitable nozzle geometry for an afterburning turbojet (or turbofan) engine is a *supersonic nozzle* with a convergent-divergent (C-D) geometry, i.e. a nozzle with a well-defined throat before a divergent cone or ramp. The throat Mach number is one, over a wide range of operating conditions, which is referred to as the *choked* throat. As the afterburner operation causes the gas temperature to rise, the gas density shall decrease in harmony with the temperature rise. Consequently, to accommodate the lower density gas at the sonic condition prevailing at the nozzle throat and satisfy continuity equation, the throat area needs to be opened. This is referred to as a *variable-geometry* nozzle requirement of the exhaust system. The continuity equation for a steady flow of a perfect gas is rewritten here to demonstrate the extent of the nozzle throat-opening requirement in a variable-geometry, convergent-divergent nozzle:

$$\dot{m} = \sqrt{\frac{\gamma}{R}} \cdot \frac{p_t}{\sqrt{T_t}} \, A.M. \left(\frac{1}{1 + \dfrac{\gamma - 1}{2} M^2} \right)^{\frac{\gamma+1}{2(\gamma-1)}} \tag{1.179}$$

Let us examine the various terms in Eq. (1.179), as a result of the afterburner operation. First, the mass flow rate increases slightly by the amount of $\dot{m}_{f,AB}$, which is but a small percentage of the gas flow with the afterburner-off condition, say 3–4%. The total pressure, p_t, drops with the afterburner operation, but this too is a small percentage, say 5–8%. The nozzle throat Mach number will remain as one; therefore, except for the gas

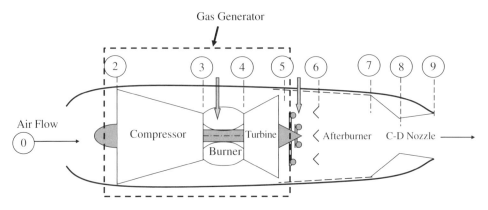

Figure 1.32 Schematic drawing of an afterburning turbojet engine.

property variation, the last parentheses in Eq. (1.179), will remain unchanged as a result of afterburner operation. Now, again neglecting small variations in the gas properties, γ and R, which are in reality a function of the gas temperature, the first term under the square root in Eq. (1.179) shall be unchanged. This brings us to the main driver for the nozzle throat area increase requirement when the afterburner is in operation, namely the gas total temperature, T_t. The effect of increasing T_t with afterburner operation is in creating lower density gas, which requires a larger area at the nozzle throat at sonic speed to accommodate the mass flow rate at lower density. Therefore, the nozzle throat area is opened directly proportional to the square root of the gas total temperature in the nozzle throat, i.e.

$$\frac{A_{8,AB-ON}}{A_{8,AB-OFF}} \approx \sqrt{\frac{T_{t8,AB-ON}}{T_{t8,AB-OFF}}} \tag{1.180}$$

The afterburner is composed of an inlet diffuser, a fuel spray bar, a flameholder to stabilize the combustion, some cooling provisions, and a screech damper, as shown in Figures 1.32 and 1.33. The inlet diffuser decelerates the gas to allow for higher efficiency combustion in the afterburner. A similar prediffuser is found at the entrance to the primary burner. The fuel spray bar is typically composed of one or more rings with distinct fuel injection heads circumferentially distributed around the ring. The V-shaped flame holder ring(s) create a fuel–air mixture recirculation region in its turbulent wake, which allows for a stable flame front to be established. There is also a perforated liner, which serves two functions. One, it serves as a cooling conduit which blankets, i.e. protects, the outer casing from the hot combustion gases. Two, it serves as an acoustic liner, which dampens the high-frequency noise, i.e. screech, which is generated as a result of combustion instability.

An afterburner is considered to be an adiabatic duct with insulated walls, which essentially neglects the small heat transfer through the casing, as compared to energy release in the AB, due to combustion, i.e.

$$_5\left(\dot{Q}_{wall}\right)_7 \cong 0 \tag{1.181}$$

Figure 1.32 shows the basic building block of an afterburning turbojet is gas generator. Additional parameters (beyond the simple turbojet engine) are needed to analyze the performance of an afterburning turbojet engine. These are related to the physical

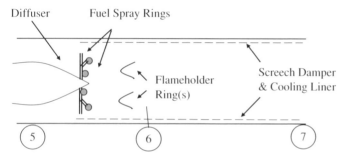

Figure 1.33 Schematic drawing of an afterburner with its station numbers.

Table 1.2 New parameters in afterburner.

1.	$Q_{R,\,AB}$	(Ideal) heating value of the fuel in the afterburner
2.	$\dot{m}_{f,AB}$ or T_{t7} or $\tau_{\lambda\text{-AB}} \equiv \dfrac{c_{pAB}T_{t7}}{c_{pc}T_0}$	Fuel flow rate or the exit temperature or $\tau_{\lambda\text{-AB}}$
3.	$\eta_{AB} \equiv \dfrac{\left(Q_{R,AB}\right)_{\text{actual}}}{\left(Q_{R,AB}\right)_{\text{ideal}}}$	Afterburner efficiency (<1)
4.	$\pi_{AB} \equiv \dfrac{p_{t7}}{p_{t5}}$	Total pressure ratio across the afterburner (<1)

characteristics of the afterburning system, e.g. the type of fuel used in the afterburner (which is typically the same as the fuel in the primary burner), either the fuel-flow setting of the afterburner fuel pump or the exit total temperature of the afterburner, the afterburner efficiency, and the afterburner total pressure loss.

Symbolically, we represent the new parameters in Table 1.2.

1.8.2 Analysis

A nondimensional T-s diagram for an ideal afterburning turbojet engine is shown in Figure 1.34. It is interesting to note the appearance of τ parameters on the nondimensional temperature axis. Also, we note that the thermodynamic process from the flight static, 0, to the turbine exit, i.e. $t5$, is unaffected by the afterburner. Although the T-s diagram depicts an ideal afterburning turbojet engine in Figure 1.34, the real engine afterburner is to operate with no *back influence* on the upstream components as well. We will continue using the *marching technique* in establishing the stagnation flow properties throughout the engine as we used in the turbojet section. The above stipulation, that the operation of an afterburner does not affect the upstream components, will bring our analysis, unchanged, to station $t5$. We will take up the analysis from station $t5$ and establish exit conditions at $t7$.

To allow marching through the afterburner, we need to know or estimate its losses in total pressure and its inefficiency in heat release in the confines of the finite volume of the afterburner. These are π_{AB} and η_{AB}, respectively. So far, the afterburner loss depiction through the total pressure and heat release capability is the same as the primary or main burner. However, unlike the primary burner, an afterburner may or may not be in operation, and this has a large influence on the total pressure loss characteristics of the afterburner. It is only obvious to expect the operating total pressure loss in the afterburner to be larger than the afterburner-off total pressure loss. One may think of it as the *compounding* of the burning losses as well as the frictional losses on the walls and the flame holder, which add up to the total pressure loss in an operating afterburner. Thus, we distinguish between $\pi_{AB\text{-Off}}$ and $\pi_{AB\text{-On}}$ by:

$$\pi_{AB\text{-Off}} \succ \pi_{AB\text{-On}} \tag{1.182}$$

We perform energy balance across the afterburner to establish the afterburner fuel-to-air ratio, f_{AB}:

$$\left(\dot{m}_0 + \dot{m}_f + \dot{m}_{f,AB}\right)h_{t7} - \left(\dot{m}_0 + \dot{m}_f\right)h_{t5} = \dot{m}_{f,AB}Q_{R,AB}\eta_{AB} \tag{1.183}$$

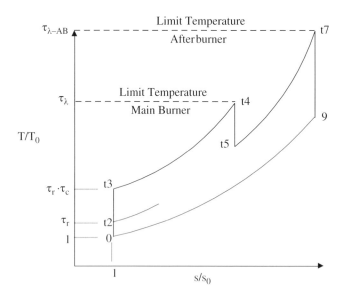

Figure 1.34 Nondimensional T-s diagram for an *ideal* afterburning turbojet engine.

Divide both sides by the air mass flow rate,$\dot{m}_0$, to get:

$$\left(1+f+f_{AB}\right)h_{t7} - \left(1+f\right)h_{t5} = f_{AB}Q_{R,AB}\eta_{AB} \tag{1.184}$$

Now, we can isolatef_{AB}, as:

$$f_{AB} = \frac{\left(1+f\right)\left(h_{t7}-h_{t5}\right)}{Q_{R,AB}\eta_{AB}-h_{t7}} \approx \frac{\left(1+f\right)\left(T_{t7}-T_{t5}\right)}{\dfrac{Q_{R,AB}\eta_{AB}}{c_{p,AB}} - T_{t7}} \tag{1.185}$$

where c_{pAB} represents an average specific heat at constant pressure in the afterburner, i.e. between the temperatures T_{t5} and T_{t7}.

The only unknown in Eq. (1.185) is the fuel-to-air ratio in the afterburner, as we had calculated all the upstream parameters, e.g. f, T_{t5}, values, and the afterburner fuel-heating value and the efficiency as well as the exit temperature are all specified. The total pressure at the afterburner exit, p_{t7}, is

$$p_{t7} = \pi_{AB} \cdot p_{t5} \tag{1.186}$$

The operation of the afterburner changes the inlet conditions to the nozzle due to a change in mass flow rate, total pressure, and temperature. The mass flow rate due to the fuel in the afterburner is calculated via Eq. (1.185), the total pressure is calculated via Eq. (1.186), and the total temperature at the exit of the afterburner constitutes another thermal limit, which is often specified. The analysis of the nozzle, therefore, remains unchanged, except the nozzle inlet values are different as a result of the afterburner operation.

The fundamental definitions of the cycle thermal and propulsive efficiency as given earlier in the turbojet section remain the same for an afterburning turbojet engine as well:

$$\eta_{th} \equiv \frac{\Delta K\dot{E}}{\wp_{thermal}} = \frac{\Delta K\dot{E}}{\dot{m}_f.Q_R + \dot{m}_{f,AB}.Q_{R,AB}} \tag{1.187a}$$

$$\eta_{th} = \frac{\left(\dot{m}_0 + \dot{m}_f + \dot{m}_{f,AB}\right)\dfrac{V_9^2}{2} - \dot{m}_0 \dfrac{V_0^2}{2}}{\dot{m}_f.Q_R + \dot{m}_{f,AB}.Q_{R,AB}} \tag{1.187b}$$

$$\eta_p \equiv \frac{F.V_0}{\Delta K\dot{E}} \tag{1.188a}$$

$$\eta_p = \frac{\left(F_n / \dot{m}_0\right)V_0}{\left(1 + f + f_{AB}\right)\dfrac{V_9^2}{2} - \dfrac{V_0^2}{2}} \tag{1.188b}$$

We may use chain rule on the nozzle total pressure ratio for a turbojet, according to:

$$\pi_n = \left\{ \left(\pi_t \pi_b \pi_c \pi_d \pi_r \frac{p_0}{p_9} \right)^{\frac{\gamma-1}{\gamma}} - \eta_n \left[\left(\pi_t \pi_b \pi_c \pi_d \pi_r \frac{p_0}{p_9} \right)^{\frac{\gamma-1}{\gamma}} - 1 \right] \right\}^{\frac{-\gamma}{\gamma-1}} \quad \text{(Turbojet)} \tag{1.189}$$

By inserting the π_{AB} in the total pressure chain, we get:

$$\pi_n = \left\{ \left(\pi_{AB}\pi_t \pi_b \pi_c \pi_d \pi_r \frac{p_0}{p_9} \right)^{\frac{\gamma-1}{\gamma}} - \eta_n \left[\left(\pi_{AB}\pi_t \pi_b \pi_c \pi_d \pi_r \frac{p_0}{p_9} \right)^{\frac{\gamma-1}{\gamma}} - 1 \right] \right\}^{\frac{-\gamma}{\gamma-1}} \quad \text{(AB TJ)} \tag{1.190}$$

The nozzle exit Mach number then becomes

$$M_9 = \sqrt{\frac{2}{\gamma-1}\left[\left(\pi_n \pi_{AB}\pi_t \pi_b \pi_c \pi_d \pi_r \frac{p_0}{p_9} \right)^{\frac{\gamma-1}{\gamma}} - 1 \right]} \,(\text{AB TJ}) \tag{1.191}$$

The nozzle exit velocity is

$$V_9 = a_0 M_9 \sqrt{\frac{\tau_{\lambda,AB}}{1 + \dfrac{\gamma-1}{2}M_9^2}} \,(\text{AB TJ}) \tag{1.192}$$

The specific thrust for an afterburning turbojet engine is:

$$\frac{F_n}{\dot{m}_0} = \left(1 + f + f_{AB}\right)V_9 \left[1 + \frac{1 - \dfrac{p_0}{p_9}}{\gamma M_9^2} \right] - V_0 \,(\text{AB TJ}) \tag{1.193}$$

TSFC for an afterburning turbojet engine is defined as:

$$TSFC \equiv \frac{\dot{m}_f + \dot{m}_{f,AB}}{F_n} = \frac{f + f_{AB}}{F_n / \dot{m}_0} \, (\text{AB TJ}) \tag{1.194}$$

1.9 Turbofan Engine

1.9.1 Introduction

To create a turbofan engine, a basic gas generator is followed by an additional turbine stage(s), which will tap into the exhaust stream thermal energy to provide shaft power to a fan. This arrangement of multiple loading demands on the turbine stages leads to a multiple shaft arrangement, referred to as *spools*. The fan stages and potentially several (low-pressure) compressor stages may be driven by a shaft, which is connected to the low-pressure turbine. The HPC stages are driven by the high-pressure turbine stage(s). The rotational speeds of the two shafts are called N_1 and N_2, respectively for the low-and high-pressure spools. An additional two new parameters enter our gas turbine vocabulary when we consider turbofan engines. The first is the *bypass ratio* and the second is the *fan pressure ratio*. The ratio of the flow rate in the fan *bypass duct* to that of the gas generator (i.e. the *core*) is called *the bypass ratio* α. The fan pressure ratio is the ratio of total pressure at the fan exit to that of the fan inlet. It is given the symbol, π_f.

$$\alpha \equiv \dot{m}_{\text{fan bypass}} / \dot{m}_{\text{core}} \tag{1.195}$$

$$\pi_f \equiv p_{t13} / p_{t2} \tag{1.196}$$

The principle behind the turbofan concept comes from sharing the power with a larger mass flow rate of air at a smaller velocity increment pays dividend at low speed flight. As we have seen earlier, the smaller the velocity increment across the engine, the higher the propulsive efficiency will be. This principle is Mach number independent; however, for supersonic flight the *installation drag* of large bypass ratio turbofan engines become excessive, and consequently, *small-frontal-area engines* are more suitable to high-speed flight. The fan exhaust stream may be separate from the *core* stream, which is then referred to as *separate exhaust turbofan engine*. Figure 1.35 shows the schematic drawing of a two-spool, separate-exhaust turbofan engine. Note that the gas generator is still the heart of this engine.

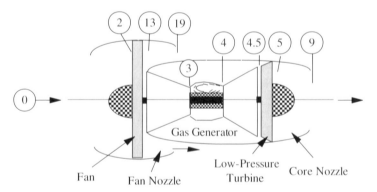

Figure 1.35 Schematic drawing of a separate-exhaust turbofan engine with two spools.

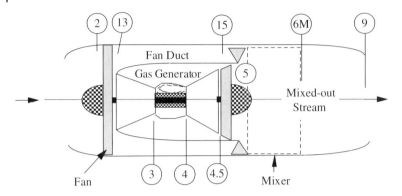

Figure 1.36 Schematic of a long-duct turbofan engine with a mixer.

It is possible to extend the fan duct and provide mixing of the fan and the core stream before entering a common exhaust nozzle. The weight penalty of the long-duct turbofan engine needs to be assessed against the thrust enhancement potential of mixing a cold and hot stream in a mixer before the exhaust nozzle. The schematic drawing of this configuration is shown in Figure 1.36.

1.9.2 Analysis of a Separate-Exhaust Turbofan Engine

In applying our marching technique developed in the analysis of a turbojet engine to a turbofan engine, we are confronted with the fan. As in a compressor, a fan is characterized by its pressure ratio and efficiency. Figure 1.37 shows the thermodynamics of fan compression on an h-s diagram.

The fan adiabatic efficiency is defined as:

$$\eta_f \equiv \frac{h_{t13s} - h_{t2}}{h_{t13} - h_{t2}} = \frac{\Delta h_{t,\text{isentropic}}}{\Delta h_{t,\text{actual}}} = \frac{\text{ideal power}}{\text{actual power}} \tag{1.197}$$

Upon dividing the numerator and denominator of Eq. (1.197) by h_{t2}, we get

$$\eta_f = \frac{T_{t13s}/T_{t2} - 1}{T_{t13}/T_{t2} - 1} = \frac{\pi_f^{\frac{\gamma-1}{\gamma}} - 1}{\tau_f - 1} \tag{1.198}$$

which relates the fan pressure and temperature ratio through adiabatic fan efficiency:

$$\tau_f = 1 + \frac{1}{\eta_f}\left(\pi_f^{\frac{\gamma-1}{\gamma}} - 1\right) \tag{1.199}$$

$$\pi_f = \left\{1 + \eta_f\left(\tau_f - 1\right)\right\}^{\frac{\gamma}{\gamma-1}} \tag{1.200}$$

The power balance between the turbine and compressor now includes the fan:

$$\eta_m \dot{m}_0\left(1 + f\right)\left(h_{t4} - h_{t5}\right) = \dot{m}_0\left(h_{t3} - h_{t2}\right) + \alpha \dot{m}_0\left(h_{t13} - h_{t2}\right) \tag{1.201}$$

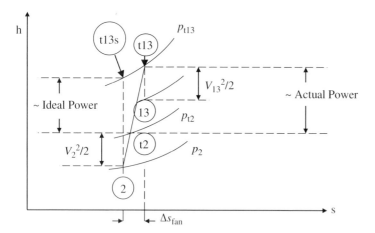

Figure 1.37 Enthalpy-entropy (h-s) diagram representing the flow process in an (adiabatic) fan.

Note that the mass flow rates in a turbofan engine are:

$\dot{m}_0$: Air mass flow rate through the engine core
$\alpha \dot{m}_0$: Air mass flow rate through the fan duct
$(1+\alpha)\dot{m}_0$: Air mass flow rate through the inlet (which later splits into core and the fan duct flow rates)
$\dot{m}_f$: The fuel flow rate in the main burner and the fuel-to-air ratio, f, is defined as $\dfrac{\dot{m}_f}{\dot{m}_0}$, as logically expected

This is depicted graphically in Figure 1.38.

From the power balance Eq. (1.201), the core mass flow rate $\dot{m}_0$ cancels out, and if we divide both sides by flight static enthalpy, h_0, to nondimensionalize the equation, we get:

$$\eta_m \left(1+f\right)\frac{h_{t4}}{h_0}\left(1-\tau_t\right)=\frac{h_{t2}}{h_0}\left[\left(\tau_c-1\right)+\alpha\left(\tau_f-1\right)\right] \tag{1.202}$$

We recognize the ratio of total enthalpy at the turbine inlet to the flight static enthalpy as τ_λ and the ratio h_{t2}/h_0 as τ_r, and we proceed to isolate τ_t in terms of known quantities, namely

$$\tau_t = 1 - \frac{\tau_r \left[\left(\tau_c-1\right)+\alpha\left(\tau_f-1\right)\right]}{\eta_m \left(1+f\right)\tau_\lambda} \tag{1.203}$$

In Eq. (1.203), all the terms on the RHS are either directly specified, e.g. α, η_m, or τ_λ, or easily calculated from the given engine design parameters. The fuel-to-air ratio in a turbofan engine is expressed the same way as the fuel-to-air ratio in a turbojet engine, i.e.

$$f = \frac{\tau_\lambda - \tau_r \tau_c}{\dfrac{Q_R \eta_b}{h_0} - \tau_\lambda} \tag{1.204}$$

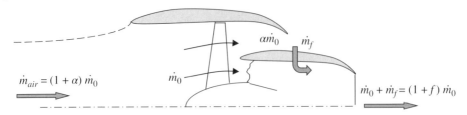

Figure 1.38 Definition sketch for various mass flow rates in a separate-exhaust turbofan engine.

The only point of caution in using Eq. (1.204) is to note that the fuel-flow rate is referenced to the airflow rate *in the core* of a turbofan engine and not the entire airflow rate through the inlet, which partially goes through the fan.

Note that the turbine expansion parameter, τ_t is a strong function of fan bypass ratio and the compressor pressure ratio, as fully expected. However, we note that under certain conditions of, say, large fan pressure ratio or the engine bypass ratio, we may end up in a negative turbine expansion! To see this, we split Eq. (1.203) into its rational constituents, namely:

$$\tau_t = 1 - \frac{\tau_r(\tau_c - 1)}{\eta_m(1+f)\tau_\lambda} - \frac{\tau_r\alpha(\tau_f - 1)}{\eta_m(1+f)\tau_\lambda} = 1 - A - B \tag{1.205}$$

where A and B are the second and third terms on the RHS of Eq. (1.205). Let us further rearrange Eq. (1.205) as follows:

$$1 - \tau_t = A + B \tag{1.206}$$

We note that $(1 - \tau_t)$ term that appears on the LHS of Eq. (1.206) is a nondimensional expression for the turbine power output:

$$1 - \tau_t = \frac{T_{t4} - T_{t5}}{T_{t4}} = \frac{\wp_t}{\dot{m}_t C_{pt} T_{t4}} \tag{1.207}$$

The quantity "A" represents the turbine expansion needed to power the compressor, as in the turbojet analysis. The magnitude of A is therefore independent of the fan bypass and pressure ratio. The last term in Eq. (1.205), namely, the B-term, represents the additional turbine expansion of gases needed to run the fan. Now, this is the term that could render Eq. (1.206) physically meaningless, i.e. lead to a negative τ_t! We further note that the maximum expansion in an aircraft gas turbine engine is produced when the turbine exit pressure is equal to the ambient pressure, i.e. $p_{t5} = p_0$. This assumes that there is no *exhaust diffuser* downstream of the turbine (in place of the exhaust nozzle) as is customary in stationary gas turbine power plants. Therefore, not only a negative τ_t is impossible but also $1 - \tau_t \succ \frac{\wp_{t,max}}{\dot{m}_t C_{pt} T_{t4}}$, based on the expansion beyond the ambient pressure, is impossible. All of these arguments are presented graphically in Figure 1.39.

In terms of fan pressure ratio, we show the α_{max} that a TF engine can support in Figure 1.40.

The question of thrust developed in a turbofan engine is best answered if we apply the momentum principles that we learned in the turbojet section to *two separate streams*

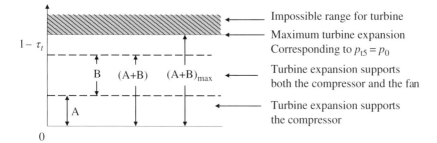

Figure 1.39 Turbine expansion possibilities, including the impossible range.

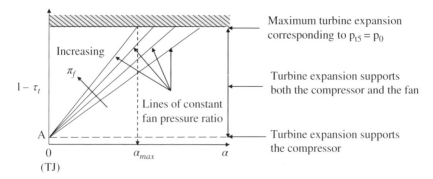

Figure 1.40 Variation of turbine power parameter $(1 - \tau_t)$ with the fan bypass and pressure ratios.

and write the gross thrust for two nozzles and account for the entire ram drag through the inlet, namely:

$$F_n = \dot{m}_0(1+f)V_9 + \alpha \dot{m}_0 V_{19} - (1+\alpha)\dot{m}_0 V_0 + (p_9 - p_0)A_9 + (p_{19} - p_0)A_{19} \qquad (1.208)$$

We recognize the terms on the RHS of Eq. (1.208) as the two-nozzle momentum thrusts followed by the ram drag and then the pressure thrust of the primary and the fan nozzles. The specific thrust for a turbofan engine is the ratio of the net (uninstalled) thrust divided by the entire airflow rate through the inlet, i.e. specific thrust is:

$$\frac{F_n}{(1+\alpha)\dot{m}_0} = \frac{1+f}{1+\alpha}V_9 + \frac{\alpha}{1+\alpha}V_{19} - V_0 + \frac{(p_9 - p_0)A_9}{(1+\alpha)\dot{m}_0} + \frac{(p_{19} - p_0)A_{19}}{(1+\alpha)\dot{m}_0} \qquad (1.209)$$

We immediately note that the specific thrust drops for a turbofan engine (see $1 + \alpha$ in the denominator of the LHS of Eq. (1.209)), in comparison to a turbojet engine that has zero bypass ratio, as more air is handled to produce the thrust in a bypass configuration. Therefore, this observation of lower specific thrust for a turbofan engine leads to the conclusion that bypass engines have by necessity larger frontal areas, again in comparison to a turbojet engine producing the same thrust. Now, we know that the larger frontal areas will result in a higher level of nacelle and installation drag. Therefore, the benefit of higher propulsive efficiency promised in a bypass engine should be carefully weighed against the penalties of higher installation drag for such configurations. The general conclusions are that for subsonic-transonic applications, the external drag penalty of a turbofan engine installation is significantly smaller than the benefits gained by higher engine propulsive efficiency. As a result, in a subsonic-transonic cruise civil transport, for example, the trend has been and continues to be in developing higher bypass ratio engines to reduce fuel burn and possibly use a gearbox on the low-pressure spool to better match the turbine efficiency. Ground clearance for an underwing installation poses the main constraint on the bypass ratio/engine envelope (especially on low-wing commercial aircraft). As the flight Mach number increases into supersonic regime, the optimum bypass ratio is then reduced. For a perfectly expanded primary and fan nozzles, the Eq. (1.209) simplifies to:

$$\frac{F_n}{(1+\alpha)\dot{m}_0} = \frac{1+f}{1+\alpha}V_9 + \frac{\alpha}{1+\alpha}V_{19} - V_0 \tag{1.210}$$

Applying the definition of specific fuel consumption to a turbofan engine, we get:

$$TSFC = \frac{\dot{m}_f}{F_n} = \frac{f}{F_n / \dot{m}_0} = \frac{f/(1+\alpha)}{F_n /(1+\alpha)\dot{m}_0} \tag{1.211}$$

1.9.3 Thermal Efficiency of a Turbofan Engine

The ideal cycle thermal efficiency η_{th} is unaffected by the placement of a fan stage in a gas generator where the fan discharge is then expanded in a nozzle. The reason is that the power utilized to compress the gas in the fan is balanced by the power turbine and therefore the net energy exchange in the bypass duct comes from the balance of the fan inlet and the fan nozzle. A perfectly expanded nozzle will recover the kinetic energy exchange of the inlet diffuser and thus the ideal thermal efficiency of a turbofan engine is unaffected. We can see this, for an ideal fan inlet and a nozzle, in the following T-s diagram shown in Figure 1.41.

Therefore, the thermal efficiency of an ideal turbofan engine is identical to a turbojet engine with the same overall cycle, pressure ratio:

$$\eta_{th} = 1 - \frac{1}{\tau_r \tau_c} = 1 - \frac{1}{\left(1 + \frac{\gamma-1}{2}M_0^2\right)\pi_c^{\frac{\gamma-1}{\gamma}}} \tag{1.212}$$

For the real, i.e. nonideal, turbofan engine, we start from the definition of thermal efficiency:

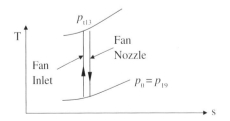

Figure 1.41 T-s diagram of a compression (fan) inlet stream followed by the expansion in a nozzle.

$$\eta_{th} \equiv \frac{\Delta K\dot{E}}{\dot{m}_f Q_R} = \frac{\alpha \dot{m}_0 \frac{V_{19}^2}{2} + (1+f)\dot{m}_0 \frac{V_9^2}{2} - (1+\alpha)\dot{m}_0 \frac{V_0^2}{2}}{\dot{m}_f Q_R} = $$

$$\frac{\alpha V_{19}^2 + (1+f)V_9^2 - (1+\alpha)V_0^2}{2fQ_R} = \frac{(1+f)V_9^2 - V_0^2}{2fQ_R} + \alpha \frac{V_{19}^2 - V_0^2}{2fQ_R} \tag{1.213a}$$

The RHS of Eq. (1.213a) identifies the bypass ratio, the two exhaust velocities, the fuel-to-air ratio, the fuel-heating value and the burner efficiency as the parameters in a turbofan engine that affect the thermal efficiency of the engine. The typical unknowns however, for the design point analysis of a turbofan engine with a given bypass ratio and a fuel type, are the two nozzle exhaust velocities and the combustor fuel-to-air ratio. We note from the last equality in Eq. (1.213a) that all the bypass contribution to engine cycle thermal efficiency is lumped in the last term and the core contribution is in the first term. Ideally, the lower kinetic energy of the core due to power drainage of the fan, i.e. the first term, appears as the kinetic energy rise in the bypass stream, i.e. the second term. Therefore, the thermal efficiency of a turbofan engine is ideally unaffected by the magnitude of the bypass stream.

The Eq. (1.213a) is correct for a turbofan engine with perfectly expanded nozzles. In case the exhaust nozzles are not perfectly expanded, we have to use *effective exhaust speeds* $V_{19\text{eff}}$ and $V_{9\text{eff}}$ in place of actual exit velocities V_{19} and V_9. We define the nozzle effective exhaust speed as:

$$V_{19\,\text{eff}} \equiv \frac{F_{g-\text{fan}}}{\dot{m}_{19}} \tag{1.213b}$$

$$V_{9\,\text{eff}} \equiv \frac{F_{g-\text{core}}}{\dot{m}_9} \tag{1.213c}$$

Example 1.1 shows the nearly constant thermal efficiency even for a TF engine with losses and underexpanded nozzles. Figure 1.42b shows the TF efficiencies.

1.9.4 Propulsive Efficiency of a Turbofan Engine

The propulsive efficiency of a turbofan engine, in contrast to the thermal efficiency of such engines, is a strong function of the engine bypass ratio. Start from the definition of propulsive efficiency:

$$\eta_p \equiv \frac{F_n V_0}{\Delta K\dot{E}} = \frac{\left[(1+f)V_9 + \alpha V_{19} - (1+\alpha)V_0 + (p_9 - p_0)A_9 + (p_{19} - p_0)A_{19}\right]V_0}{(1+f)\frac{V_{9\text{eff}}^2}{2} + \alpha \frac{V_{19\text{eff}}^2}{2} - (1+\alpha)\frac{V_0^2}{2}} \tag{1.214}$$

Assuming the nozzles are perfectly expanded and $f \ll 1$, Eq. (1.214) simplifies to:

$$\eta_p \approx \frac{2V_0\left[V_9 + \alpha V_{19} - (1+\alpha)V_0\right]}{V_9^2 + \alpha V_{19}^2 - (1+\alpha)V_0^2} \tag{1.215}$$

By dividing the numerator and denominator of Eq. (1.215) by V_0^2, we get

$$\eta_p \approx \frac{2\left[\left(\dfrac{V_9}{V_0}\right)+\alpha\left(\dfrac{V_{19}}{V_0}\right)-(1+\alpha)\right]}{\left(\dfrac{V_9}{V_0}\right)^2+\alpha\left(\dfrac{V_{19}}{V_0}\right)^2-(1+\alpha)} \tag{1.216}$$

In a turbofan engine, the propulsive efficiency seems to be influenced by three parameters, as shown in Eq. (1.216), the bypass ratio, α, the primary jet-to-flight velocity ratio, V_9/V_0, and the fan jet-to-flight velocity ratio, V_{19}/V_0. In case the fan nozzle and the primary nozzle velocities are equal, Eq. (1.216) reduces to

$$\eta_p \approx 2/\left(1+V_9/V_0\right) \tag{1.217}$$

Interestingly, this does not explicitly depend on the bypass ratio, α! The effect of bypass ratio is implicit however in the exhaust velocity reduction that appears in the denominator of Eq. (1.217). Again as noted earlier, if the nozzles are not perfectly expanded, we have to use the *effective exhaust speeds* instead of the "actual" velocities.

Example 1.1
For a parametric study of the effect of bypass ratio on the thermal, propulsive, and overall efficiencies of a separate-flow turbofan engine with convergent nozzles, we performed design-point cycle analysis using the following parameters:

Cycle overall pressure ratio, OPR, p_{t3}/p_0 ~61

Cruise altitude: 12 km ($p_0 = 19.19$ kPa, $T_0 = 216$ K)
Cruise Mach number of 0.80
Fan pressure ratio, p_{t13}/p_{t2}, of 1.6
Compressor pressure ratio (including inner fan), p_{t3}/p_{t2}, of 40
Turbine entry temperature (TET): 2000 K, which is high for cruise in current practice
Fuel LHV: 43.3 MJ kg^{-1}

Component efficiencies were assumed to be: $\pi_d = 0.995$, $e_f = 0.90$, $e_c = 0.90$, $\pi_{fn} = 0.98$, $\pi_b = 0.95$, $\eta_b = 0.99$, $e_t = 0.85$, $\eta_m = 0.99$, $\pi_n = 0.98$
Gas thermal properties: $\gamma_c = 1.4$, $c_{pc} = 1004$ J kgK^{-1}, $\gamma_t = 1.33$, $c_{pt} = 1156$ J kgK^{-1}
The range of bypass ratio (BPR) was: 5–17.
First, we look at the effect of bypass ratio on core and fan nozzle (effective) jet speeds in Figure 1.42a. The fan jet speed remains constant, as it is independent of bypass ratio. However, core effective jet speed is reduced with increasing bypass ratio. On energy basis, we are draining the energy from the core stream, as we supply more shaft power to a larger fan. The thermal, propulsive, and overall efficiencies of the turbofan engine are shown in Figure 1.42b. The propulsive efficiency continuously improves with BPR (with $\eta_{p,\text{max}}$~76%), whereas thermal efficiency remains nearly constant (at ~45%), as the overall cycle pressure ratio is kept constant in this example. The overall engine efficiency improves with bypass ratio in harmony with propulsive efficiency.

(a) Separate-Flow Turbofan Engine with Convergent Nozzles
M_{Cruise}=0.80, Altitude: 12 km

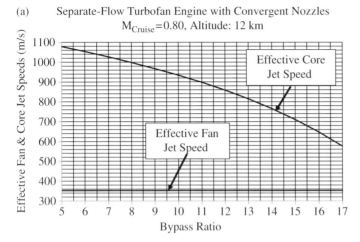

(b) Separate-Flow Turbofan Engine with Convergent Nozzles
M_{Cruise}=0.80, Altitude: 12 km

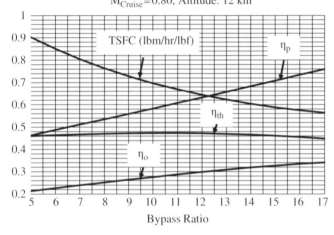

(c) Separate-Flow Turbofan Engine with Convergent Nozzles
M_{Cruise}=0.80, Altitude: 12 km

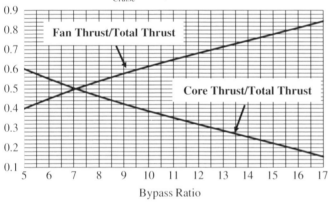

Figure 1.42 The effect of bypass ratio on TF-efficiency and fuel burn.

The fuel burn (TSFC) drops by nearly 40% as we increase the BPR. Finally, in Figure 1.42 (c) we have plotted the fraction of total thrust provided by fan and core as a function of BPR. The fan's share of thrust production steadily increases from nearly 40% to 84% as BPR increases from 5 to 17. The core loses thrust, as we drain the power from core stream and provide it to the fan; consequently, the core share of thrust drops from nearly 60% to about 15%.

Based on this simple example, we can deduce the direction that industry is pursuing in achieving fuel burn improvements; namely, push toward higher BPR, i.e. ultra-high BPR or UHB. By performing similar parametric studies for OPR or TET, we note the direction that industry must follow in those parameters, namely push toward higher OPR and higher TET. There are other parameters that we did not examine, e.g. the use of renewable fuel instead of fossil fuel, hybrid gas-electric propulsion, and finally, electric propulsion.

Comments

We note that the peak overall efficiency is still (dismal) 34%. The implication for this example of GT-propulsor, with $\eta_{o,\,max} \sim 0.34$, is that $\sim66\%$ of the thermal power invested in the combustor by the fuel $\left(\dot{m}_f Q_R\right)$ was not converted to thrust power $(F_n V_0)$. This wasted energy appears as the heat in the (hot) exhaust jet from the core! For the best case, i.e. BPR = 17, we calculated the disparity between the core and the fan nozzle exit temperatures:

Core (primary nozzle) exit temperature: $T_8 \sim 840\,K$ $T_{t8} \sim 978\,K$
Fan nozzle exit temperature: $T_{18} \sim 236\,K$ $T_{t18} \sim 283\,K$

The jet velocities (for BPR = 17) are comparable and are listed here for reference.

Core exhaust jet speed: $V_8 \sim 566\,m\,s^{-1}$ $V_{8,eff} \sim 577\,m\,s^{-1}$
Fan nozzle exhaust jet speed: $V_{18} \sim 308\,m\,s^{-1}$ $V_{18,eff} \sim 353\,m\,s^{-1}$

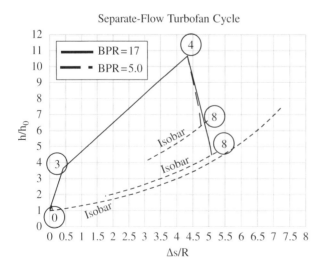

Figure 1.43 The effect of bypass ratio on thermodynamics of a turbofan cycle.

Finally, we note that the low overall efficiency in this turbofan engine is not caused by its low component efficiencies; rather, it is due to the thermal efficiency limitations of Brayton cycle. We ran the same engine at max-BPR (in this example, it was 17), and assumed ideal components throughout the engine and we produced the whopping $\eta_{o, \max} \sim 38\%$. It is instructive to plot the h-s diagram of the TF cycle. Figure 1.43 shows Brayton cycle for two BPRs of 5 and 17. Note the core nozzle in both cycles ($\alpha = 5$ and 17) is underexpanded. Turbine in $\alpha = 17$ produces more power; thus, its exit temperature is lower than BPR 5.0 case. Propulsive efficiency improved by $\sim 64\%$ and the overall efficiency by $\sim 60\%$ with BPR.

1.9.5   Ultra-High Bypass (UHB) Geared Turbofan Engines

There is a new class of turbofan engines, called ultra-high bypass, or UHB, that promises to reduce fuel consumption (by nearly –16% as compared to International Aero Engines V2500 as baseline), cut noise (by –20 dB margin to FAA Stage 4 Aircraft Noise Standards), and produce less pollutants/emissions. Typical bypass ratio in this class starts around 12. To make a comparison to the advanced turboprops (ATPs) that were developed in 1980s and 1990s (also known as Propfan or Unducted Fan); they had bypass ratios of 35–70 (GE-UDF had a bypass ratio of 35 and the P&W/AGT had a bypass ratio of 70). Therefore, the UHB turbofan engines of 2010 fall between the current conventional turbofans with bypass ratio in 6–8 range and the ATP of 35–70. We had learned that propulsive efficiency improved with bypass ratio in a turbofan engine in subsonic aircraft. We derived an expression for the thrust-specific fuel consumption in a jet engine (see Farokhi 2014) that was inversely proportional to propulsive efficiency, η_p, namely:

$$TSFC = \frac{V_0 / Q_R}{\eta_p \eta_{th}} \tag{1.218}$$

Therefore, UHB development mainly targets propulsive efficiency improvements to reduce fuel consumption as well as offering other attractive features (e.g. reduced noise and pollution). Airline industry and commercial cargo business/transporters are most affected by the fuel consumption performance as it represents the lion's share of their direct operating cost (DOC). So, it stands to reason to increase the bypass ratio as long as we can innovate our way through the technology challenges associated with the UHB engines. The practical problems of developing/implementing a high-efficiency large bypass ratio turbofan engine are inherent in operational disparity between a large-diameter fan and a very small engine core diameter, i.e. LPT and intermediate pressure compressor (IPC) have much smaller diameters than the fan, yet driven by the same shaft (i.e. the low-pressure spool).

To improve the efficiency of the fan, IPC, and LPT, we need a suitable (i.e. lightweight, reliable) fan drive gear system.

The UHB turbofan installation concerns on aircraft are mainly (i) wing/fuselage integration, clearance/envelope, limits on fan diameter; (ii) transonic nacelle drag calling for a "slim line" nacelle design with natural laminar flow or hybrid laminar flow (i.e. with boundary layer suction); (iii) nacelle weight concerns demanding the use of advanced composite materials; and (iv) innovative fan reverser design and integration into nacelle. Pratt & Whitney and NASA have worked closely to develop the geared turbofan (GTF)

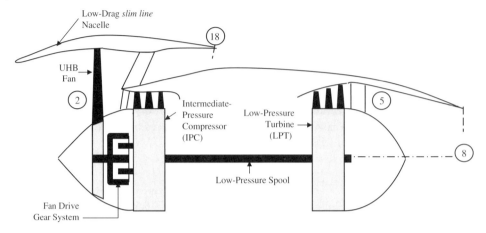

Figure 1.44 Schematic drawing of the low-pressure spool in a UHB with fan drive gear system.

engine technology in connection with the UHB requirements. The first GTF entry into service, PW1000G family, was in 2013. The engine uses a patented lightweight "floating gearbox" system that separates the fan from the IPC and LPT. The gear ratio is 3:1 (for more details, see Duong, McCune, Dobek, 2009).

The weight reduction in the nacelle is accomplished through a composite fan nacelle design. The schematic drawing of the low-pressure spool in a GTF engine is shown in Figure 1.44.

To appreciate the cycle parameters of typical UHB engines, for example its fan low-pressure ratio, and their thrust-specific fuel consumption performance, we solve an example problem.

Example 1.2

A GTF is used in a UHB engine with a single-stage, low-pressure ratio, fan. The engine design point is at the standard sea level static condition, according to:

$M_0 = 0$, $p_0 = 101\,\text{kPa}$, $T_0 = 288\,\text{K}$, $\gamma_c = 1.4$, and $c_{pc} = 1004\,\text{J kgK}^{-1}$ and the engine at the design point is sized for the air mass flow rate (into the inlet) of $600\,\text{kg s}^{-1}$.

The inlet at the design point is assumed to have $\pi_d = 0.995$ total pressure recovery.

The fan bypass ratio at the design point is $\alpha = 12$, its design pressure ratio is $\pi_f = 1.36$ (which is considered low as compared to 1.6 fan pressure ratio), and its polytropic efficiency, $e_f = 0.90$. The reduction gearbox is characterized by its efficiency, which is the ratio of power transmission across the gearbox, namely:

$$\eta_{gb} \equiv \frac{\left(\wp_{\text{gearbox}}\right)_{\text{out}}}{\left(\wp_{\text{gearbox}}\right)_{\text{in}}} \tag{1.219}$$

We assume $\eta_{gb} = 0.998$. The core compressor pressure ratio (i.e. excluding the inner fan section) is $p_{t3}/p_{t13} = 22$ with polytropic efficiency, $e_c = 0.90$. Note that the overall compressor (total) pressure ratio that includes the inner fan section is then $p_{t3}/p_{t2} = (1.36)(22) = 29.92$.

The combustor at takeoff is characterized by:

$T_{t4} = 1600\,\text{K}$, $Q_R = 42\,800\,\text{kJ kg}^{-1}$ (lower heating value of fuel), $\eta_b = 0.99$, $\pi_b = 0.96$

The turbine has a polytropic efficiency of $e_t = 0.85$ and power transmission (or mechanical) efficiency, $\eta_m = 0.975$. The gas in the hot section may be modeled as $\gamma_t = 1.33$ and $c_{pt} = 1156\,\text{J kgK}^{-1}$.

Both the fan and the core nozzles are of convergent design with $\pi_{fn} = 0.985$ and $\pi_{cn} = 0.990$.

We simplify our analysis in this example by treating the gas properties as specified in two "cold" and "hot" zones as described.

Calculate:

a) Total pressures and temperatures at every station inside the engine
b) Combustor fuel flow rate in kg s^{-1} and lbm s^{-1}
c) Shaft power delivered to the gearbox in MW and hp.
d) Shaft power delivered to the fan in MW and hp.
e) Fan nozzle exit Mach number, M_{18}
f) Core nozzle exit Mach number, M_8
g) Fan nozzle gross thrust, F_{gf} in kN and lbf
h) Core nozzle gross thrust, F_{gc} in kN and lbf
i) Thrust-specific fuel consumption, TSFC in mg/s/N and lbm/h/lbf

Estimate the gearbox mass if its mass is proportional to the shaft horsepower according to the following gearbox mass density rule: gearbox mass density = 0.008 lbm hp^{-1}.

Solution

Since $M_0 = 0$, $p_{t0} = p_0 = 101\,\text{kPa}$ and $T_{t0} = T_0 = 288\,\text{K}$. The total temperature is the inlet remains constant (adiabatic flow), which gives, $T_{t2} = 288\,\text{K}$. The total pressure in the inlet suffers some loss due to friction, which results in $p_{t2} = \pi_d \cdot p_{t0} = 100.5\,\text{kPa}$

Based on fan total pressure ratio, we get the total pressure in station 13 as

$$p_{t13} = \pi_f \cdot p_{t2} = 136.7\,\text{kPa}.$$ We use $\tau_c = \pi_c^{\frac{\gamma-1}{\gamma e_c}}$ to get $T_{t13} = 317.5\,\text{K}$. To march toward the fan nozzle exit, we use the fan nozzle total pressure ratio to get $p_{t18} = p_{t13} \cdot \pi_{fn} = 134.6\,\text{kPa}$. Since the fan nozzle flow is adiabatic, the total temperature remains constant, therefore, $T_{t18} = T_{t13} = 317.5\,\text{K}$.

We calculate the fan NPR p_{t13}/p_0 to check it against the "critical" NPR where the nozzle chokes. In this problem, NPR$_{fn} = 1.353$ which is far less than the critical value (of ~ 2). Therefore, we conclude that the exit of our convergent fan nozzle is unchoked and thus subsonic. We also conclude that the static pressure at the nozzle exit is the same as the ambient pressure; therefore, $p_{18} = p_0 = 101\,\text{kPa}$. From p_{18} and p_{t18} we calculate M_{18}, which produces $M_{18} = 0.654$. From T_{t18} and M_{18}, we calculate $T_{18} = 292.5\,\text{K}$, with the corresponding speed of sound, $a_{18} = 342.7\,\text{ms}^{-1}$, and fan nozzle exit velocity is $V_{18} = M_{18} \cdot a_{18} = 224.2\,\text{ms}^{-1}$. Since the nozzle is perfectly expanded, $V_{18,\text{eff}} = V_{18} = 224.2\,\text{ms}^{-1}$. Therefore, the fan nozzle gross thrust is:

$$F_{gf} = \alpha \dot{m}_0 V_{18} = 124.16\,\text{kN}\left(\text{or}\,27{,}913\,\text{lbf}\right)$$

Now, we do the core calculations. The compressor OPR, including the fan inner section is the product of the fan and the core pressure ratio, i.e. $\pi_c = \pi_f \cdot \pi_{core} = 29.92$ (this is the cycle pressure ratio). Based on π_c and p_{t2}, we get $p_{t3} = 3006.8\,\text{kPa}$. From the polytropic efficiency of the compressor, we calculate the compressor total temperature ratio, $\tau_c = 2.9414$. Therefore the compressor exit total temperature is $T_{t3} = 847$ K. The combustor exit total temperature is a given design value at takeoff of $T_{t4} = 1600$ K; therefore, through energy balance across the combustor, we calculate the fuel-to-air ratio to be $f = 0.0247$. From total pressure ratio across the burner, π_b, we get the exit total pressure, $p_{t4} = 2886.5$ kPa. The turbine shaft power is delivered to the core compressor and the fan gearbox. The gearbox delivers its input power times the efficiency of gearbox to the fan. We calculate fan power through energy balance across the fan, namely,

$$
\begin{aligned}
\wp_f &= (1+\alpha)\dot{m}_0 c_{pc}(T_{t13} - T_{t2}) = 600\frac{\text{kg}}{\text{s}}\left(1004\frac{\text{J}}{\text{kgK}}\right)(317.5 - 288)\text{K} \\
&= 17.789\,\text{MW}\left(\text{or}\,23859\,\text{hp}\right)
\end{aligned}
$$

The power delivered to the gearbox is $\wp_{\text{gb-in}} = \dfrac{\wp_f}{\eta_{\text{gb}}} = 17.825\,\text{MW}$ (or 23 907 hp).

The core compressor shaft power is calculated from the power balance across the core compressor:

$$
\wp_{\text{core}} = \dot{m}_0 c_{pc}(T_{t3} - T_{t13}) = 46.15\frac{\text{kg}}{\text{s}}\left(1004\frac{\text{J}}{\text{kgK}}\right)(847 - 317.5)\text{K} = 24.54\,\text{MW}
$$

The turbine shaft power produced is

$$
\wp_t = \frac{\wp_f + \wp_{\text{core}}}{\eta_m} = 43.452\,\text{MW}
$$

Therefore, the turbine exit total temperature is calculated from the power balance across the turbine to be: $T_{t5} = 805.2$ K. From the polytropic efficiency of the turbine and its τ_t, we calculate $\pi_t = 0.0385$ and thus $p_{t5} = 111.2$ kPa. The core nozzle exit total pressure is the product of π_{cn} and p_{t5}; therefore, $p_{t8} = 110.14$ kPa. The total temperature in the adiabatic nozzle remains constant; therefore $T_{t8} = T_{t5} = 805.2$ K.

The core nozzle pressure ratio, $NPR_{\text{core-nozzle}} = 1.10$, which is far below the critical NPR of ~2. Therefore, the core nozzle exit is subsonic and it is perfectly expanded, i.e. $p_8 = p_0 = 101$ kPa, and we calculate M_8 from p_{t8} and p_8 to be $M_8 = 0.363$. In the core nozzle calculations we use the hot gas properties. From the T_{t8} and M_8, we calculate $T_8 = 788$ K and thus $a_8 = 548.3\,\text{m s}^{-1}$ and $V_8 = 199\,\text{m s}^{-1}$. The core nozzle gross thrust is:

$$
F_{\text{gc}} = (1+f)\dot{m}_0 V_8 = 9.411\,\text{kN}\left(\text{or}\,2116\,\text{lbf}\right)
$$

The ratio of fan-to-core thrust is thus 27 913/2116 = 13.2

The engine thrust is the sum of the two gross thrusts (since there is no ram drag at takeoff),

$F_{\text{total}} = 133.6$ kN (or 30 000 lbf).

The thrust-specific fuel consumption is TSFC = 8.519 mg/s/N, which is equal to 0.301 lbm/h/lbf, which is about 20% below a conventional turbofan engine (of bypass ratio 6) that delivers a TSFC of about 0.367 lbm/h/lbf.

When we calculate the specific impulse, I_s, for this engine, we get $I_s \approx 12\,000\,\text{sec}$, which is again another indicator for a tremendous takeoff performance, more in line with turboprops.

The mass of the gearbox is directly proportional to the power delivered to the gearbox through this correlation: $0.008\,\text{lbm}\,\text{hp}^{-1}$ (as suggested by Ian Halliwell 2013). Consequently, since the power into the gearbox was calculated to be 23\,907 hp, the mass of the gearbox is estimated at 191 lbm or 86.6 kg, which is insignificant as a key component in a 30\,000-pound thrust class engine.

Comments

In this example, we performed a one-dimensional cycle analysis with a simple gas model (i.e., as either cold or hot). Our turbine was uncooled, but we specified a lower turbine efficiency to partially offset or account for the effect of cooling. Also, we did not extract any auxiliary shaft power, as it is always done in aircraft gas turbine engines. However, more accurate engine simulators are available (e.g. see GasTurb by Joachim Kurzke 2017) that will properly model the gas and perform cooling bleed and power extraction, which must be used for higher accuracy. In our simple analysis, we demonstrated a 20% reduction in TSFC whereas in real engine testing a 16% reduction in TSFC is measured compared to the baseline International Aero Engines V2500. Additional information on GTF can be found on the P&W website, Asbury and Yetter (2000), which may be consulted for thrust reversers, and Guynn et al. (2013) have treated advanced cycles for single-aisle transport aircraft. CFM website provides performance parameters of other modern turbofan engines.

A new book on gas turbine performance modeling, by Kurzke and Halliwell (2018), is highly recommended for students and practicing engineers in propulsion and power industry.

1.9.6 Analysis of Mixed-Exhaust Turbofan Engines with Afterburners

The schematic diagram in Figure 1.45 of a mixed-exhaust turbofan engine shows the station numbers for this engine and identifies a new component, namely the mixer. The engine core discharges a hot gas into the mixer whereas the fan duct injects a "cold" gas into the mixer. Upon mixing of the two streams, the gas will attain a "mixed-out" state at station 6 M. Now, what about the pressure of the two incoming streams? On the question of static pressure between the two streams, as they enter the mixer, we rely on our understanding of *Kutta trailing-edge condition* applied to airfoils with sharp trailing edge. We learned this

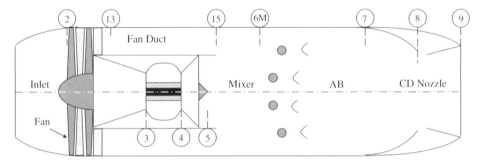

Figure 1.45 Long-duct turbofan engine with a mixer and an afterburner.

principle in aerodynamics. In simple terms, Kutta condition demands the continuity of static pressure at the (sharp) trailing edge, i.e. $p_{upper} = p_{lower}$ on an airfoil at its trailing edge. It physically suggests that the two merging streams at the trailing edge of an airfoil, or fan duct and engine core streams, cannot support a *static pressure jump*. Therefore, we demand:

$$p_{15} = p_5 \tag{1.220}$$

We can think of Eq. (1.220) as that the static pressure is *communicated* between the core and the fan duct, which is a physical principle. Therefore, the fan pressure ratio and hence the fan mass flow rate are basically set by the *engine backpressure*. Here, we note that the two parameters of a fan, namely, its pressure ratio and the bypass ratio, cannot be independently set in this configuration. Here, we can only stipulate one parameter, and the second parameter *falls out* of the common back-pressure requirement.

1.9.6.1 Mixer

To analyze a constant-area mixer, we employ the conservation principles of mass, momentum, and energy. Also, the mixture gas laws establish the mixed-out gas properties at the mixer exit. Figure 1.46 shows a constant-area mixer where mixed-out gas properties c_{p6M} and c_{v6M} are mass averaged and γ_{6M} is the ratio of the two.

The thermodynamic state of gas before and after the mixing process is shown on a T-s diagram in Figure 1.47. The physical mixing process takes place in the shear layer, where large-scale vortical structures are formed, which entrain cold flow into the central hot jet and cause mixing (on the macro-scale).

The application of the law of conservation of energy to an insulated mixer gives:

$$\dot{m}_5 h_{t5} + \dot{m}_{15} h_{t15} = \dot{m}_{6M} h_{t6M} \tag{1.221}$$

Assuming the inlet conditions are known for the mixer, the only unknown in the above equation is the mixed-out total enthalpy, h_{t6M}, namely

$$h_{t6M} = \frac{(1+f) h_{t5} + \alpha h_{t15}}{1 + \alpha + f} \tag{1.222}$$

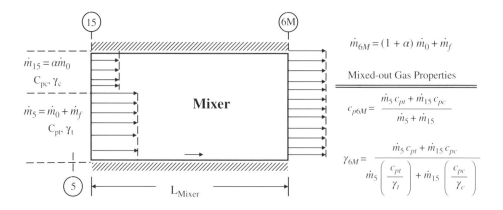

Figure 1.46 Mixer control volume showing nonuniform inlet and *mixed-out* exit velocity.

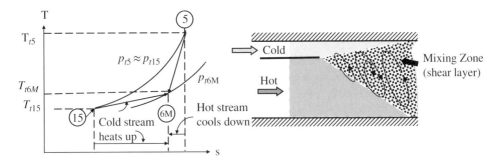

Figure 1.47 Mixing of two streams at different temperatures on a T-s diagram and in physical domain.

Here, we introduce the nondimensional parameters in the above equation as follows:

$$h_{t6M} = h_0 \cdot \frac{(1+f)\tau_t.\tau_\lambda + \alpha\tau_r.\tau_f}{1+\alpha+f} \qquad (1.223)$$

In order to establish the mixed-out pressure, p_{6M}, at the exit of the mixer, we apply the conservation of momentum in the streamwise direction, i.e. x as shown in Figure 1.46. A simple expression for the x-momentum is just $\dot{m}V_x$, which holds for a steady uniform flow and may be applied to the mixer control volume as follows:

$$\dot{m}_{6M}V_{6M} - [\dot{m}_5 V_5 + \dot{m}_{15} V_{15}] = \sum F_x \qquad (1.224)$$

In Eq. (1.224), the right-hand side is the sum of all external forces acting on the control volume in the x-direction. The external forces are:

$$\sum F_x = p_5 A_5 + p_{15} A_{15} - p_{6M} A_{6M} - \int \tau_w dA_{\text{wetted}} \qquad (1.225)$$

where the last term is the wall friction drag force acting on the mixer wall, $D_{f\text{mixer}}$, associated with the mixer. We also note that the static pressures at the entrance to the mixer from the core and the fan duct side are equal according to the Kutta condition:

$$p_5 = p_{15} \qquad (1.226)$$

For a constant-area mixer

$$A_{6M} = A_5 + A_{15} \qquad (1.227)$$

Substitution of Eqs. (1.226) and (1.227) in Eq. (1.224) gives:

$$(\dot{m}_{6M}V_{6M} + p_{6M}A_{6M}) - (\dot{m}_5 V_5 + p_5 A_5) - (\dot{m}_{15}V_{15} + p_{15}A_{15}) = -D_f \qquad (1.228)$$

The grouped terms in the above equation are the *impulse*, given a symbol I, corresponding stations, namely 6 M, 5 and 15, respectively. We note that the fluid impulse out of the mixer is somewhat less than the (sum of the) fluid impulse entering the mixer and this difference or loss is caused by the friction drag on the mixer wall. Therefore, had we assumed frictionless flow in the mixer, we would have concluded that the impulse is conserved, i.e.

$$I_{6M} = I_5 + I_{15} \qquad (1.229)$$

We proceed with our analysis using the assumption of zero boundary layer frictional losses on the walls of the mixer in order to derive a closed form solution for the unknown mixer exit conditions. We then incorporate a correction factor due to friction to account for the viscous flow losses on the total pressure.

The impulse function I attains a very simple form when we express it in terms of flow Mach number:

$$I \equiv \dot{m}V + pA = A\left(\rho V^2 + p\right) = Ap\left(\frac{\rho V^2}{p} + 1\right) = Ap\left(1 + \gamma M^2\right) \qquad (1.230)$$

We had used the equation for the speed of sound, $a^2 = \gamma p/\rho$, in the above derivation. Now, in light of this expression for impulse, let us recast Eq. (1.228) in terms of flow Mach number and pressure area terms:

$$p_{6M}A_{6M}\left(1 + \gamma_{6M}M_{6M}^2\right) - p_5 A_5\left(1 + \gamma_5 M_5^2\right) - p_{15}A_{15}\left(1 + \gamma_c M_{15}^2\right) = 0 \qquad (1.231)$$

By dividing Eq. (1.231) by A_5, we get:

$$p_{6M}1 + \gamma_{6M}M_{6M}^2\left(1 + A_{15}/A_5\right) - p_5\left(1 + \gamma_5 M_5^2\right) + \left(A_{15}/A_5\right)\left(1 + \gamma_c M_{15}^2\right) = 0 \qquad (1.232)$$

Eq. (1.232) has two unknowns, namely p_{6M} and M_{6M}. The continuity equation is expressed in terms of pressure and Mach number according to:

$$\dot{m} \equiv \rho AV = AVp/RT = \frac{\gamma pAM}{\sqrt{\gamma RT}} = \frac{\gamma pAM}{\sqrt{\gamma RT_t}}\sqrt{1 + \frac{\gamma - 1}{2}M^2} \qquad (1.233)$$

Therefore, applying the continuity equation to the mixer, we get:

$$\frac{p_{6M}M_{6M}\sqrt{1 + \left(\gamma_{6M} - 1\right)M_{6M}^2/2}}{\sqrt{\gamma_{6M}R_{6M}T_{t6M}}}\left(1 + A_{15}/A_5\right) =$$
$$\frac{p_5 M_5\sqrt{1 + \left(\gamma_t - 1\right)M_5^2/2}}{\sqrt{\gamma_t R_t T_{t5}}} + \frac{p_{15}M_{15}\sqrt{1 + \left(\gamma_c - 1\right)M_{15}^2/2}}{\sqrt{\gamma_c R_c T_{t15}}}\left(A_{15}/A_5\right) \qquad (1.234)$$

Equation (1.234) and the impulse equation, (1.232), form two equations, two unknown system in terms of the unknowns, p_{6M} and M_{6M}. We may solve the equations in closed form as they reduce to a quadratic equation in M_{6M}^2.

1.9.6.2 Mixed-Turbofan Cycle Analysis
To reinforce the powerful *marching technique* that we have learned so far in cycle analysis, let us apply it to a mixed-flow turbofan engine at this time.

First, we will identify a *mix of parameters* that we need to specify in order to start the cycle analysis in a mixed-flow turbofan engine.

Flight Condition:
We need the flight Mach number (or speed), static temperature, and pressure at the flight altitude, i.e. M_0 (or V_0), T_0, and p_0, as well as air properties (γ, c_p, humidity level), in this case we treat air as *dry*.

Inlet:
We need the inlet adiabatic efficiency or its total pressure recovery parameter, i.e. η_d or π_d, only one figure-of-merit is needed.

Fan:
We need the fan's pressure ratio, $\pi_f \equiv (p_{t13}/p_{t2})$, *or the fan bypass ratio, α, and its polytropic efficiency, e_f or its adiabatic efficiency, η_f.*

Compressor:
We need compressor's total pressure ratio, $\pi_c \equiv p_{t3}/p_{t2}$, which is the cycle OPR and its polytropic efficiency e_c or its adiabatic efficiency, η_c.

Combustor:
We need combustor's total pressure ratio, π_b, combustion efficiency, η_b, and fuel heating value, Q_R and either turbine inlet temperature, T_{t4} or the fuel-to-air ratio, f.

Turbine:
We need turbine's polytropic efficiency, e_t, or adiabatic efficiency, η_t, mechanical efficiency, η_m, exit Mach number, M_5, and an extra assumption about the pressure matching condition between the core and fan ducts at their merging boundary, typically $p_{t5} = p_{t15}$ or $p_5 = p_{15}$.

Mixer:
We need a viscous loss parameter that accounts for the mixer wall friction on the total pressure loss across the mixer, namely, $\pi_{M,f}$

Afterburner
We need the afterburner efficiency, η_{AB}, the maximum temperature T_{t7} (or f_{AB}), afterburner total pressure ratio, π_{AB}, and the heating value of the fuel, $Q_{R,AB}$ in the afterburner.

Nozzle:
We need adiabatic efficiency of the nozzle, η_n or its total pressure loss parameter, π_n, and the nozzle exit pressure, p_9, or instead of exit pressure, we may specify nozzle's (exit-to-throat) area ratio, A_9/A_8.

1.9.6.3 Solution Procedure

We first calculate the total temperature and pressure distribution throughout the engine. We follow a streamline that passes through the core and a streamline that passes through the fan. Once we establish the total fluid properties after each interaction along the streamline path, we have the 1-D information for the core and the fan streams. Then we allow the Kutta condition to establish the communication between the two streams before the mixer takes over and achieves a mixed-out state of the gas prior to discharge into the nozzle or afterburner. These steps are summarized as follows.

Station 0

$$p_{t0} = p_0 \left(1 + \frac{\gamma_c - 1}{2} M_0^2 \right)^{\frac{\gamma_c}{\gamma_c - 1}}$$

$$\pi_r = \left(1 + \frac{\gamma_c - 1}{2} M_0^2 \right)^{\frac{\gamma_c}{\gamma_c - 1}} \tag{1.235}$$

$$T_{t0} = T_0 \left(1 + \frac{\gamma_c - 1}{2} M_0{}^2 \right)$$

$$\tau_r = \left(1 + \frac{\gamma_c - 1}{2} M_0{}^2 \right) \tag{1.236}$$

$$a_0 = \sqrt{\gamma_c R_c T_0}$$

$$V_0 = a_0 M_0$$

Station 2

$$p_{t2} = \pi_d \cdot p_{t0} \text{ or in terms of adiabatic efficiency, } p_{t2} = p_0 \left(1 + \eta_d \frac{\gamma_c - 1}{2} M_0{}^2 \right)^{\frac{\gamma_c}{\gamma_c - 1}}$$

$$\tag{1.237}$$

$$T_{t2} = T_{t0}$$

Now, following a streamline that enters the core, we proceed to station 3.

Station 3

$$p_{t3} = p_{t2} . \pi_c$$

$$T_{t3} = T_{t2} . \pi_c^{\frac{\gamma - 1}{\gamma . e_c}}$$

$$\tau_c = \pi_c^{\frac{\gamma_c - 1}{\gamma_c e_c}} \tag{1.238}$$

Station 4

$$p_{t4} = p_{t3} . \pi_b$$

T_{t4} is either explicitly given as the engine maximum temperature or can be calculated from the fuel-to-air ratio, f, according to:

$$h_{t4} = C_{pt} T_{t4} = \frac{1}{1 + f} \left(h_{t3} + f Q_R \eta_b \right)$$

We also calculate the fuel-to-air ratio in the burner, if only T_{t4} is specified, i.e.

$$f = \frac{h_{t4} - h_{t3}}{Q_R \eta_b - h_{t4}} = \frac{\tau_\lambda - \tau_r \left(\tau_c - 1 \right)}{\dfrac{Q_R \eta_b}{h_0} - \tau_\lambda} \tag{1.239}$$

where $\tau_\lambda \equiv \dfrac{h_{t4}}{h_0}$

Before marching to station 5, let us follow a streamline that passes through the fan.

Station 13

$$p_{t13} = p_{t2}.\pi_f$$

$$T_{t13} = T_{t2}.\pi_f^{\frac{\gamma_c-1}{\gamma_c.e_f}}$$

$$or\,\tau_f = \pi_f^{\frac{\gamma_c-1}{\gamma_c e_f}} \tag{1.240}$$

Station 15
Here, we assume the fan duct to be frictional but *adiabatic* and we characterize the frictional aspect of it by a total pressure ratio π_{fd} parameter,

$$p_{t15} = p_{t13}.\pi_{fd} \tag{1.241}$$

where π_{fd} is given/assumed. Also, our adiabatic flow assumption in the fan duct gives:

$$T_{t15} = T_{t13} \quad \Big[\text{for a calorically perfect gas, otherwise,} h_{t15} = h_{t13} \Big]$$

or

$$\tau_{fd} = 1$$

Now, let us proceed through the turbine by a representative core streamline.

Station 5
Power balance between the turbine and the compression system (fan and compressor) gives:

$$\eta_m \left(1+f\right)\left(h_{t4} - h_{t5}\right) = \left(h_{t3} - h_{t2}\right) + \alpha\left(h_{t13} - h_{t2}\right) \tag{1.242}$$

We note that in the above equation; there are two unknowns, namely the bypass ratio, α and the turbine exit (total) enthalpy, h_{t5}. Let us divide the above equation by the flight static enthalpy, h_0, to get the nondimensional power balance equation:

$$\eta_m \left(1+f\right)\tau_\lambda\left(1-\tau_t\right) = \tau_r\left(\tau_c - 1\right) + \alpha\tau_r\left(\tau_f - 1\right) \tag{1.243}$$

Here, we still have the two unknowns in terms of the bypass ratio and the turbine expansion ratio, τ_t. We can make a reasonable assumption that the total pressure in the fan duct and turbine exit are nearly equal, since static pressures were equal by Kutta condition and we may choose the Mach number (i.e. sizing parameter) in the two streams to yield:

$$p_{t5} \approx p_{t15} \tag{1.244}$$

This immediately ties the turbine pressure ratio to the fan pressure ratio:

$$\frac{p_{t5}}{p_{t2}} = \pi_t.\pi_b.\pi_c = \frac{p_{t15}}{p_{t2}} = \pi_{fd}.\pi_f \tag{1.245}$$

Consequently, turbine total pressure ratio may be expressed in terms of all known parameters of the cycle as:

$$\pi_t = \frac{\pi_{fd}.\pi_f}{\pi_b.\pi_c} \tag{1.246}$$

The turbine temperature expansion ratio, τ_t may be linked to π_t via the polytropic efficiency, namely

$$\tau_t = \pi_t^{\frac{\gamma_t-1}{\gamma_t}e_t} = \left(\frac{\pi_{fd}.\pi_f}{\pi_b.\pi_c} \right)^{\frac{\gamma_t-1}{\gamma_t}e_t} \tag{1.247}$$

Now, we see that the power balance equation written in Eq. (1.243), fan and compressor contains only one unknown, and that is the bypass ratio, α, which is expressed as:

$$\alpha = \frac{\eta_m\left(1+f\right)\tau_\lambda\left(1-\tau_t\right)-\tau_r\left(\tau_c-1\right)}{\tau_r\left(\tau_f-1\right)} \tag{1.248}$$

The assumption of constant total pressure at the exit of the turbine and the fan duct is a reasonable assumption for our design-point analysis. This has enabled us to calculate the turbine pressure ratio and hence the temperature ratio. The power balance then produced the bypass ratio.

Station 6M

From the energy balance across an adiabatic mixer, we obtain:

$$h_{t6M} = h_0.\frac{\left(1+f\right)\tau_t.\tau_\lambda+\alpha\tau_f.\tau_r}{1+\alpha+f} \tag{1.249}$$

All parameters in the above equation are either given as design/limit parameters or have been calculated in the previous steps, thus enabling us to march through the gas turbine. Now, we are ready to calculate the flow Mach number at the fan duct exit, M_{15}, in terms of turbine exit Mach number, M_5, according to:

$$p_{t15} = p_{15}.\left(1+\frac{\gamma_c-1}{2}M_{15}^2\right)^{\frac{\gamma_c}{\gamma_c-1}} = p_5.\left(1+\frac{\gamma_t-1}{2}M_5^2\right)^{\frac{\gamma_t}{\gamma_t-1}} \tag{1.250}$$

Equation (1.250) reduces to the trivial solution, $M_{15} = M_5$, except for different γ_c and γ_t:

$$M_{15}^2 = \frac{2}{\gamma_c-1}\left\{ \left[\left(1+\frac{\gamma_t-1}{2}M_5^2\right)^{\frac{\gamma_t}{\gamma_t-1}} \right]^{\frac{\gamma_c-1}{\gamma_c}} -1 \right\} \tag{1.251}$$

We use the mass flow rate expressions:

$$\dot{m}_{15} = \frac{\gamma_c p_{15}A_{15}M_{15}}{a_{15}} = \alpha\dot{m}_0 \tag{1.252}$$

$$\dot{m}_5 = \frac{\gamma_t p_5 A_5 M_5}{a_5} = \left(1+f\right)\dot{m}_0 \tag{1.253}$$

Now, if we take the ratio of the above two expressions and note that static pressures are identical (a' la Kutta condition), we get an expression for the unknown area ratio, i.e.

$$\frac{A_{15}}{A_5} = \frac{\alpha}{1+f} \cdot \left(\frac{\gamma_t}{\gamma_c}\right) \frac{a_{15}}{a_5} \frac{M_5}{M_{15}} \tag{1.254}$$

So far, we have established the mixed-out total enthalpy, h_{t6M} and all the upstream parameters to the mixer, such as M_{15} and the area ratio A_{15}/A_5. In the mixer analysis, we had derived two equations and two unknowns in Eqs. (1.232) and (1.234):

$$p_{6M} 1 + \gamma_{6M} M_{6M}{}^2 \left(1 + A_{15}/A_5\right) - p_5\left(1 + \gamma_t M_5{}^2\right) + \left(A_{15}/A_5\right)\left(1 + \gamma_c M_{15}{}^2\right) = 0 \tag{1.255}$$

$$\frac{\gamma_{6M} p_{6M} M_{6M} \sqrt{1 + \left(\gamma_{6M}-1\right)M_{6M}{}^2/2}}{\sqrt{\left(\gamma_{6M}-1\right)c_{p6M}T_{t6M}}} \left(1 + A_{15}/A_5\right) =$$
$$\frac{\gamma_t p_5 M_5}{a_5} + \frac{\gamma_c p_{15} M_{15}}{a_{15}} \left(A_{15}/A_5\right) \tag{1.256}$$

Let us nondimensionalize Eq. (1.255) by dividing through by p_5 and rewriting it in the following form:

$$\frac{p_{6M}}{p_5}\left[1 + \gamma_{6M} M_{6M}{}^2\right] = \left[\left(1 + \gamma_t M_5{}^2\right) + \left(A_{15}/A_5\right)\left(1 + \gamma_c M_{15}{}^2\right)\right]/\left(1 + A_{15}/A_5\right) = C_1 \tag{1.257}$$

The right-hand side of the above equation is known and we have given it a shorthand notation of C_1. Now, we can isolate the same unknowns in Eq. (1.256) to get:

$$\frac{p_{6M}}{P_5}\left(M_{6M}\sqrt{1 + \left(\gamma_{6M}-1\right)M_{6M}{}^2/2}\right) = \left[\left(\frac{\gamma_t}{\gamma_{6M}}\right)\frac{M_5}{a_5} + \left(\frac{\gamma_c}{\gamma_{6M}}\right)\frac{M_{15}\left(A_{15}/A_5\right)}{a_{15}}\right]$$
$$\frac{\sqrt{\left(\gamma_{6M}-1\right)c_{p6M}T_{t6M}}}{\left(1 + A_{15}/A_5\right)} = C_2 \tag{1.258}$$

Again, the RHS of the above equation is known and it is labeled C_2. By dividing Eq. (1.257) by (1.258), we eliminate the mixer static pressure ratio and get M_{6M} via:

$$\frac{1 + \gamma_{6M} M_{6M}{}^2}{M_{6M}\sqrt{1 + \left(\gamma_{6M}-1\right)M_{6M}{}^2/2}} = \frac{C_1}{C_2} = \sqrt{C} \tag{1.259}$$

Let us cross multiply and square both sides of the Eq. (1.259) to arrive at a quadratic equation in $M_{6M}{}^2$, as follows:

$$\left(1 + \gamma_{6M} M_{6M}{}^2\right)^2 = C.M_{6M}{}^2\left(1 + \left(\gamma_{6M}-1\right)M_{6M}{}^2/2\right) \tag{1.260}$$

$$\gamma_{6M}{}^2 - C\left(\gamma_{6M}-1\right)/2M_{6M}{}^4 + \left(2\gamma_{6M}-C\right)M_{6M}{}^2 + 1 = 0 \tag{1.261}$$

$$M_{6M}^2 = \frac{C - 2\gamma_{6M} - \sqrt{\left(C - 2\gamma_{6M}\right)^2 - 4\left[\gamma_{6M}^2 - C\left(\gamma_{6M} - 1\right)/2\right]}}{2\gamma_{6M}^2 - C\left(\gamma_{6M} - 1\right)} \tag{1.262}$$

where C is defined in Eq. (1.259) and is expressible in terms of earlier established parameters C_1 and C_2. Now, the mixer static pressure ratio can be determined from:

$$\frac{p_{6M}}{p_5} = \frac{C_1}{1 + \gamma_{6M} M_{6M}^2} \tag{1.263}$$

The *ideal* (i.e. inviscid) total pressure ratio across the mixer is p_{t6M}/p_{t5}, which is expressible in terms of the static pressure ratio of Eq. (1.263) and Mach number according to:

$$\pi_{M_i} \equiv \frac{p_{t6M}}{p_{t5}} = \frac{p_{6M}}{p_5} \frac{\left[1 + \left(\gamma_{6M} - 1\right) M_{6M}^2 / 2\right]^{\frac{\gamma_{6M}}{\gamma_{6M} - 1}}}{\left[1 + \left(\gamma_t - 1\right) M_5^2 / 2\right]^{\frac{\gamma_t}{\gamma_t - 1}}} \tag{1.264}$$

The *actual* total pressure ratio across the mixer should also account for the mixer wall frictional losses, which may be expressed as a multiplicative factor of the ideal parameter:

$$\pi_M \equiv \pi_{M_i}.\pi_{M_f} \tag{1.265}$$

where the frictional loss parameter, π_{M_f}, needs to be specified or assumed.

Station 7

Application of the energy equation to the afterburner yields an expression for the fuel-to-air ratio in the afterburner in terms of known parameters as follows:

$$\left(\dot{m}_{6M} + \dot{m}_{fAB}\right)h_{t7} - \dot{m}_{6M}h_{t6M} = \dot{m}_{fAB}Q_{R,AB}\eta_{AB} \tag{1.266}$$

$$f_{AB} \equiv \frac{\dot{m}_{fAB}}{\dot{m}_{6M}} = \frac{h_{t7} - h_{t6M}}{Q_{R,AB}\eta_{AB} - h_{t7}} = \frac{\tau_{\lambda AB} - h_{t6M}/h_0}{\dfrac{Q_{R,AB}\eta_{AB}}{h_0} - \tau_{\lambda AB}} \tag{1.267}$$

The total pressure at the afterburner exit is known via the loss parameter according to:

$$p_{t7} = p_{t6M}.\pi_{AB} \tag{1.268}$$

Note that we need two afterburner total pressure loss parameters, one for the *afterburner on* and the second for the *afterburner off*, i.e. $\pi_{\text{AB-On}}$ and $\pi_{\text{AB-Off}}$. We expect $\pi_{\text{AB-On}} < \pi_{\text{AB-Off}}$.

Station 9

For adiabatic flow and constant gas properties in the (CD) nozzle, we have

$$T_{t9} = T_{t7} \tag{1.269}$$

The total pressure at the nozzle exit, p_{t9} is expressible in terms of the nozzle total pressure ratio parameter, according to:

$$p_{t9} = p_{t7}.\pi_n \tag{1.270}$$

or in case of an adiabatic nozzle efficiency, η_n, we have:

$$\eta_n = \left[\left(\frac{p_{t7}}{p_9}\right)^{\frac{\gamma_{AB}-1}{\gamma_{AB}}} - \pi_n^{\frac{\gamma_{AB}-1}{-\gamma_{AB}}}\right] \bigg/ \left[\left(\frac{p_{t7}}{p_9}\right)^{\frac{\gamma_{AB}-1}{\gamma_{AB}}} - 1\right] \tag{1.271}$$

Now, with the total pressure calculated at the nozzle exit and the static pressure, p_9, as a given in the cycle analysis, we establish the nozzle exit Mach number, according to:

$$M_9{}^2 = \frac{2}{\gamma_{AB}-1}\left[\left(\frac{p_{t9}}{p_9}\right)^{\frac{\gamma_{AB}-1}{\gamma_{AB}}} - 1\right] \tag{1.272}$$

The exit speed of sound from the exit Mach number and total temperature is:

$$a_9{}^2 = \frac{\gamma_{AB}R_{AB}T_{t9}}{1+\left(\gamma_{AB}-1\right)M_9{}^2/2} \tag{1.273}$$

The nozzle exit velocity, as the product of $M_9.a_9$, is:

$$V_9 = M_9.a_9 \tag{1.274}$$

Example 1.3
We calculate the performance of a mixed-flow TF engine cruising at Mach 2.0 at an altitude where $p_0 = 10\,\text{kPa}$, $T_0 = 223\,\text{K}$, and dry air is characterized by $\gamma_c = 1.4$, $c_{pc} = 1004\,\text{J kgK}^{-1}$. The fan pressure ratio was chosen as $\pi_f = 2.0$ and the compressor pressure ratio (including the inner fan) varies between 6 and 16 (for a parametric study). The main burner is set at $T_{t4} = 1600\,\text{K}$ and the afterburner was on with $T_{t7} = 2000\,\text{K}$. The turbine exit (axial) Mach number, M_5 is set at 0.50. The fuel LHV in both primary and the afterburner is $43\,\text{MJ kg}^{-1}$ and the nozzle exit pressure is assumed to be $p_9/p_0 = 3.8$. The component efficiencies and losses are assumed to be: $\pi_d = 0.90$, $\pi_{fd} = 0.99$, $\pi_b = 0.95$, $\pi_{AB} = 0.92$, $\pi_n = 0.95$, $e_f = e_c = 0.90$, $e_t = 0.80$, $\eta_b = 0.99$, $\eta_m = 0.98$, $\Pi_{Mf} = 0.98$. The gas properties in the turbine and afterburner are assumed to be: $\gamma_t = 1.33$, $c_{pt} = 1156\,\text{J kgK}^{-1}$, and $\gamma_{AB} = 1.30$ and $c_{pAB} = 1243\,\text{J kgK}^{-1}$, respectively.

Figure 1.48a shows the variation of mixed-turbofan engine efficiency with compressor pressure ratio. Propulsive efficiency remains constant (at nearly 62%) whereas thermal efficiency shows a peak (with $\eta_{th}\sim49.4\%$, at an *optimal* OPR ≈ 11.0. Note that the peak overall engine efficiency is still dismal, i.e. $\eta_{o,max}\sim31\%$.

In part (b) of Figure 1.48, we note that the engine bypass ratio varies between 0.44 and 0.63. The engine fuel burn parameter (TSFC in lbm/h/lbf) and specific thrust is shown in Figure 1.48c. The specific fuel consumption is at its minimum, $\text{TSFC}_{min}\sim1.61$ lbm/h/ lbf, where the compressor pressure ratio is ~11, i.e. the optimal π_c for this engine at this Mach number and altitude. The specific thrust is shown in the nondimensional form as $F_n/(1+\alpha)\dot{m}_0 a_0$

And shows a nearly constant behavior (~2.77) with compressor pressure ratio.

Comment
The low overall efficiency of the mixed-exhaust TF engine is due to the thermal efficiency limitations in heat engines, such as the Brayton cycle. The slightly higher thermal

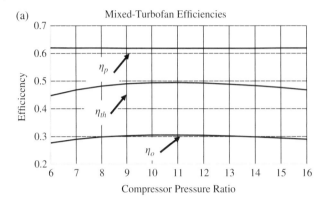

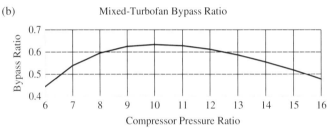

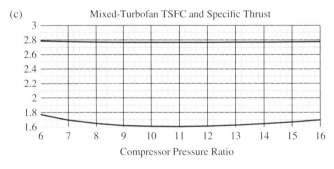

Figure 1.48 Performance characteristics of a mixed TF engine with AB in cruise.

efficiency in this engine as compared to the separate-flow TF engine (Example 1.1), i.e. ~50% vs. 45%, is due to the higher ram compression, π_r, in this engine at Mach 2 flight that produces the cycle OPR of ~86 versus ~61 in Example 1.1.

1.10 Turboprop Engine

1.10.1 Introduction

To construct a turboprop engine, we start with a gas generator.

The turbine in the gas generator provides the shaft power to the compressor. However, we recognize that the gas at the turbine exit is still highly energetic (i.e. high p_t and T_t) and capable of producing shaft power, similar to a turbofan engine.

Once this shaft power is produced in a follow-on turbine stage that we is called "power" or "free" turbine, we can supply the shaft power to a propeller. A schematic drawing of this arrangement is shown in Figure 1.49.

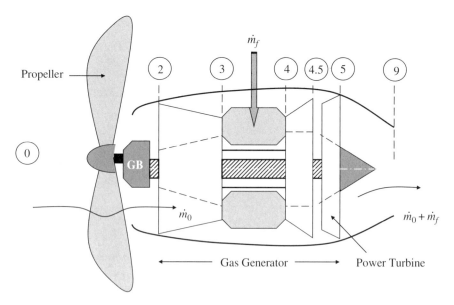

Figure 1.49 Schematic drawing of a turboprop engine with station numbers identified.

The attractiveness of a turbo-propeller engine as compared to a turbo-fan engine lies in its ability to offer a very large bypass ratio, which may be between 30 and 100. The large bypass ratio, by necessity, will then cut back on the exhaust velocities of the pro-pulsor, thereby attaining higher propulsive efficiencies for the engine. The high propul-sive efficiency, however, comes at a price. The limitation on the tip Mach number of a rotating propeller, say to less than 1.3, leads to a cruise Mach number in the 0.7 to ~0.8 range for advanced turboprops and to 0.4–0.6 for conventional propellers. We do not have *this limitation* on cruise Mach number in a turbofan engine. Also, the large diam-eter of the propeller often requires a reduction gearbox.

1.10.2 Turboprop Cycle Analysis

1.10.2.1 The New Parameters
Let us identify the new parameters that we have introduced by inserting a propeller in the gas turbine engine. We will examine the turboprop from the power distribution point of view as well as its thrust producing capabilities namely the propeller thrust contribution to the overall thrust, which includes the core thrust.

From the standpoint of power, the low-pressure turbine power is supplied to a gear-box, which somewhat diminishes it in its frictional loss mechanism in the gearing and then delivers the remaining power to the propeller. We will call the fractional delivery of shaft power through the gearbox, the *gearbox efficiency*, η_{gb}, which symbolically is defined as:

$$\eta_{gb} \equiv \wp_{prop} / \wp_{LPT} \tag{1.275}$$

where the numerator is the power supplied to the propeller and the denominator is the shaft power provided by the power turbine to the gearbox. Also, we define the fraction of *propeller shaft power* that is converted in the *propeller thrust power* as the *propeller efficiency*, η_{prop}, as

$$\eta_{\text{prop}} \equiv F_{\text{prop}} \cdot V_0 / \wp_{\text{prop}} \qquad (1.276)$$

We recognize that the propeller and the engine core both contribute to thrust production:

$$F_{\text{total}} = F_{\text{prop}} + F_{\text{core}} \qquad (1.277)$$

The contribution of the engine core to the overall thrust, which we have called as the *core thrust*, takes on the familiar form of the gross thrust of the nozzle minus the ram drag, i.e.

$$F_{\text{core}} = \left(\dot{m}_0 + \dot{m}_f\right)V_9 - \dot{m}_0 V_0 + \left(p_9 - p_0\right)A_9 \qquad (1.278)$$

The pressure thrust contribution of the nozzle, i.e. the last term in Eq. (1.278), for a turboprop engine is often zero due to perfectly expended exhaust, i.e. $p_9 = p_0$. So, for all practical purposes, the engine core of a turboprop produces a thrust based solely on the momentum balance between the exhaust and the intake of the engine, namely,

$$F_{\text{core}} \cong \left(\dot{m}_0 + \dot{m}_f\right)V_9 - \dot{m}_0 V_0 \qquad (1.279)$$

1.10.2.2 Design-Point Analysis

We require the following set of input parameters in order to estimate the performance of a turboprop engine. The following list, which sequentially proceeds from the flight condition to the nozzle exit, summarizes the input parameters per component. In this section, we will practice the powerful marching technique that we have learned so far in this book.

Station 0

The flight Mach number, M_0, the ambient pressure and temperature, p_0 and T_0, and air properties γ and R, are needed to characterize the flight environment. We can calculate the flight total pressure and temperature, p_{t0} and T_{t0}, the speed of sound at the flight altitude, a_0 and the flight speed V_0, based on the input.

Station 2

At the engine face, we need to establish the total pressure and temperature, p_{t2} and T_{t2}. From adiabatic flow assumption in the inlet, we conclude that:

$$T_{t2} = T_{t0}$$

We need to define an inlet total pressure ratio parameter, π_{d}, or adiabatic diffuser efficiency, η_{d}. This results in establishing p_{t2}, similar to our earlier cycle analysis; for example,

$$p_{t2} = \pi_{\text{d}} \cdot p_{t0}$$

Station 3

To continue our march through the engine, we need to know the compressor pressure ratio, π_{c}, which again is treated as a design choice, and the compressor polytropic efficiency, e_c. This allows us to calculate the compressor total temperature ratio in terms of compressor total pressure ratio using the polytropic exponent, i.e.

$$\tau_c = \pi_c^{\frac{\gamma_c - 1}{\gamma_c e_c}}$$

Now, the compressor discharge total pressure and temperature, p_{t3} and T_{t3} are determined.

Station 4

To establish the burner exit conditions, we need to know the loss parameters, π_b and η_b, as well as the limiting temperature T_{t4}. The fuel type with its energy content, that we had called the *heating value* of the fuel, Q_R, needs to be specified. Again, we establish the fuel-to-air ratio, f, by energy balance across the burner and the total pressure at the exit, p_{t4}, by loss parameter π_b.

Station 4.5

For the upstream turbine, or the so-called *high-pressure turbine*, HPT, we need to know the mechanical efficiency, η_{mHPT}, which is a power transmission efficiency, and the turbine polytropic efficiency, e_{tHPT}, which measures the internal efficiency of the turbine. The power balance between the compressor and high-pressure turbine is

$$\eta_{mHPT}\left(1+f\right)\left(h_{t5} - h_{t4.5}\right) = h_{t3} - h_{t2} \tag{1.280a}$$

$$h_{t4.5} = h_{t4} - \frac{h_{t3} - h_{t2}}{\eta_{mHPT}\left(1+f\right)} \tag{1.280b}$$

leads to the only unknown in the above equation, which is $h_{t4.5}$. The total pressure at station 4.5 may be linked to the turbine total temperature ratio according to:

$$\frac{p_{t4.5}}{p_{t4}} = \left(\frac{T_{t4.5}}{T_{t4}}\right)^{\frac{\gamma_t}{(\gamma_t - 1)e_{tHPT}}} \tag{1.281}$$

$$\pi_{HPT} = \tau_{HPT}^{\frac{\gamma_t}{(\gamma_t - 1)e_{tHPT}}} \tag{1.282}$$

Stations 5 and 9

Since the power turbine drives the *load*, i.e. the propeller, we need to specify the turbine expansion ratio that supports this *load*. In this sense, we consider the propeller as an external load to the cycle and hence as an input parameter to the turboprop analysis. It serves a purpose to put this and the following station, i.e. 9, together, as both are responsible for the thrust production. Another view of stations 5 and 9 downstream of 4.5 points to the *power split*, decision made by the designer, between the propeller and the exhaust jet. The following T-s diagram best demonstrates this principle.

In the T-s diagram (Figure 1.50), both the actual and the ideal expansion processes are shown. We will use this diagram to define the component efficiencies as well as the

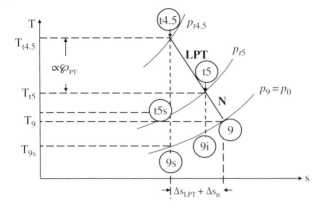

Figure 1.50 Thermodynamic states of an expansion process in free turbine and nozzle of a turboprop engine.

power split choice. For example, we define the power turbine (i.e. LPT) adiabatic efficiency as:

$$\eta_{LPT} \equiv \frac{h_{t4.5} - h_{t5}}{h_{t4.5} - h_{t5s}} = \frac{h_{t4.5}\left(1 - \tau_{LPT}\right)}{h_{t4.5}\left[1 - \pi_{LPT}^{\frac{\gamma_t - 1}{\gamma_t}}\right]} = \frac{1 - \tau_{LPT}}{1 - \pi_{LPT}^{\frac{\gamma_t - 1}{\gamma_t}}} \tag{1.283}$$

Also, we may define the nozzle adiabatic efficiency, η_n, as:

$$\eta_n \equiv \frac{h_{t5} - h_9}{h_{t5s} - h_{9s}} \tag{1.284}$$

We note that the above definition for the nozzle adiabatic efficiency deviates slightly from our earlier definition in that we have assumed

$$h_{t5} - h_{9i} \approx h_{t5s} - h_{9s} \tag{1.285}$$

which, in light of small expansions in the nozzle and hence near parallel isobars, is considered reasonable.

The total ideal power available at station 4.5, per unit mass flow rate, may be written as:

$$\frac{\wp}{\dot{m}_0\left(1 + f\right)} = h_{t4.5} - h_{9s} = h_{t4.5}\left[1 - \left(\frac{p_9}{p_{t4.5}}\right)^{\frac{\gamma_t - 1}{\gamma_t}}\right] = \wp_{i,total} / \dot{m}_9 \tag{1.286}$$

If we examine the RHS of Eq. (1.286), we note that all terms on the RHS are known. Therefore, the total ideal power is known to us *a priori*. Now, let us assume that the power split between the free turbine and the nozzle is, say α and $1 - \alpha$, respectively (see Figure 1.51).

We can define the power split as:

$$\alpha \equiv \frac{h_{t4.5} - h_{t5s}}{h_{t4.5} - h_{9s}} = \frac{\wp_{LPT} / \dot{m}_9}{\wp_{i,total} / \dot{m}_9} = \frac{\eta_{LPT}}{\ } \tag{1.287}$$

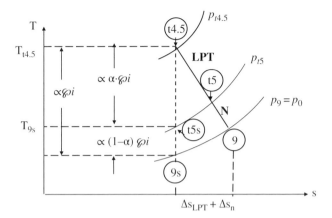

Figure 1.51 Definition of power split, α, in a turboprop engine.

which renders the following equation for the free turbine (LPT) power in terms of a given α,

$$\wp_{LPT} = \dot{m}_9 \eta_{LPT} \cdot \alpha \cdot h_{t4.5} \left[1 - \left(\frac{p_9}{p_{t4.5}} \right)^{\frac{\gamma_t - 1}{\gamma_t}} \right] = \dot{m}_9 \left(h_{t4.5} - h_{t5} \right) \tag{1.288}$$

This expression for the power turbine (LPT) can be applied to the propeller through gearbox and propeller efficiency in order to arrive at the thrust power produced by the propeller, namely

$$F_{prop} \cdot V_0 = \eta_{prop} \cdot \eta_{gb} \cdot \eta_{mLPT} \cdot \wp_{LPT} = \dot{m}_0 \left(1 + f \right) \left[\alpha \eta_{LPT} \eta_{mLPT} \eta_{gb} \eta_{prop} \right] h_{t4.5} \left[1 - \left(\frac{p_9}{p_{t4.5}} \right)^{\frac{\gamma_t - 1}{\gamma_t}} \right] \tag{1.289}$$

We note that the R.H.S. of the above equation, per unit mass flow rate, is known. Now, let us examine the nozzle thrust. The kinetic energy per unit mass at the nozzle exit is linked to:

$$V_9^2 / 2 \cong \eta_n \left(h_{t5s} - h_{9s} \right) = \eta_n \left(1 - \alpha \right) \wp_{i,total} / \dot{m}_9 = \left(1 - \alpha \right) \eta_n h_{t4.5} \left[1 - \left(\frac{p_9}{p_{t4.5}} \right)^{\frac{\gamma_t - 1}{\gamma_t}} \right] \tag{1.290}$$

Therefore, the exhaust velocity is now approximated by the power split parameter, α, and the total ideal power available after the gas generator, i.e. station 4.5, according to:

$$V_9 \approx \sqrt{2 \left(1 - \alpha \right) \eta_n h_{t4.5} \left[1 - \left(\frac{p_9}{p_{t4.5}} \right)^{\frac{\gamma_t - 1}{\gamma_t}} \right]} \tag{1.291}$$

Now, the turboprop thrust per unit air mass flow rate (through the core) can be expressed in terms of the propeller thrust, the core thrust using Eq. (1.291) incorporated for the exhaust velocity.

$$\frac{F_{total}}{\dot{m}_0} = (1+f)V_9 - V_0 + \frac{(1+f)\left[\alpha\eta_{PT}\eta_{mLPT}\eta_{gb}\eta_{prop}\right]h_{t4.5}\left[1-\left(\frac{p_9}{p_{t4.5}}\right)^{\frac{\gamma_t-1}{\gamma_t}}\right]}{V_0} \quad (1.292)$$

The fuel efficiency of a turboprop engine is often expressed in terms of the fraction of the fuel consumption in the engine to produce a unit shaft/mechanical power, according to:

$$PSFC \equiv \frac{\dot{m}_f}{\wp_{prop} + \wp_{core}} = \frac{f}{\left(\frac{\wp_{prop}}{\dot{m}_0} + \frac{\wp_{core}}{\dot{m}_0}\right)} \quad (1.293)$$

$$\wp_{prop} = \eta_{gb}\eta_{mLPT}\wp_{LPT} = \dot{m}_0(1+f)\eta_{gb}\eta_{mLPT}\eta_{LPT}.\alpha.h_{t4.5}\left[1-\left(\frac{p_9}{p_{t4.5}}\right)^{\frac{\gamma_t-1}{\gamma_t}}\right] \quad (1.294)$$

$$\wp_{core} = \frac{\dot{m}_0}{2}\left[(1+f)V_9^2 - V_0^2\right] \quad \text{for a perfectly expanded nozzle, otherwise use } V_{9eff}$$
$$\quad (1.295)$$

We can define the thermal efficiency of a turboprop engine as:

$$\eta_{th} \equiv \frac{\wp_{prop} + \wp_{core}}{\dot{m}_f Q_R} = \frac{\left(\frac{\wp_{prop}}{\dot{m}_0} + \frac{\wp_{core}}{\dot{m}_0}\right)}{f Q_R} \quad (1.296)$$

The propulsive efficiency, η_p, may be defined as:

$$\eta_p \equiv \frac{F_{total}.V_0}{\wp_{prop} + \wp_{core}} = \frac{\frac{F_{total}}{\dot{m}_0}.V_0}{\left(\frac{\wp_{prop}}{\dot{m}_0} + \frac{\wp_{core}}{\dot{m}_0}\right)} \quad (1.297)$$

where all the terms in the above efficiency definitions have been calculated in earlier steps and the overall efficiency is again the product of the thermal and propulsive efficiencies

1.10.2.3 Optimum Power Split between the Propeller and the Jet

For a given fuel flow rate, flight speed, compressor pressure ratio, and all internal component efficiencies, we may ask a very important question, which is, "At what power split, α, would the total thrust be maximized?" This is a simple mathematics question. What we need first is to express the total thrust in terms of all independent parameters, i.e. f, V_0, π_c etc., and then differentiate it with respect to α and set the derivative equal to zero. From that equation, obtain the solution(s) for α that satisfies the equation. We use Eq. (1.292)

$$\frac{F_{\text{total}}}{\dot{m}_0} = (1+f)V_9 - V_0 + \frac{(1+f)\left[\alpha\eta_{\text{LPT}}\eta_{mLPT}\eta_{\text{gb}}\eta_{\text{prop}}\right]h_{t4.5}\left[1-\left(\dfrac{p_9}{p_{t4.5}}\right)^{\frac{\gamma_t-1}{\gamma_t}}\right]}{V_0}$$

We express the exhaust velocity, V_9, as Eq. (1.291):

$$V_9 \approx \sqrt{2(1-\alpha)\eta_n h_{t4.5}\left[1-\left(\frac{p_9}{p_{t4.5}}\right)^{\frac{\gamma_t-1}{\gamma_t}}\right]}$$

We note that the bracketed term in the above equations, i.e.

$$1-\left(\frac{p_9}{p_{t4.5}}\right)^{\frac{\gamma_t-1}{\gamma_t}} = 1-\left(\frac{p_9}{p_0}\frac{p_0}{p_{t0}}\frac{p_{t0}}{p_{t2}}\frac{p_{t2}}{p_{t3}}\frac{p_{t3}}{p_{t4}}\frac{p_{t4}}{p_{t4.5}}\right)^{\frac{\gamma_t-1}{\gamma_t}} = 1-\left(\frac{p_9}{p_0}\left(\pi_r\pi_d\pi_c\pi_b\pi_{\text{HPT}}\right)^{-1}\right)^{\frac{\gamma_t-1}{\gamma_t}}$$

$$(1.298)$$

which is a constant. Also let us examine the total enthalpy at station 4.5, $h_{t4.5}$,

$$h_{t4.5} = \frac{h_{t4.5}}{h_{t4}}\frac{h_{t4}}{h_0}h_0 = \tau_{\text{HPT}}.\tau_\lambda.h_0 \qquad (1.299)$$

which is a constant, as well. Therefore, the expression for the total thrust of the engine (per unit mass flow rate in the engine nozzle) is essentially composed of a series of constants and the dependence on power split parameter, α takes on the following simplified form:

$$\frac{F_{\text{total}}}{(1+f)\dot{m}_0} = C_1\sqrt{1-\alpha} + C_2\alpha + C_3 \qquad (1.300)$$

where C_1, C_2, and C_3 are all constants. Now, let us differentiate the above function with respect to α and set the derivative equal to zero, i.e.

$$\frac{d}{d\alpha}\left[\frac{F_{\text{total}}}{(1+f)\dot{m}_0}\right] = \frac{-C_1}{2\sqrt{1-\alpha}} + C_2 = 0 \qquad (1.301)$$

which produces a solution for the power split parameter, α, that maximizes the total thrust of a turboprop engine. Hence, we may call this special value of α, the "optimum" α, namely

$$\alpha_{\text{opt}} = 1-\left(\frac{C_1}{2C_2}\right)^2 \qquad (1.302)$$

Now, upon substitution for the constants C_1 and C_2 and some simplification, we get:

$$\alpha_{opt} = 1 - \frac{\eta_n}{\left(\eta_{PT}\eta_{mLPT}\eta_{gb}\eta_{prop}\right)^2} \cdot \frac{\gamma_c - 1}{2} \cdot \frac{M_0^2}{\tau_{HPT}\tau_\lambda\left[1 - \left(\dfrac{p_9/p_0}{\pi_r\pi_d\pi_c\pi_b\pi_{HPT}}\right)^{\frac{\gamma_t-1}{\gamma_t}}\right]} \tag{1.303}$$

This expression for the optimum power split between the propeller and the jet involves all component and transmission (of power) efficiencies, as expected. However, let us assume that all efficiencies were 100% and further assume that the exhaust nozzle was perfectly expanded, i.e. $p_9 = p_0$. What does the above expression tell us about the optimum power split in a perfect turboprop engine? Let us proceed with the simplifications.

$$\alpha_{opt_{ideal}} = 1 - \frac{\dfrac{\gamma_c - 1}{2} M_0^2}{\tau_{HPT}\tau_\lambda - \dfrac{\tau_\lambda}{\tau_r\tau_c}} \tag{1.304}$$

From power balance between the compressor and the high-pressure turbine, we can express the following:

$$\tau_r\left(\tau_c - 1\right) = \left(1 + f\right)\tau_\lambda\left(1 - \tau_{HPT}\right) \approx \tau_\lambda\left(1 - \tau_{HPT}\right) \tag{1.305}$$

This simplifies to

$$\tau_\lambda\tau_{HPT} \cong \tau_\lambda - \tau_r\left(\tau_c - 1\right) \tag{1.306}$$

Substitute the above equation in the optimum power split in an ideal turboprop engine, Eq. (1.304), to get:

$$\alpha_{opt_{ideal}} = 1 - \frac{\dfrac{\gamma_c - 1}{2} M_0^2}{\tau_\lambda - \tau_r\left(\tau_c - 1\right) - \dfrac{\tau_\lambda}{\tau_r\tau_c}} = 1 - \frac{\tau_r - 1}{\tau_\lambda - \tau_r\left(\tau_c - 1\right) - \dfrac{\tau_\lambda}{\tau_r\tau_c}} \tag{1.307}$$

At takeoff condition and low-speed climb/descent ($\tau_r \to 1$), the optimum power split approaches 1, as expected. The propeller is the most efficient propulsor at low speeds, as it attains the highest propulsive efficiency. As flight Mach number increases, the power split term α becomes less than one.

Example 1.4

For the turboprop, we chose Mach 0.6 cruise at 12 km (standard) altitude. Compressor pressure ratio is $\pi_c = 30$, turbine entry temperature TET = 1600 K. To examine the effect of power drainage in the exhaust stream, we vary the power split parameter, α, between 0.78 and 0.98. We choose propeller efficiency at cruise $\eta_{prop} = 0.80$, fuel LHV is 43 MJ kg^{-1}, and component efficiency and loss parameters: $\pi_d = 0.98$, $\pi_b = 0.96$, $e_c = 0.92$, $e_{tHPT} = 0.82$, $\eta_{tLPT} = 0.88$, $\eta_n = 0.96$, $\eta_{m,H} = 0.99$, $\eta_{m,L} = 0.99$, $\eta_b = 0.99$ and the nozzle was perfectly expanded, with $p_9/p_0 = 1.0$. Gas properties were assumed to be: $\gamma_c = 1.4$, $c_{pc} = 1004$ J kgK^{-1} and $\gamma_t = 1.33$, $c_{pt} = 1156$ J kgK^{-1}.

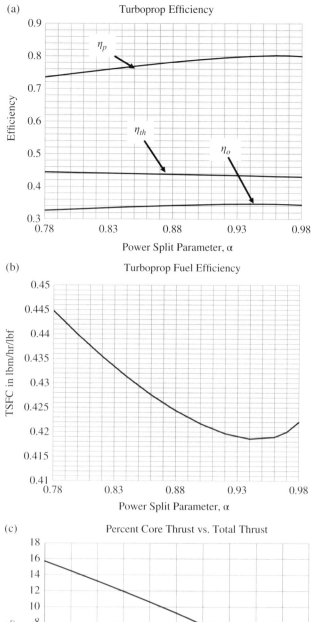

(a) Turboprop Efficiency

(b) Turboprop Fuel Efficiency

(c) Percent Core Thrust vs. Total Thrust

Figure 1.52 Performance of a TP engine in cruise with variable power split, α.

Figure 1.52a shows that the propulsive efficiency increases with power split in favor of the propeller, until it plateaus at ~0.94, whereas thermal efficiency remains nearly constant at ~43%. The peak overall efficiency is slightly better than TF, but it is dismal at ~ 35%.

Figure 1.52b shows that the TP fuel burn continually improves until it reaches a plateau at α~0.94 power split. The core starts to produce a net drag, instead of thrust as the nozzle jet speed falls below flight speed when we drain the exhaust stream at levels of $\alpha \gtrsim 0.97$.

The behavior of core thrust as percent total shows energy drainage with α. Core ceases to produce thrust at $\alpha \gtrsim 0.97$, as shown in Figure 1.52c.

1.10.2.4 Advanced Propeller: Prop-Fan

Conventional propellers lose their thrust production capability when their tip operates in supersonic flow and stall, also known as compressibility losses. In the United States, Pratt & Whitney/Allison Gas Turbine, GE Aviation, and NASA collaborated in developing the technology of advanced turboprop engines in the 1970s and 1980s (Hager and Vrabel 1988). These engines represent open-rotor architecture, which are at times called Propfan, while GE's gearless, direct-drive ATP is called the unducted fan (UDF). The advanced propellers operate with relative (or helical) supersonic tip Mach number (M_T~1.1–1.15) without stalling! With increasing capability in relative tip Mach number of the advanced propeller, the cruise flight Mach number is increased to M_0~0.8–0.85. Several configurations in co- and counter-rotating propeller sets and pusher versus tractor configurations were developed and flight-tested. The advanced propellers are thin (with 3–5% thickness-to-chord ratio) and highly swept at the tip (between 30–40°) to improve tip efficiency at high relative Mach numbers. This is akin to reducing effective Mach number on a transonic aircraft wing based on sweep angle and thickness. Maximum tip sweep, within the aeroelastic stability constraints of the fan, has resulted in 30°–40° angle in advanced propellers. The subjects of sweep and aeroelasticity are detailed in a modern transonic aerodynamic textbook by Vos and Farokhi (2015) that may be consulted for review.

Figure 1.53 shows the first generation of ATP tested at NASA (from NASA SP-495 1988).

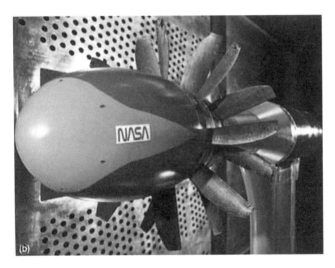

Figure 1.53 Unducted-Fan (UDF) counter-rotating model at NASA-Lewis (now Glenn) wind tunnel. *Source:* Courtesy of NASA.

The critical design areas in an ATP powered aircraft are:

- ATP integration issues, e.g. tractor vs. pusher, wing vs. aft-fuselage
- Fan blade aero-structural design for max aero-efficiency and dynamic stability
- Fan actuation/pitch control
- Blade retention system
- Blade-out condition
- Nacelle cowl system design
- Spinner area-ruling to prevent blade root choking (in tractor configuration)

1.11 High-Speed Air-Breathing Engines

Ramjets offer the highest (fuel) specific impulse at flight Mach numbers above ~4, as depicted in Figure 1.54. However, they produce no static thrust. The analysis of conventional, i.e. subsonic combustion, ramjets is the same as other air-breathing engines such as turbojets (see Kerrebrock (1992) or Hill and Peterson (1992)). A typical flow path in a conventional ramjet engine is shown in Figure 1.55.

The T-s diagram of the ramjet cycle is shown in Figure 1.56. The ram compression in supersonic-hypersonic flow includes shock compression that causes an increase in gas temperature and a reduction in the Mach number. In the conventional ramjet, there is a normal shock in the inlet that transitions the flow into subsonic regime. The typical combustor Mach numbers in a conventional ramjet is ~0.2–0.3. The combustor exit temperature, T_{t4}, may reach stoichiometric levels of ~2000–2500 K.

The equations that are used in the calculation of conventional ramjet performance are listed in Tables 1.3 and 1.4. The equations in these tables are listed sequentially;

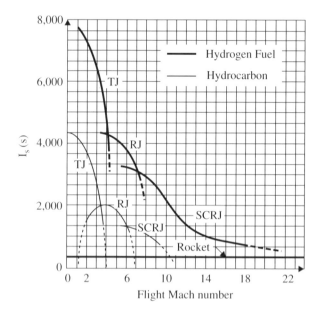

Figure 1.54 Approximate variation of specific impulse with flight Mach number for different air-breathing engines (TJ: turbojet, RJ: ramjet and SCRJ: scramjet) and a typical chemical rocket. *Source:* Adapted from Kerrebrock 1992.

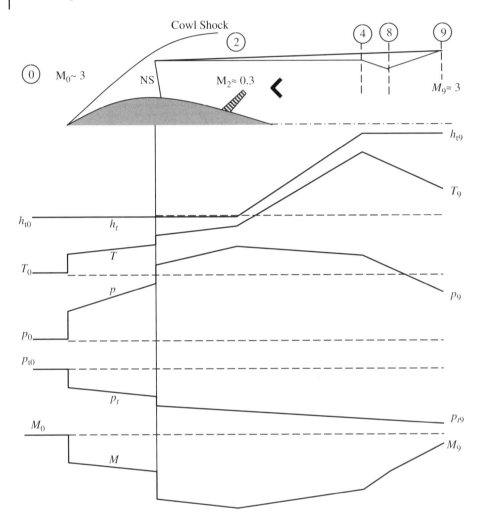

Figure 1.55 Flow parameter trends in a typical subsonic-combustion ramjet.

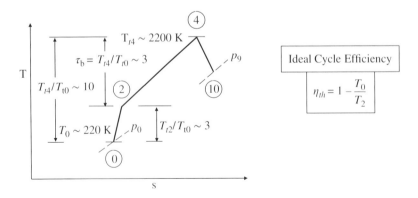

Figure 1.56 Conventional (subsonic-combustion) ramjet cycle and its ideal thermal efficiency.

Table 1.3 Summary of ramjet equations in terms of dimensional parameters (for calorically perfect gas).

Given:

M_0, p_0, T_0 (or altitude), γ and R

η_d or π_d

π_b, η_b, Q_R and T_{t4} or τ_λ

π_n or η_n and p_9

Calculate: $f, F_n / \dot{m}_0, TSFC, I_s, \eta_{th}, \eta_p$ and η_o

---

$$T_{t0} = T_0 \left(1 + \frac{\gamma - 1}{2} M_0^2\right) = T_{t2}$$
Get flight total temperature

$$p_{t0} = p_0 \left(1 + \frac{\gamma - 1}{2} M_0^2\right)^{\frac{\gamma}{\gamma - 1}}$$
Get flight total pressure

$$V_0 \equiv M_0 . a_0 = M_0 \sqrt{\gamma R T_0}$$
Get flight velocity

$$\frac{p_{t2}}{p_0} = \left\{1 + \eta_d \frac{\gamma - 1}{2} M_0^2\right\}^{\frac{\gamma}{\gamma - 1}} \text{ or } p_{t2} = \pi_d . p_{t0}$$
Get total pressure at diffuser exit

$$p_{t4} = p_{t2} . \pi_b$$
Get burner exit total pressure

$$f = \frac{h_{t4} - h_{t0}}{Q_R \eta_b - h_{t4}} = \frac{T_{t4} - T_{t0}}{Q_R \eta_b / c_p - T_{t4}}$$
Get fuel-to-air ratio in the burner

$$\pi_n = \left\{\left(\frac{p_{t4}}{p_9}\right)^{\frac{\gamma - 1}{\gamma}} - \eta_n \left[\left(\frac{p_{t4}}{p_9}\right)^{\frac{\gamma - 1}{\gamma}} - 1\right]\right\}^{\frac{-\gamma}{\gamma - 1}}$$
Get nozzle total pressure ratio

$$p_{t9} = p_{t4} . \pi_n$$
Get total pressure at nozzle exit

$$M_9 = \sqrt{\frac{2}{\gamma - 1}\left[\left(\frac{p_{t9}}{p_9}\right)^{\frac{\gamma - 1}{\gamma}} - 1\right]}$$
Get nozzle exit Mach number

$$T_9 = \frac{T_{t9}}{1 + \frac{\gamma - 1}{2} M_9^2}$$
Get nozzle exit static temperature

$$a_9 = \sqrt{\gamma R T_9}$$
Get speed of sound at nozzle exit

$$V_9 = a_9 . M_9$$
Get nozzle exit velocity

$$\frac{F_n}{\dot{m}_0} = (1 + f) V_9 \left(1 + \frac{1}{\gamma M_9^2}\left(1 - \frac{p_0}{p_9}\right)\right) - V_0$$
Get specific thrust

$$TSFC \equiv \frac{\dot{m}_f}{F_n} = \frac{f}{F_n / \dot{m}_0}$$
Get thrust-specific fuel consumption

$$I_s = 1 / (g_0 TSFC)$$
Get fuel-specific impulse

$$\eta_p = \frac{2 (F_n / \dot{m}_0) V_0}{(1 + f) V_9^2 - V_0^2}$$
Get propulsive efficiency

$$\eta_{th} = \frac{(1 + f) V_9^2 - V_0^2}{2 f Q_R \eta_b}$$
Get cycle thermal efficiency

$$\eta_o = \eta_p . \eta_{th} = \frac{(F_n / \dot{m}_0) V_0}{f Q_R \eta_b} = \frac{V_0 / Q_R \eta_b}{TSFC}$$
Get overall efficiency

Table 1.4 Summary of ramjet equations in terms of nondimensional parameters (for calorically perfect gas).

Given:

M_0, p_0, T_0 (or altitude), γ and R

η_d or π_d

π_b, η_b, Q_R and T_{t4} or τ_λ

π_n or η_n and p_9

Calculate: $f, F_n / \dot{m}_0$, TSFC, I_s, η_{th}, η_p, and η_o

---

$$\tau_r = 1 + \frac{\gamma - 1}{2} M_0^2 \qquad\qquad \text{Get ram temperature ratio}$$

$$\pi_r = \left(1 + \frac{\gamma - 1}{2} M_0^2\right)^{\frac{\gamma}{\gamma-1}} = \tau_r^{\frac{\gamma}{\gamma-1}} \qquad\qquad \text{Get ram pressure ratio}$$

$$\pi_d = \frac{p_{t2}}{p_{t0}} = \left(\frac{1 + \eta_d \dfrac{\gamma - 1}{2} M_0^2}{1 + \dfrac{\gamma - 1}{2} M_0^2}\right)^{\frac{\gamma}{\gamma-1}} \qquad\qquad \text{Get inlet total pressure ratio}$$

$$f = \frac{\tau_\lambda - \tau_r}{\dfrac{Q_R \eta_b}{c_p T_0} - \tau_\lambda} \qquad\qquad \text{Get fuel-to-air ratio}$$

$$\pi_n = \left\{\left(\pi_b \pi_d \pi_r \frac{p_0}{p_9}\right)^{\frac{\gamma-1}{\gamma}} - \eta_n \left[\left(\pi_b \pi_d \pi_r \frac{p_0}{p_9}\right)^{\frac{\gamma-1}{\gamma}} - 1\right]\right\}^{\frac{-\gamma}{\gamma-1}} \qquad\qquad \text{Get nozzle pressure ratio}$$

$$M_9 = \sqrt{\frac{2}{\gamma - 1}\left[\left(\pi_n \pi_b \pi_d \pi_r \frac{p_0}{p_9}\right)^{\frac{\gamma-1}{\gamma}} - 1\right]} \qquad\qquad \text{Get nozzle exit Mach number}$$

$$V_9 = M_9 \sqrt{\gamma R T_0 \frac{\tau_\lambda}{1 + \dfrac{\gamma - 1}{2} M_0^2}} = a_0 M_9 \sqrt{\frac{\tau_\lambda}{1 + \dfrac{\gamma - 1}{2} M_9^2}} \qquad\qquad \text{Get exhaust velocity}$$

$$\frac{F_n}{\dot{m}_0} = (1 + f) V_9 \left(1 + \frac{1}{\gamma M_9^2}\left(1 - \frac{p_0}{p_9}\right)\right) - V_0 \qquad\qquad \text{Get specific thrust}$$

$$TSFC \equiv \frac{\dot{m}_f}{F_n} = \frac{f}{F_n / \dot{m}_0} \qquad\qquad \text{Get thrust-specific fuel consumption}$$

$$I_s = 1/(g_0 . TSFC) \qquad\qquad \text{Get fuel-specific impulse}$$

$$\eta_p = \frac{2(F_n / \dot{m}_0) V_0}{(1 + f) V_9^2 - V_0^2} \qquad\qquad \text{Get propulsive efficiency } \eta_p$$

$$\eta_{th} = \frac{(1 + f) V_9^2 - V_0^2}{2 f Q_R \eta_b} \qquad\qquad \text{Get thermal efficiency } \eta_{th}$$

$$\eta_o = \eta_p . \eta_{th} = \frac{(F_n / \dot{m}_0) V_0}{f Q_R \eta_b} = \frac{V_0 / Q_R \eta_b}{TSFC} \qquad\qquad \text{Get overall efficiency } \eta_o$$

---

therefore, they are useful in computer-based calculations/simulation of ramjets. These are the same equations as in turbojets except we have set $\pi_c = 1$ to simulate a ramjet.

1.11.1 Supersonic Combustion Ramjet

Ramjets cannot produce static thrust. They require ram compression, i.e. some forward flight speed, before these simple air-breathing engines (or ducts with burners) can produce thrust. It is seemingly ironic that the same engines (i.e. ramjets) *cease* to produce thrust at very high ram compressions! Two aspects of ram compression are detrimental to thrust production at high speeds:

1) The inlet total pressure recovery is exponentially deteriorated with flight Mach number.
2) The rising gas temperature in the inlet cuts back $(\Delta T_t)_{\text{burner}}$ to eventually zero.

The worst offender in total pressure recovery of supersonic/hypersonic inlets is the normal shock, which incidentally is also responsible for the rising gas temperatures ahead of the burner. Therefore, if we could only do away with the normal shock in the inlet, we should be getting a super-efficient ramjet. But the flow in a supersonic inlet without a (terminal) normal shock would still be supersonic in the combustor! Despite the obvious challenges, supersonic combustion ramjets, or scramjets are born out of necessity. Scramjets hold out the promise of being the most efficient air-breathing engines, i.e. with the highest (fuel) specific impulse, at flight Mach numbers above ~6 to suborbital Mach numbers (i.e. $M \sim 20$); although the upper limits in scramjet operational Mach number is still unknown.

Scramjet engines, more than any other air-breathing engine, need to be integrated with the aircraft. The need for integration stem from a long forebody that is needed at hypersonic Mach numbers to efficiently compress the air. In addition, a long forebody offers the largest capture area possible for the engine, A_0. Aft-integration with the aircraft allows for a large area ratio nozzle suitable for high-altitude hypersonic vehicles. Figure 1.57 shows a generic scramjet engine that is integrated in a hypersonic aircraft. The inlet achieves compression both externally and internally through oblique shocks. The series of oblique shock reflections inside the inlet create a shock train, known as the *isolator*. The flow that emerges from the isolator is supersonic, say Mach 3.0, as it enters the combustor. Achieving efficient combustion at supersonic speeds is a challenge.

Fuel injection, atomization, vaporization, mixing, and chemical reaction time scales should, by necessity, be short. If air is moving at Mach-3 where speed of sound is ~333 m s^{-1}, it traverses 1 m in ~1 millisecond. This is just an indication of the convective or residence time scale in the combustor. Hydrogen offers these qualities. For example, hydrogen offers about 1/10 of the chemical reaction time scale of hydrocarbon fuels. Additionally, hydrogen is in a cryogenic state as a liquid; therefore, it offers aircraft and engine structure regenerative cooling opportunities.

By accepting high Mach numbers inside the scramjet engine; the static temperature of the gas will be lower throughout the engine and thus the burner is allowed to release heat in the combustor that is needed for propulsion. The h-s diagram of the scramjet cycle is shown in Figure 1.58.

1.11.1.1 Inlet Analysis
Scramjet inlets are integrated with the aircraft forebody, involving several external and internal oblique shocks. In addition, the vehicle's angle of attack, α, impacts the inlet

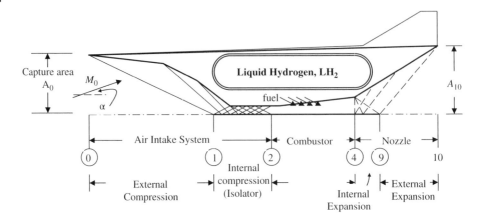

Figure 1.57 Typical arrangement of a generic scramjet engine on a hypersonic aircraft.

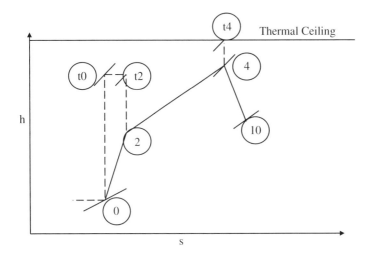

Figure 1.58 Static and stagnation states of gas in a scramjet engine.

recovery through oblique shock waves' angles. For a given forebody shape, we may use shock-expansion theory to calculate the waves' orientations and the associated total pressure recovery. In the absence of the detailed aircraft forebody geometry, or in the preliminary phase, we may assume an inlet recovery or resort to standards such as AIA or MIL-E-5008B (MIL-STD). We listed these here for convenience:

$$\pi_d = 1 - 0.1(M_0 - 1)^{1.5} \qquad 1 < M_0 \qquad \text{AIA – Standard} \tag{1.308}$$

$$\pi_d = 1 - 0.075(M_0 - 1)^{1.35} \qquad 1 < M_0 < 5 \qquad \text{MIL – E – 5008B} \tag{1.309}$$

$$\pi_d = 800 / (M_0^4 + 935) \qquad 5 < M_0 \qquad \text{MIL – E – 5008B} \tag{1.310}$$

1.11.1.2 Scramjet Combustor

Although the detail design and analysis of supersonic combustion is beyond the scope of this book, we can still present global models and approaches that are amenable to closed-form solution. Supersonic branch of Rayleigh flow, i.e. a frictionless, constant-area duct with heating, is a possible (and the simplest) model for a scramjet combustor. The fact that the analysis was derived for a constant-area duct is not a major limitation, since we can divide a combustor with area change into a series of combustors with constant areas. Rayleigh flow is detailed in Chapter 2 of Farokhi (2014), and we will not review it here. However, we may offer to analyze a variable-area combustor that maintains a constant pressure. From cycle analysis viewpoint, constant-pressure combustion is advantageous to cycle efficiency, and thus we consider it here further. Figure 1.59 shows a frictionless duct with heat exchange but with constant static pressure. We assume the gas is calorically perfect and the flow is steady and one-dimensional. The contribution of fuel mass flow rate to the gas in the duct is small; therefore, combustion is treated as heat transfer through the wall. The problem statement specifies the inlet condition and heat transfer rate (or fuel-flow rate or fuel-to-air ratio) and assumes a constant static pressure in a duct with a variable area. The purpose of the analysis is to calculate the duct exit flow condition as well as the area, A_4.

We apply conservation principles to the slab of fluid, as shown. The continuity requires:

$$\rho A V = (\rho + d\rho)(V + dV)(A + dA) \quad \frac{dV}{V} + \frac{d\rho}{\rho} + \frac{dA}{A} = 0 \tag{1.311}$$

The balance of x-momentum gives

$$\dot{m}(V + dV) - \dot{m}V = pA - p(A + dA) + pdA \equiv 0 \quad dV = 0 \tag{1.312}$$

Equation (1.312) signifies a constant-velocity flow, i.e.

$$V_4 = V_2 \tag{1.313}$$

Therefore, continuity Eq. (1.311) demands ρA = constant; or area is inversely proportional to density.

The energy balance written for the duct gives:

$$h_{t4} - h_{t2} = \frac{\dot{Q}}{\dot{m}} = q = fQ_R\eta_b \tag{1.314a}$$

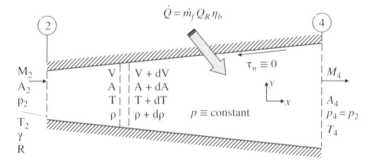

Figure 1.59 A frictionless, constant-pressure combustor.

Since with constant velocity, the kinetic energy (per unit mass) remains constant, then

$$h_{t4} - h_{t2} = h_4 - h_2 \tag{1.314b}$$

$$T_4 = T_2 + q / c_p = T_2 + f Q_R \eta_b / c_p \tag{1.315}$$

The RHS of Eq. (1.315) is known (as an input to the problem), which then establishes the exit static temperature, T_4. Since static pressure is constant, we can calculate the exit density, ρ_4. Also, the area ratio follows the inverse of the density ratio, i.e.

$$A_4 = A_2 \left(\rho_2 / \rho_4 \right) \tag{1.316}$$

which establishes the exit flow area. The exit static temperature calculates the exit speed of sound and Mach number (since V = constant).

We may calculate a critical heat flux, q^* (or critical fuel-to-air ratio, f^*), that will choke the duct at its exit. Eq. (1.315) is written as

$$\frac{T_4}{T_2} = 1 + \frac{q}{c_p T_2} \tag{1.317}$$

Also, Mach number ratio, M_4 / M_2 is related to the ratio of speeds of sound following:

$$\frac{M_4}{M_2} = \frac{V_4 a_2}{V_2 a_4} = \frac{a_2}{a_4} = \sqrt{\frac{T_2}{T_4}} = \frac{1}{\sqrt{1 + \dfrac{q}{c_p T_2}}} \tag{1.318}$$

For a choked exit, $M_4 = 1$, therefore the critical heat flux, q^*, in terms of duct inlet flow conditions is expressed as:

$$\frac{q^*}{c_p T_2} = M_2{}^2 - 1 \tag{1.319a}$$

In terms of fuel-to-air ratio, we may express q^* as $f^* Q_R \eta_b$ to get:

$$f^* \approx \frac{c_p T_2}{Q_R \eta_b} \left(M_2{}^2 - 1 \right) \tag{1.319b}$$

With calculated exit Mach number, we may arrive at the total pressure at the exit, following

$$p_{t4} = p_2 \left(1 + \frac{\gamma - 1}{2} M_4{}^2 \right)^{\frac{\gamma}{\gamma - 1}}$$

An input to this analysis is the burner efficiency, η_b. However, combustion efficiency in supersonic streams is not as well established or understood as the conventional low-speed burners. Mixing efficiency in supersonic shear layers, without chemical reaction, forms the foundation of scramjet combustion efficiency. A supersonic stream mixes with a lower speed stream (in our case the fuel) along a shear layer by vortex formations and supersonic wave interactions. These interactions require space along the (shear layer) or flow direction. Consequently, burner efficiency is a function of the combustor length and continually grows with distance along the scramjet combustor. Burrows and Kurkov (1973) may be consulted for some data and analysis related to supersonic combustion of hydrogen in vitiated air. Heiser et al. (1994) should be consulted for detailed discussion of hypersonic air-breathing propulsion.

1.11.1.3 Scramjet Nozzle

The large area ratios needed for high-speed, high-altitude flight prompted the use of aircraft aft-underbody as the expansion ramp for the scramjet nozzle (shown in Figure 1.57). For perfect expansion, we may approximate the area ratio requirements, A_{10}/A_9, by one-dimensional gas dynamic equations. Severe overexpansion at lower altitudes causes a shock to appear on the aft underbody of the vehicle. The complicating factors in the analysis of hypersonic air-breathing nozzles are similar to those in the rocket nozzles, i.e. continuing chemical reaction in the nozzle, flow separation, cooling, and flow unsteadiness, among others. Simple nozzle design based on the method of characteristics is a classical approach that students in gas dynamics learn. However, developing robust, high-fidelity computational fluid dynamics codes for direct numerical simulation of viscous reacting flows is the most powerful tool that researchers and industry are undertaking.

1.12 Rocket-Based Airbreathing Propulsion

Integrating rockets with ramjets has long been the solution of overcoming ramjets lack of takeoff capability. A ram-rocket may be a configuration similar to the one shown in Figure 1.60.

A fuel-rich solid propellant rocket provides the takeoff thrust. The air intake is sealed off until forward speed of the aircraft can produce the needed ram compression for the ramjet combustor to sustain stable thrust. The rocket motor serves as the *gas generator* for the ramjet, i.e. combustion gases from the fuel-rich solid propellant in the rocket motor mix with the air and the mixture is combusted in the ramjet burner to produce thrust. Variations of this scheme, such as a separate ramjet fuel, may be used in a gas generator ram-rocket. There is no new theory to be presented here, at least at the preliminary level, i.e. we have developed basic tools to analyze both components individually and combined. A ram-rocket configuration where the rocket provides takeoff thrust is shown in Figure 1.61. The ramjet fuel is injected in the air stream for sustained thrust, as shown.

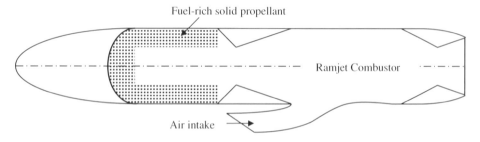

Figure 1.60 Schematic drawing of a gas generator ram-rocket.

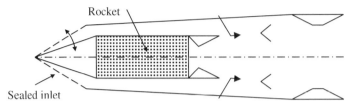

Figure 1.61 Ram-rocket configuration.

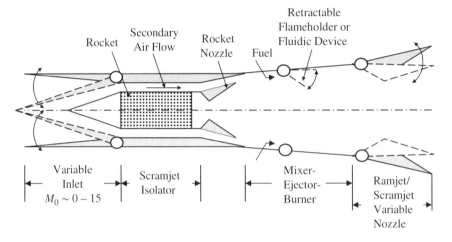

Figure 1.62 Concept in rocket-based-combined-cycle (RBCC) propulsion.

A rocket placed in a duct will draw air in and through mixing of the cold and hot gases will enhance its thrust level. This is the ejector principle. Since the propulsion is based on both the rocket and assisted by the secondary air mixing through an inlet, this may be referred to as an *air-breathing rocket*. The duct that follows the rocket nozzle is then the mixer-ejector, which may also serve as the combustor of a conventional or scramjet engine for combined operation. Now, we have clearly passed the air-breathing and rocket propulsion boundaries and entered into the realm of RBCC propulsion (see Figure 1.62). This is a promising approach to the future of hypersonic flight from takeoff to orbit. It clearly is an exciting time to be a propulsion engineer/researcher.

1.13 Summary

In this chapter, we reviewed air-breathing engine cycles and developed 1-D, steady solution methodology based on thermodynamics and conservation laws for their performance parameters. These were identified to be: specific thrust, specific fuel consumption, thermal, propulsive, and overall efficiencies. We had limited our approach to steady, one-dimensional flow and the "design-point" analysis. The gas was treated as perfect, but its properties were described in cold-hot-very-hot zones. Namely, the cold described inlet-fan-compressor, the hot described turbine and nozzle (if no afterburner was used), and the afterburner gas properties were used in the wet mode and in the nozzle. We included the effect of fluid viscosity and thermal conductivity in our cycle analysis empirically, i.e. through the introduction of component efficiency. The knowledge of component efficiency at different operating conditions is critical to their design and optimization. We always treated that knowledge, i.e. the component efficiency, as a "given" in our analysis, e.g. based on published data. RBCC propulsion system is introduced with the promise for hypersonic air-breathing propulsion.

We intentionally left the discussion of combustion-generated pollutants out of the current chapter, which is focused on cycle analysis. In three examples on separate-flow TF engine, mixed-flow TF engine, and a TP, we calculated cycle efficiencies, η_{th}, η_p, and η_o. All three examples revealed an inherent weakness in thermal efficiency of heat engines, or in our application, the Brayton cycle. The best thermal efficiency was about 50% and when combined with propulsive efficiency to produce the overall efficiency, we could only reach the peak overall efficiency of ~35%. This *inefficiency* to convert the thermal power in fuel–air combustion to thrust power in an aircraft is the *Achilles' heel* of the heat engines. We also learned that higher BPR, as well as higher OPR and TET, improve the cycle efficiency and reduce fuel burn. In addition to improved efficiency, higher TET reduces the core size; thus, engine weight is reduced. In the near-to-medium term (i.e. 2020–2035), these are the directions that the aircraft propulsion industry explores while developing practical hybrid and electric propulsion options for sustainable aviation. MacKay (2009) has addressed sustainable energy and energy conversion pathways/efficiencies, which is relevant to our discussion. David MacKay's book is on my highly recommended reading list.

Gordon and McBride (1996) address the composition of combustion gases in equilibrium reactions, which complement our discussions in primary combustor and afterburner. Archer and Saarlas 1998), Cumpsty (2003), Flack and Rycroft (2005), Hesse and Mumford (1964), Mattingly (1996), Mattingly et al. (2018), Oates (1988), Oates (1985), Oates (1989), Shepherd (1972), El-Sayed (2008), and Greitzer et al. (2004) are excellent aircraft propulsion textbooks and references that are recommended for further reading. On rocket propulsion and SSTO concepts, Sutton and Biblarz (2001), Varvill and Bond (2003), and Huzel and Huang (1992) may be consulted. Rolls-Royce (2015) and Pratt and Whitney (1980) have produced valuable reference manuals that are highly recommended (Anon. 2015, 1980).

References

Pratt & Whitney (1980). The Aircraft Gas Turbine Engine and Its Operation. Hartford, CT: Pratt & Whitney Aircraft Group/United Technologies.

Rolls-Royce (2015). The Jet Engine, 5e. New York: Wiley.

Anderson, J.D. (2003). Modern Compressible Flow, 3e. New York: McGraw-Hill & Co.

Archer, R.D. and Saarlas, M. (1998). An Introduction to Aerospace Propulsion. New York: Prentice Hall.

Asbury, S.C., and Yetter, J.A. 2000. Static performance of six innovative thrust reverser concepts for subsonic transport applications. Summary of the NASA Langley Innovative Thrust Reverser Test Program. NASA/TM-2000-210300, July.

Burrows, M.C. and Kurkov, A.P. (1973). Analytical and Experimental Study of Supersonic Combustion of Hydrogen in a Vitiated Airstream, *NASA TMX-2628*. Washington, D.C.

Cumpsty, N.A. (2003). Jet Propulsion: A Simple Guide to the Aerodynamic and Thermodynamic Design and Performance of Jet Engines, 2e. Cambridge, UK: Cambridge University Press.

Duong, L., McCune, M. and Dobek, L., 2009. US Patent 3, 0293278, December 2009.

El-Sayed, A. (2008). Aircraft Propulsion and Gas Turbine Engines. New York: CRC Press.

Farokhi, S. (2014). Aircraft Propulsion, 2e. Chichester, UK: Wiley.

Flack, R.D. and Rycroft, M.J. (2005). Fundamentals of Jet Propulsion with Applications. Cambridge, UK: Cambridge University Press.

Gordon, S. and McBride, S., "NASA Reference Publication 1311, Computer Program for Calculation of Complex Chemical Equilibrium Compositions and Applications, Vol. 1: Analysis," October 1994 and "Vol. 2: User Manual and Program Description," National Aeronautics and Space Administration, Washington, D.C, 1996.

Greitzer, E.M., Tan, C.S., and Graf, M.B. (2004). Internal Flow: Concepts and Applications. Cambridge, UK: Cambridge University Press.

Guynn, M.D., Berton, J.J., Tong, M.T., and Haller, W.J. August 2013. Advanced single-aisle transport propulsion design options revisited. AIAA Paper Number 2013-4330.

Hager, R.D. and Vrabel, D. 1988. Advanced Turboprop Project. NASA SP-495, Washington, DC.

Halliwell, I.2013. AIAA-IGTI Undergraduate Team Engine Design Competition Request for Proposal, 2010–2011. www.aiaa.org. Last accessed 24 November.

Heiser, H.W., Pratt, D.T., Daley, D.H., and Mehta, U.B. (1994). Hypersonic Airbreathing Propulsion. Reston, VA: AIAA, Inc.

Hesse, W.J. and Mumford, N.V.S. (1964). Jet Propulsion for Aerospace Applications, 2e. New York: Pittman Publishing Corporation.

Hill, P.G. and Peterson, C.R. (1992). Mechanics and Thermodynamics of Propulsion, 2e. Reading, MA: Addison-Wesley.

Huzel, D.K. and Huang, D.H. (1992). Design of Liquid Propellant Rocket Engines. Reston, VA: AIAA, Inc.

Kerrebrock, J.L. (1992). Gas Turbines and Aircraft Engines, 2e. Cambridge, Massachusetts: MIT Press.

Kurzke, J. (2017). A design and off-design performance program for gas turbines. *GasTurb* www.gasturb.de. Last accessed 19 December 2017.

Kurzke, J. and Halliwell, I. (2018). Propulsion and Power: An Exploration of Gas Turbine Performance Modelling. Cham, Switzerland: Springer International Publishing, AG.

Lotter, K. 1977. Aerodynamische Probleme der Integration von Triebwerk und Zelle beim Kampfflugzeugen. Proceedings of the 85th Wehrtechnischen Symposium, Mannheim, Germany.

MacKay, D.J.C. (2009). Sustainable Energy – Without the Hot Air. Cambridge (UK): UIT Cambridge Ltd.

Mattingly, J.D. (1996). Elements of Gas Turbine Propulsion. New York: McGraw-Hill.

Mattingly, J.D., Heiser, W.H., Pratt, D.T. et al. (2018). Aircraft Engine Design, 3e. Washington, DC: AIAA.

Oates, G.C. (ed.) (1985). Aerothermodynamics of Aircraft Engine Components. Washington, D.C.: AIAA Education Series.

Oates, G.C. (1988). Aerothermodynamics of Gas Turbine and Rocket Propulsion. Washington, DC: AIAA.

Oates, G.C. (ed.) (1989). Aircraft Propulsion Systems Technology and Design. Washington, D.C.: AIAA Education Series.

Shepherd, D.G. (1972). Aerospace Propulsion. New York: American Elsevier Publication.

Sutton, G.P. and Biblarz, O. (2001). Rocket Propulsion Elements, 7e. New York: Wiley.

Varvill, R. and Bond, A. (2003). A Comparison of Propulsion Concepts for SSTO Reusable Launchers. *Journal of British Interplanetary Society* 56: 108–117.

Vos, R. and Farokhi, S. (2015). Introduction to Transonic Aerodynamics. Heidelberg, Germany: Springer Dordrecht.

2

Aircraft Aerodynamics – A Review

2.1 Introduction

The purpose of this chapter is to review the principles of compressible flows as they apply to aircraft and cover the traditional areas of undergraduate education in aerodynamics. It is intended to serve as a companion chapter to the main topics in the book on future propulsion systems and energy sources for sustainable aviation. By necessity, detailed step-by-step mathematical derivations are mostly avoided, with an emphasis placed on physics of fluids.

Aerodynamic forces and moments on an airplane stem from shear and pressure distribution on the body that integrate into a resultant force and moment. The resultant aerodynamic force has a component in the flight direction, known as *drag*. It also has a component normal to the flight direction, known as *lift*. The aerodynamic drag is a retarding force that acts on the vehicle, which needs to be overcome, or balanced by engine(s) thrust. In level steady flight, which is unaccelerated, thrust and drag are equal in magnitude. Since the dawn of flight, aircraft drag reduction has occupied a special place on aerodynamics research landscape that always involves deeper understanding of the underlying flow physics and system design optimization. Interestingly, lift also contributes to aircraft drag, known as the *induced drag*. In this chapter, we review briefly the underlying principles of aircraft aerodynamics, with emphasis on drag, but without lengthy derivations and proofs of the adopted correlations that are beyond the scope of the present book.

There are 10 main contributors to aircraft drag:

1) Friction drag based on the integral of wall shear stress, with its origin in fluid viscosity, no-slip boundary condition, and boundary-layer formation
2) Pressure(or form) drag due to boundary-layer formation that causes displacement thickness and possibly separation and wake formation
3) Drag due to lift, or *induced drag* on wings with finite aspect ratio (AR) caused by the trailing vortices in the wake
4) Compressibility drag due to Mach number influence on density and pressure distribution

Future Propulsion Systems and Energy Sources in Sustainable Aviation, First Edition. Saeed Farokhi.
© 2020 John Wiley & Sons Ltd. Published 2020 by John Wiley & Sons Ltd.
Companion website: www.wiley.com/go/farokhi/power

5) Wave drag due to wave formation in transonic and supersonic speeds that radiates energy away from the body
6) Propulsion installation drag due to nacelle aerodynamics, inlet, and nozzle integration, which are called *throttle-dependent drag*
7) Interference drag due to wing-pylon-nacelle flow interactions
8) Trim drag, e.g. caused by the deflection of elevator and/or rudder trim tab(s) to maintain steady flight
9) Landing gear and spoiler drag in landing–takeoff (LTO) cycle
10) Environmentally induced drag, e.g. rain or particulate flow, icing, surface roughness

Let's go through the vocabulary of aerodynamic drag first. Integrating shear stress on the wetted surface of a body in the flow direction yields *friction drag*. Integrating static pressure around a closed body, in the direction of flight, produces *pressure drag*. The sum of the friction and pressure drag in 2D (e.g. airfoil) is called *profile drag*. *Induced drag* is caused by the vortical wake downstream of a lifting body, which induces downwash on the wing. Downwash effectively *tilts* the plane of local flow on the wing and creates induced drag. *Compressibility drag* in subsonic flow is caused by fluid density variation that affects pressure and shear distribution on a body. Once the flight Mach number exceeds a "critical" value, $M > M_{crit}$, Mach and shock waves are formed on the body. These waves are oblique with respect to flight direction and radiate energy away from the body. Upon integration of the pressure distribution on such bodies, even in the inviscid limit, a drag force appears. Since the source of this drag force is rooted in wave formation, the force is called *wave drag*. In transonic flow, shock waves interact with the boundary layer on the body and often cause flow separation and thus a rapid rise in the vehicle drag. This phenomenon is known as *drag divergence*. In air-breathing propulsion, inlet(s) and exhaust nozzle(s), pylons, open rotors and propellers, as well as other sources of interference with aircraft wing and fuselage contribute to and alter the basic aircraft drag. Due to its origin, this is called *propulsion installation drag*.

Interference drag is due to the altered flowfield that is caused by the integration of wing, fuselage, and empennage. It is customary to include interference drag in vehicle *zero-lift drag*, which is then supplemented by the drag due to lift to represent vehicle drag.

An additional form/source of drag is called *trim drag*. To achieve longitudinal stability in level, constant altitude flight, or in perturbed flight conditions, deflection of control surfaces, e.g. elevators and/or flaps, are required. The increment of drag due to deflected control surfaces (elevator and rudder) to achieve trim condition is the trim drag. As this subject relates to stability and control of aircraft, it is not treated here directly. However, aerodynamic drag due to deflected control surfaces is presented.

For additional information on thrust and drag of aircraft in flight, Covert (1985) should be consulted. The classical fluid-dynamic drag book by Hoerner (1965) is a valuable resource for students of aerodynamics. The companion book on fluid-dynamic lift (also a classic) by Hoerner and Borst (1985) should be consulted. More detailed studies of aerodynamics are found in textbooks such as Anderson (2016), Kuethe and Chow (2009), Bertin and Cummings (2013), and Vos and Farokhi (2015), among many others. Figure 2.1 identifies the elements of aircraft drag on a *drag tree*.

Similarity parameters and boundary conditions in viscous, thermally conducting compressible flows are fundamental to the study of aircraft aerodynamics and is presented briefly before we take on the boundary-layer discussion.

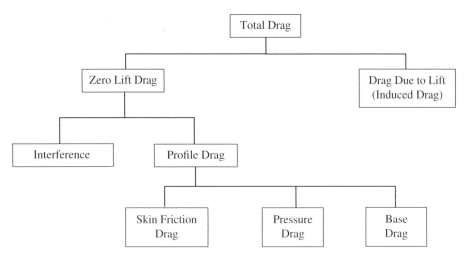

Figure 2.1 Aircraft drag tree.

2.2 Similarity Parameters in Compressible Flow: Flight vs. Wind Tunnel

Using the principles of dimensional analysis and the Buckingham Pi Theorem, a group of nondimensional parameters appear that govern the dynamics of fluid flow. In addition to Mach number, M, and the ratio of specific heats, γ, to simulate compressibility effects, we have Reynolds number to simulate the viscous effects and the Prandtl number to simulate thermal diffusivity (or conduction heat transfer) in a fluid. Discussion and derivation of similarity parameters may be found in fluid mechanics/aerodynamics books, e.g. Anderson (2012, 2016).

The main similarity parameters in a viscous fluid flow that are unaffected by gravity (i.e. buoyancy-driven effects), with reference to freestream conditions, are:

1) $\gamma_\infty \equiv \dfrac{c_{p\infty}}{c_{v\infty}}$. Ratio of specific heats in a gas is a gas property, which is related to the number of the degrees of freedom in a gas molecule

2) $M_\infty \equiv \dfrac{V_\infty}{a_\infty}$. Mach number is the ratio of local speed of gas to the speed of sound

3) $\mathrm{Re}_{\infty l} \equiv \dfrac{\rho_\infty V_\infty l}{\mu_\infty}$. Reynolds number based on characteristic length, l, is the ratio of inertia-to-viscous forces in a viscous fluid flow

4) $\mathrm{Pr}_\infty \equiv \dfrac{\mu_\infty c_{p\infty}}{k_\infty}$. Prandtl number is a fluid property, which is also the ratio of viscous-to-thermal diffusivity in a thermally conducting fluid

Note that we left out the Knudsen number, Kn, which is defined as the ratio of the molecular mean-free-path in the gas to a characteristic length scale of the problem. Unless we are flying in rarefied gas (i.e. at very high altitudes) where the mean-free-path of the gas molecules is of the same order of magnitude as the vehicle characteristic length, air is treated as continuum. For reference, we note that the mean-free path at 80-mi altitude is about 1 ft (i.e. about 30.5 cm at 128 km altitude). Thus, the Knudsen number does not impact our results in dynamic similitude.

Upon matching of the similarity parameters between the flight condition and the (scale model in the) wind tunnel, the two flows are called *dynamically similar*. The mathematical implication is that the shape of the streamlines is similar and the pressure coefficients at corresponding points in the two flows are also equal. Therefore, the non-dimensional aerodynamic force and moment coefficients are the same between the scale model in the wind tunnel and the flight.

However, another complexity arises that stems from boundary-layer transition point in flight that is different from the scale model in the wind tunnel. The use of trip strips on scale models in wind tunnels offers a practical, but crude "fix" to the problem of transition mismatch. In addition, transonic speeds, i.e. $M_\infty \approx 1$, pose a different kind of complexity, namely test section choking and reflected waves from the solid wind tunnel walls that impinge on the aircraft model and contaminate the measurements. These reflected waves in wind tunnels, as well as choking, have no counterparts in flight (as there are no walls to reflect the waves or choke the flow). As a result, there are dedicated transonic wind tunnel facilities in the United States and in Europe with slotted test section walls to prevent choking and cancel wave reflections. Figure 2.2 shows the

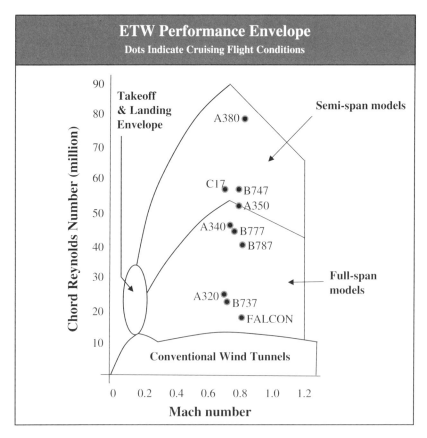

Figure 2.2 European Transonic Windtunnel (ETW) Mach-Reynolds number envelope. *Source:* Data from www.etw.de.

performance envelope of the European Transonic Windtunnel (ETW), i.e. Mach number versus chord Reynolds number.

To assist with Reynolds number simulation in wind tunnels, cryogenic gas (e.g. nitrogen) is used. For example, the use of nitrogen in cryogenic wind tunnels produces a *fivefold increase in Reynolds number* simulation as compared to a conventional (air) wind tunnel. Nitrogen is a good choice since it is a diatomic gas and constitutes 79% of air (by volume). Comprehensive catalogs on wind tunnels are compiled and edited by Peñaranda and Freda (1985a,b). These are valuable resources on facilities, capabilities, and their instrumentation. Handbook on subsonic wind tunnel testing by Alexander (1991) and the classical book by Rae and Pope (1984) are all valuable references.

At hypersonic speeds, aircraft create strong shock waves and extreme thermal loads. The gas mixture downstream of strong shocks becomes hot and thus reacting. This is where dissociation (of O_2, N_2) and ionization (of O and N) occurs, and the aerodynamic simulation in ground test takes on new challenges. To combat high temperature effects, cryogenic wind tunnels are developed that simulate flows with high Reynolds number and high Mach number. Finally, the static aeroelastic effects of the scaled models in wind tunnel and in flight, as well as dynamic aeroelastic effects, e.g. flutter, are important subjects but are beyond the scope of this book. The hypersonic aerodynamic books by Anderson (1989) and Hayes and Probstein (1959) are highly recommended for high-temperature gas dynamics. For lessons learned in reaching hypersonic speeds, read the *History of Hypersonics* by Heppenheimer (2007).

2.3 Physical Boundary Conditions on a Solid Wall (in Continuum Mechanics)

The viscous flow adheres to solid walls and thus assumes a *no-slip boundary condition* on the surface. Thus, the three components of fluid velocity, u, v, and w, identically vanish at the wall:

$$u = v = w = 0 \text{ at the wall } \left(\text{i.e. at a solid wall all relative velocity components vanish}\right)$$

The general statement of no-slip boundary condition is that the fluid tangential (and normal) velocity at the wall is the same as the tangential (and normal) velocity of the wall. This allows the wall to be moving and not necessarily be stationary. In addition, if the wall is porous, the velocity component normal to the wall is finite (specified by the porosity at the wall). The slope of the velocity profile at the wall is proportional to the wall shear stress, i.e. in 2-D flows, using Cartesian coordinates, (x, y), with velocity components (u, v), we have:

$$\left(\tau_{yx}\right)_w = \mu\left(\frac{\partial u}{\partial y} + \frac{\partial v}{\partial x}\right)_w \approx \mu\left(\frac{\partial u}{\partial y}\right)_w \tag{2.1}$$

We used Newton's law of friction in Eq. (2.1), which applies to *Newtonian fluids*. The latter approximation in Eq. (2.1) is often valid because:

$$\partial v / \partial x \ll \partial u / \partial y \tag{2.2}$$

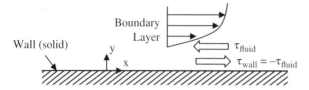

Figure 2.3 The wall shear stress in a viscous fluid causes friction drag on the wall.

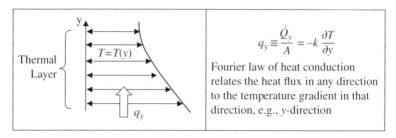

Figure 2.4 Temperature distribution in a viscous fluid (in the *y*-direction).

where the length scale in *x*-direction is proportional to the length of the body, e.g. chord length, c, whereas the lateral length scale, i.e. *y*, scales with the boundary-layer thickness, δ, which is much smaller than c. This suggests that there is a larger gradient in the *lateral* direction to a boundary layer than in the *streamwise* direction along a boundary layer on a flat plate. This assertion is correct as long as the boundary layers are thin and attached. The wall shear stress is equal and opposite of the shear stress acting on the fluid (at the wall). This is graphically shown in Figure 2.3.

Thermal boundary conditions at a solid wall are also governed by *no-slip*, which states that the gas adjacent to (or in contact with) the wall attains the wall temperature, T_w, in equilibrium state. The governing equation for heat conduction at the wall is (the Fourier law of heat conduction):

$$\left(q_y\right)_{wall} = -k\left(\frac{\partial T}{\partial y}\right)_{wall} \tag{2.3}$$

where q_y *is the heat flux in* *y*-*direction,* *k* *is the thermal conductivity of the fluid.* Figure 2.4 shows the Fourier law of heat conduction, in the *y*-direction in a thermal layer. In this example $\partial T/\partial y$ is negative and q_y is positive, consistent with the Fourier Law.

An *adiabatic wall* is defined when $q_w = 0$, and thus Eq. (2.3) implies that the temperature gradient at the wall must vanish. The adiabatic wall temperature, T_{aw}, is an important reference parameter in heat transfer studies. If the wall temperature is higher than the adiabatic wall temperature, i.e. $T_w > T_{aw}$, the fluid heats the wall. For $T_w < T_{aw}$, the fluid is cooled by the wall. In the two cases, heat transfer is in the opposite direction to the temperature gradient. Figure 2.5 shows the three thermal boundary conditions at the wall, namely wall is heated, cooled, or is adiabatic.

We will discuss adiabatic wall temperature, T_{aw}, in compressible boundary layer, Section 2.4.1.

(a) (b) (c)

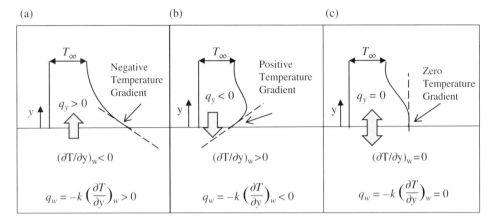

Figure 2.5 Thermal boundary conditions on a (solid) wall: (a) heated wall ($T_w > T_{aw}$); (b) cooled wall ($T_w < T_{aw}$); (c) adiabatic wall, $T_w = T_{aw}$.

2.4 Profile and Parasite Drag

Viscous, pressure, and induced drags are initially treated at the incompressible limit, i.e. $M = 0$. In addition, in 2-D cases, such as airfoils, the sum of the friction and pressure drag is combined to form the *profile drag*. On three-dimensional objects, such as aircraft, the sum of friction and pressure drag is called *parasite drag*.

In flows with a high Reynolds number, boundary-layer formation and its behavior on bodies in motion hold the key to all viscous aerodynamic studies. These include profile and parasite drag. Therefore, we treat briefly boundary layers first, for reference.

2.4.1 Boundary Layers

In the limit of very large Reynolds number for streamlined bodies, the viscous-dominated region becomes confined to a very thin layer next to the body, which Prandtl (1904) called *Grentzschicht* (German). *Grentzschicht* translates into *boundary layer*. The boundary-layer equations are based on the Navier-Stokes equations by applying order-of-magnitude arguments to a thin layer where viscous effects dominate. Consider a flat plate in a uniform two-dimensional viscous flow. The x-velocity, u, scales as the velocity at the *edge* of the boundary layer, U_e, the y-velocity, v is small, the x-coordinate scales as the length of the body, e.g. chord length of an airfoil, the y-coordinate scales as the boundary-layer thickness, δ.

The result of such order of magnitude study is the following four equations governing the continuity, two momentum equations and the energy equation in a compressible boundary layer. Viscous and thermal transport coefficients are conventionally called μ and k, where μ is the coefficient of viscosity and k is the fluid thermal conductivity coefficient.

Steady, two-dimensional continuity is satisfied when:

$$\frac{\partial(\rho u)}{\partial u} + \frac{\partial(\rho v)}{\partial y} = 0 \tag{2.4}$$

Steady, x-momentum equation in 2-D demands:

$$\rho u \frac{\partial u}{\partial u} + \rho v \frac{\partial u}{\partial y} = -\frac{dp_e}{dx} + \frac{\partial}{\partial y}\left(\mu \frac{\partial u}{\partial y}\right) \tag{2.5}$$

$\underbrace{\hspace{3cm}}$ \quad $\underbrace{\hspace{3cm}}$

Fluid inertia force/volume \qquad Viscous force/volume

Y-momentum equation simply yields the condition of constant pressure normal to the wall in the boundary layer, i.e.

$$\frac{\partial p}{\partial y} = 0 \tag{2.6}$$

Steady, 2-D, energy equation in the absence of external heat and shaft work produces:

$$\rho u \frac{\partial h}{\partial x} + \rho v \frac{\partial h}{\partial y} = \frac{\partial}{\partial y}\left(k\frac{\partial T}{\partial y}\right) + u\frac{dp_e}{dx} + \mu\left(\frac{\partial u}{\partial y}\right)^2 \tag{2.7}$$

Subscript "e" stands for the *edge* of the boundary layer, h is the enthalpy and ρ is density.

2.4.1.1 Case 1: Incompressible Laminar Flow

Here, we impose the incompressibility condition, i.e. $\rho =$ constant, in the flow and thus the continuity Eq. (2.4) reduces to:

$$\frac{\partial u}{\partial x} + \frac{\partial v}{\partial y} = 0 \tag{2.8}$$

Within the boundary layer, the fluid inertia force and the viscous force are in balance. Using Eq. (2.5), we get:

$$\rho U_e^2/x \sim \mu U_e/\delta^2 \tag{2.9}$$

This shows that the laminar boundary-layer thickness, δ, scales as inverse of the square root of Reynolds number:

$$\delta \sim \sqrt{\frac{\mu x}{\rho U_e}} \sim \frac{x}{\sqrt{Re_x}} \tag{2.10a}$$

Therefore,

$$\frac{\delta}{x} \sim \frac{1}{\sqrt{Re_x}} \sim \sqrt{v/U_e x} \tag{2.10b}$$

We define a stream function, ψ, according to $u = \partial\psi/\partial y$ and $v = -\partial\psi/dx$, which satisfies the (incompressible) continuity equation identically. Here, we also invoke the principle of self-similarity in nondimensional velocity profile, u/U_e, expressed in nondimensional lateral coordinate, or similarity variable η, namely, we define

$$\eta \sim \frac{y}{\delta} \quad \text{which gives} \quad \eta \equiv \frac{y}{\sqrt{\dfrac{vx}{U_e}}} \tag{2.11}$$

where v is the kinematic viscosity (μ/ρ) and since the (incompressible) stream function, ψ, scales as $U_e\delta$, we get

$$\psi \sim U_e\delta \sim \frac{U_e x}{\sqrt{\rho U_e x / \mu}} \sim \sqrt{v U_e x} \tag{2.12}$$

This allows for the general form of the stream function to be expressed in terms of an unknown dimensionless function of lateral nondimensional coordinate η according to:

$$\psi \sim \sqrt{v U_e x}\, F(\eta) \tag{2.13}$$

Note that self-similarity assumption resulted in the separation of variables, i.e. in x and y. In Eq. (2.13), $F(\eta)$ is called the nondimensional stream function and the dimensions of ψ are in the square root preceding $F(\eta)$. The streamwise velocity, u, is

$$u = \frac{\partial \psi}{\partial y} = \frac{\partial \psi}{\partial \eta}\frac{\partial \eta}{\partial y} = U_e F'(\eta) \tag{2.14}$$

The lateral, or y-component of, velocity, v, is

$$v = -\frac{\partial \psi}{\partial x} = \frac{\partial \psi}{\partial \eta}\frac{\partial \eta}{\partial x} = \frac{1}{2}\sqrt{\frac{v U_e}{x}}\left(\eta F' - F\right) \tag{2.15}$$

Now, if we substitute for these and their derivatives in x-momentum equation, and after some simplification, we get the *Blasius equation* for laminar boundary layers on a flat plate:

$$FF'' + 2F''' = 0 \tag{2.16}$$

This is a third-order nonlinear ordinary differential equation in $F(\eta)$, or equivalently, the nondimensional stream function, ψ, which requires three boundary conditions, namely F, F', and F'' to be specified at the wall and/or freestream. These are the no-slip velocity condition, which sets u and v equal to zero at the wall as well as the merging velocity magnitude to the freestream value at large distances from the wall, i.e.

1) $u = 0$ sets $F' = 0$ at the wall (from Eq. (2.14)).
2) $v = 0$ sets $F = 0$ at the wall (from Eq. (2.15)).
3) Outside the boundary layer, $u = U_e$, therefore, $F' = 1$ as $\eta \to \infty$.

Note that the velocity merges smoothly with the freestream, asymptotically, at large distances from the wall. Conventionally, it is taken at the *edge* of the boundary layer where the fluid attains 99% of the freestream velocity. The slope of u at large distances vanishes forcing F'' to be zero as $\eta \to \infty$ (from the η-derivative of Eq. (2.14)).

The Blasius equation is a nonlinear ordinary differential equation (ODE) and it has to be solved numerically. Since it is stated as a boundary-value problem (BVP), we need to use the so-called *shooting method* (or the *predictor-corrector method*) to convert it to an IVP (initial-value problem). This is accomplished with a guessed value of F'' at $\eta = 0$, which allows the use of a fourth-order accurate Runge-Kutta method to solve the IVP in marching away from the wall. We then check for the convergence of the boundary condition at large distances (i.e. $F'' = 0$ *and* $F' = 1$ as $\eta \to \infty$), to accept the initial condition F'' at the wall that we had assumed. Upon convergence, the function $F(\eta)$, $F'(\eta)$, and $F''(\eta)$ are tabulated and related to the velocity profiles as well as the shear stress at the wall. The numerical solution of the Blasius boundary-layer equation is tabulated in Table 2.1.

As noted earlier, we define the boundary-layer edge, δ, arbitrarily as the location normal to the wall where the local velocity is at $U_e = 0.99\,U$, where U is the freestream velocity:

$$\frac{\delta(x)}{x} = \frac{5.0}{\sqrt{Re_x}} \tag{2.17}$$

Note that in Table 2.1, at $\eta = 5.0$, $F'(\eta)$, which is u/U_e, approaches 0.9915. Also, skin friction coefficient, c_f is related to $F''(\eta = 0)$, according to:

$$c_f = \frac{\tau_w}{\rho U_e^2 / 2} = \frac{\mu\left(\frac{\partial u}{\partial y}\right)_{y=0}}{\rho U_e^2 / 2} = \frac{2\nu U_e \left(\frac{\partial\left(\frac{u}{U_e}\right)}{\partial\eta}\right)_{\eta=0}\left(\frac{\partial\eta}{\partial y}\right)}{U_e^2} \tag{2.18}$$

From Eq. (2.11),

$$\frac{\partial\eta}{\partial y} = \frac{1}{\sqrt{\dfrac{\nu x}{U_e}}} \tag{2.19}$$

From Eq. (2.14), we relate the slope of velocity profile at the wall to $F''(0)$, according to

$$\left(\frac{\partial\left(\frac{u}{U_e}\right)}{\partial\eta}\right)_{\eta=0} = F''(0) \tag{2.20}$$

Substituting Eqs. (2.19) and (2.20) into (2.18), we get the local skin friction coefficient:

$$c_f(x) = \frac{2F''(0)}{\sqrt{Re_x}} = \frac{0.664}{\sqrt{Re_x}} \tag{2.21}$$

Note that $F''(0) \cong 0.332$ in Table 2.1.

Table 2.1 Numerical solution of the Blasius equation for incompressible laminar boundary layer on a flat plate with zero pressure gradient.

η	F	F'	F''	η	F	F'	F''	η	F	F'	F''
0	0	0	0.33206	3.5	1.837712	0.913046	0.107773	7	5.279272	0.999927	0.00022
0.1	0.00166	0.033206	0.332051	3.6	1.929539	0.923335	0.098086	7.1	5.379265	0.999946	0.000169
0.2	0.006641	0.066408	0.331986	3.7	2.022348	0.932679	0.088859	7.2	5.479261	0.999961	0.000129
0.3	0.014942	0.099599	0.331812	3.8	2.116045	0.941124	0.080126	7.3	5.579257	0.999972	9.75E-05
0.4	0.02656	0.132765	0.331472	3.9	2.210544	0.948721	0.071911	7.4	5.679255	0.999981	7.36E-05
0.5	0.041493	0.165887	0.330914	4	2.305763	0.955524	0.064234	7.5	5.779254	0.999987	5.53E-05
0.6	0.059735	0.198939	0.330082	4.1	2.401624	0.961586	0.057103	7.6	5.879253	0.999992	4.13E-05
0.7	0.081278	0.231892	0.328925	4.2	2.498057	0.966963	0.05052	7.7	5.979252	0.999996	3.07E-05
0.8	0.106109	0.264711	0.327392	4.3	2.594996	0.971708	0.04448	7.8	6.079252	0.999998	2.27E-05
0.9	0.134214	0.297356	0.325435	4.4	2.69238	0.975876	0.038972	7.9	6.179252	1	1.67E-05
1	0.165573	0.329783	0.32301	4.5	2.790154	0.97952	0.033981	8	6.279252	1.000002	1.22E-05
1.1	0.200162	0.361941	0.320074	4.6	2.888268	0.982689	0.029484	8.1	6.379252	1.000003	8.92E-06
1.2	0.237951	0.393779	0.316592	4.7	2.986677	0.985432	0.025456	8.2	6.479252	1.000003	6.47E-06
1.3	0.278905	0.42524	0.312531	4.8	3.085342	0.987795	0.021871	8.3	6.579253	1.000004	4.67E-06
1.4	0.322984	0.456265	0.307868	4.9	3.184225	0.98982	0.018698	8.4	6.679253	1.000004	3.35E-06
1.5	0.370142	0.486793	0.302583	5	3.283296	0.991547	0.015907	8.5	6.779253	1.000005	2.39E-06
1.6	0.420324	0.516761	0.296666	5.1	3.382526	0.993013	0.013465	8.6	6.879254	1.000005	1.7E-06
1.7	0.473473	0.546105	0.290114	5.2	3.481891	0.994251	0.011342	8.7	6.979254	1.000005	1.2E-06
1.8	0.529522	0.574763	0.282933	5.3	3.581369	0.995291	0.009506	8.8	7.079255	1.000005	8.47E-07
1.9	0.5884	0.602671	0.275138	5.4	3.680943	0.996161	0.007928	8.9	7.179255	1.000005	5.93E-07
2	0.65003	0.629771	0.266753	5.5	3.780597	0.996884	0.006579	9	7.279256	1.000005	4.13E-07

(Continued)

Table 2.1 (Continued)

η	F	F'	F''	η	F	F'	F''	η	F	F'	F''
2.1	0.714326	0.656003	0.257811	5.6	3.880316	0.997483	0.005432	9.1	7.379256	1.000005	2.86E-07
2.2	0.7812	0.681315	0.248352	5.7	3.98009	0.997976	0.004463	9.2	7.479257	1.000005	1.98E-07
2.3	0.850556	0.705658	0.238427	5.8	4.079908	0.998381	0.003648	9.3	7.579257	1.000005	1.36E-07
2.4	0.922297	0.728987	0.228093	5.9	4.179764	0.998711	0.002968	9.4	7.679258	1.000005	9.26E-08
2.5	0.996319	0.751265	0.217413	6	4.279648	0.998978	0.002402	9.5	7.779259	1.000005	6.29E-08
2.6	1.072514	0.772461	0.206456	6.1	4.379557	0.999194	0.001934	9.6	7.879259	1.000005	4.25E-08
2.7	1.150774	0.792549	0.195295	6.2	4.479486	0.999368	0.00155	9.7	7.97926	1.000005	2.86E-08
2.8	1.230987	0.811515	0.184007	6.3	4.57943	0.999507	0.001236	9.8	8.07926	1.000005	1.92E-08
2.9	1.313039	0.829349	0.17267	6.4	4.679386	0.999617	0.000981	9.9	8.179261	1.000005	1.28E-08
3	1.396819	0.84605	0.161361	6.5	4.779353	0.999704	0.000774	10	8.279261	1.000005	8.46E-09
3.1	1.482212	0.861625	0.150156	6.6	4.879327	0.999773	0.000608	10.1	8.379262	1.000005	5.58E-09
3.2	1.569107	0.876087	0.139128	6.7	4.979307	0.999827	0.000475	10.2	8.479262	1.000005	3.66E-09
3.3	1.657393	0.889459	0.128347	6.8	5.079292	0.999869	0.00037	10.3	8.579263	1.000005	2.39E-09
3.4	1.746963	0.901767	0.117876	6.9	5.17928	0.999902	0.000286	10.4	8.679263	1.000005	1.55E-09

We also define the three different characteristic deficit thicknesses in the boundary layer, namely mass deficit or *displacement thickness* δ^*, the momentum deficit thickness θ, and (kinetic) energy deficit thickness δ^{**} as follows:

$$\rho_e U_e \delta^* \equiv \int_0^\infty \left(\rho_e U_e - \rho u \right) dy \tag{2.22a}$$

From Eq. (2.22a), we get the displacement thickness in a compressible boundary layer as

$$\delta^* \equiv \int_0^\infty \left(1 - \frac{\rho u}{\rho_e U_e} \right) dy \tag{2.22b}$$

Assuming the fluid may be treated as incompressible, we have the simplified form of the displacement thickness as:

$$\delta^* \equiv \int_0^\infty \left(1 - \frac{u}{U_e} \right) dy \tag{2.22c}$$

By defining a new nondimensional variable η, we may express the nondimensional displacement thickness, δ^*, as

$$\delta^* = \sqrt{\frac{\nu x}{U_e}} \int_0^\infty \left(1 - \frac{u}{U_e} \right) d\eta \quad \text{where } \eta \equiv \frac{y}{\sqrt{\nu x / U_e}} \tag{2.22d}$$

$$\frac{\delta^*}{x} = \frac{1}{\sqrt{Re_x}} \int_0^\infty \left(1 - F' \right) d\eta \tag{2.22e}$$

The *momentum deficit thickness*, θ, in a boundary layer is defined as:

$$\rho_e U_e^2 \theta \equiv \int_0^\infty \rho u \left(U_e - u \right) dy \tag{2.23a}$$

We may isolate θ in Eq. (2.23a) to get:

$$\theta \equiv \int_0^\infty \frac{\rho u}{\rho_e U_e} \left(1 - \frac{u}{U_e} \right) dy \tag{2.23b}$$

The momentum deficit thickness, θ, for an incompressible boundary-layer flow is:

$$\theta \equiv \int_0^\infty \frac{u}{U_e} \left(1 - \frac{u}{U_e} \right) dy \tag{2.23c}$$

Using the nondimensional coordinate, η, the momentum deficit thickness is written as:

$$\theta = \sqrt{\nu x / U_e} \int_0^\infty \frac{u}{U_e} \left(1 - \frac{u}{U_e} \right) d\eta \tag{2.23d}$$

$$\frac{\theta}{x} = \frac{1}{\sqrt{Re_x}} \int_0^\infty F'(1-F')d\eta \tag{2.23e}$$

The *kinetic energy deficit thickness*, δ^{**}, in the boundary layer is based on the definition:

$$\rho_e U_e^3 \delta^{**} \equiv \int_0^\infty \rho u \left(U_e^2 - u \right) dy \tag{2.24a}$$

The (kinetic) energy deficit thickness, δ^{**}, may be isolated in Eq. (2.24a) and expressed as:

$$\delta^{**} \equiv \int_0^\infty \frac{\rho u}{\rho_e U_e} \left(1 - \frac{u^2}{U_e^2} \right) dy \tag{2.24b}$$

The incompressible boundary layer, i.e. $\rho = $ constant, simplifies Eq. (2.24b) to:

$$\delta^{**} \equiv \int_0^\infty \frac{u}{U_e} \left(1 - \frac{u^2}{U_e^2} \right) dy \tag{2.24c}$$

Finally, using nondimensional coordinate η, we express the energy deficit thickness for an incompressible boundary layer on a flat plate as:

$$\delta^{**} = \sqrt{vx/U_e} \int_0^\infty \frac{u}{U_e} \left(1 - \frac{u^2}{U_e^2} \right) d\eta \tag{2.24d}$$

$$\frac{\delta^{**}}{x} = \frac{1}{\sqrt{Re_x}} \int_0^\infty F'\left(1 - F'^2 \right) d\eta \tag{2.24e}$$

Therefore, by integrating $(1 - F')$ in Eq. (2.22e), across the boundary layer, we get the nondimensional displacement thickness in laminar boundary layer on a flat plate as:

$$\frac{\delta^*(x)}{x} = \frac{1.721}{\sqrt{Re_x}} \tag{2.25}$$

By integrating $F'(1 - F')$ across the boundary layer, the nondimensional momentum deficit thickness in Blasius boundary layer is calculated to be:

$$\frac{\theta(x)}{x} = \frac{0.664}{\sqrt{Re_x}} \tag{2.26}$$

Similarly, by integrating $F'(1 - F'^2)$ across the boundary layer, the nondimensional energy deficit thickness in laminar incompressible boundary layer is calculates to be:

$$\frac{\delta^{**}(x)}{x} = \frac{1.044}{\sqrt{Re_x}} \tag{2.27}$$

The friction drag on (one side of) a flat plate is the integral of wall shear stress along the surface, i.e.

$$D_f' = \int_0^c \tau_w dx = \frac{\rho_e U_e^2}{2} \int_0^c c_f dx \qquad (2.28)$$

where c_f is the local skin friction coefficient. Defining the overall friction drag coefficient C_f as the *average* of the local skin friction coefficient, we get:

$$C_f \equiv \frac{1}{c} \int_0^c c_f dx \qquad (2.29)$$

By comparing Eqs. (2.28) and (2.29), we conclude that the average skin friction drag coefficient, C_f, is indeed the nondimensional friction drag coefficient on the wall, namely

$$C_f = \frac{D_f'}{\frac{1}{2}\rho_e U_e^2 c} \qquad (2.30)$$

Since friction drag on a flat plate may be calculated from the integral of shear stress along the wall, or equivalently, through the momentum deficit in the wake, we may relate the skin friction drag coefficient C_f to the nondimensional momentum deficit thickness, θ/c at the plate's trailing edge (TE). A definition sketch, with exaggerated boundary-layer thickness, is shown in Figure 2.6 that will help the derivation. Consider a (2-D) control volume ABCD on one side of the plate, as shown in Figure 2.6. AC is the plate and BD is the streamline outside the boundary layer. The mass flow rate between the plate and any streamline remains constant, which reveals a relationship between the height AB and CD in Figure 2.6, namely

$$\rho_e U_e (AB) = \int_C^D \rho u \, dy \qquad (2.31)$$

The (time rate of) change of fluid momentum between the exit and the entrance stations to the flat plate is then:

$$\int_C^D \rho u^2 dy - \rho_e U_e (AB) U_e \qquad (2.32)$$

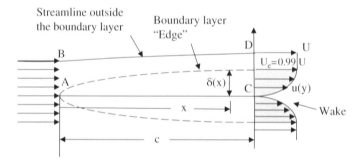

Figure 2.6 Flat plate of length c, at zero incidence, in viscous fluid flow.

This momentum change is caused by the net forces acting on the fluid in the x-direction, namely, the negative of the friction drag force felt by the plate, $-D'_f$. Therefore, we have:

$$-D'_f = \int_C^D \rho u^2 dy - \rho_e U_e (AB) U_e = \int_C^D \rho u^2 dy - \int_C^D \rho u U_e dy = \int_C^D \rho u (u - U_e) dy \qquad (2.33)$$

Note that in deriving Eq. (2.33), we used the continuity equation between the wall and streamline, i.e.

$$\rho_e U_e (AB) = \int_C^D \rho u dy \qquad (2.34)$$

The friction drag coefficient is then expressed as:

$$C_f = \frac{D'_f}{\frac{1}{2}\rho_e U_e^2 c} = \frac{2}{c} \int_C^D \frac{\rho u}{\rho_e U_e} \left(1 - \frac{u}{U_e}\right) dy = 2\left(\frac{\theta}{c}\right) \qquad (2.35)$$

Therefore, using Eq. (2.26), we relate the friction drag coefficient for a laminar boundary layer on a flat plate with zero pressure gradient to the momentum deficit thickness divided by the length of the plate, i.e.

$$C_f = 2\frac{\theta_c}{c} \cong \frac{1.328}{\sqrt{Re_c}} \qquad (2.36)$$

The graph of the velocity profile based on the Blasius solution representing a laminar boundary layer on a flat plate with zero pressure gradient is shown in Figure 2.7. Due to

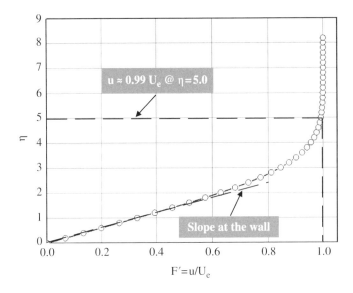

Figure 2.7 Laminar boundary-layer velocity profile on a flat plate based on Blasius solution.

Table 2.2 Summary of laminar boundary-layer characteristic parameters on a flat plate with zero pressure gradient for an incompressible fluid.

Incompressible laminar boundary layer on a flat plate

$\dfrac{\delta(x)}{x} \cong \dfrac{5.0}{\sqrt{Re_x}}$	Local boundary-layer thickness, $\delta(x) \sim x^{1/2}$
$c_f(x) \cong \dfrac{0.664}{\sqrt{Re_x}}$	Local skin friction coefficient, $c_f(x) \sim x^{-1/2}$
$\dfrac{\delta^*(x)}{x} \cong \dfrac{1.721}{\sqrt{Re_x}}$	Local displacement thickness, $\delta^*(x) \sim x^{1/2}$
$\dfrac{\theta(x)}{x} \cong \dfrac{0.664}{\sqrt{Re_x}}$	Local momentum deficit thickness, $\theta(x) \sim x^{1/2}$
$\dfrac{\delta^{**}(x)}{x} \cong \dfrac{1.044}{\sqrt{Re_x}}$	Local energy deficit thickness, $\delta^{**}(x) \sim x^{1/2}$
$C_f \cong \dfrac{1.328}{\sqrt{Re_c}}$	Friction drag coefficient (flat plate)

---

asymptotic nature of the boundary-layer solution, there is no physical "edge" where the velocity reaches the 100% freestream value. Therefore, it is agreed that the "edge" of the viscous layer is reached when the local flow speed is 99% of the freestream velocity. Here the 99% is reached when $\eta \approx 5$.

Table 2.2 is the summary of incompressible laminar boundary-layer characteristic length scales as well as nondimensional frictional characteristics, following Blasius.

2.4.1.2 Case 2: Laminar Compressible Boundary Layers

As the flight Mach number increases, viscous dissipation within the boundary layer becomes significant and thus the fluid experiences aerodynamic heating. Under intense aerodynamic heating conditions, structural cooling, i.e. *thermal management*, becomes necessary and dominates the aerodynamic design, as in hypersonic vehicles (see Heppenheimer 2007). With increasing temperature within the boundary layer, the fluid density decreases and thus the thickness of the boundary layer increases to satisfy continuity. Skin friction is also reduced in harmony with a thickened boundary layer. Figure 2.8 shows the variation of skin friction drag coefficient with flow Mach number and Reynolds number for an adiabatic wall (on a log-log scale). We note that the skin friction drag coefficient decreases with increasing Mach number, both in the laminar as well as the turbulent regimes of the boundary layer. Adiabatic wall temperature on a flat plate is close to the freestream stagnation temperature, except for the appearance of a *recovery factor, r*, in front of the kinetic energy term, namely

$$T_{aw} = T_\infty + r\frac{V_\infty^2}{2c_p} \tag{2.37}$$

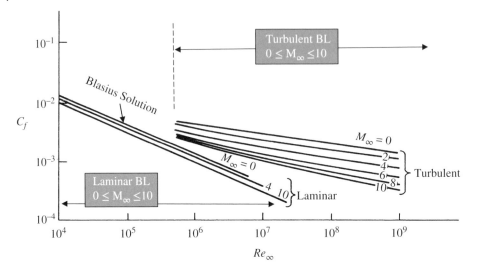

Figure 2.8 Friction drag coefficient in compressible boundary layer With adiabatic wall boundary condition, $\gamma = 1.4$ and $Pr = 1.0$. *Source:* Adapted from van Driest (1951).

For laminar boundary layers on a flat plate, the recovery factor is:

$$r = \sqrt{pr} \tag{2.38}$$

Since Prandtl number of gases is nearly 1 (e.g. for air is 0.71 for a large temperature range), we note that the adiabatic wall temperature is very close to the flow stagnation temperature. The diffusion in thermal boundary layer depends on the fluid Prandtl number as well as the wall boundary condition, e.g. adiabatic wall, cooled/heated wall, or constant temperature wall. The convective heat transfer follows Newton's law of cooling:

$$q = \frac{\dot{Q}}{A} = h(T_{\mathrm{w}} - T_{\mathrm{aw}}) \tag{2.39}$$

where q is the heat flux, A is the area, h is the convective heat transfer coefficient, T_{w} is the wall temperature, and T_{aw} is the adiabatic wall temperature. The parenthesis in Eq. (2.39) is seen as the *driving* mechanism for the heat transfer. However, the heat transfer coefficient h in Eq. (2.39) is a dimensional parameter that is to be established experimentally. Through application of Reynolds analogy between skin friction and heat transfer, we relate the nondimensional local heat transfer coefficient, known as the *Stanton number*, to the local skin friction coefficient and Prandtl number, Pr, according to:

$$St(x) = \frac{1}{2}c_f(x)Pr^{-2/3} \tag{2.40}$$

where

$$St \equiv \frac{q}{\rho u c_p (T_{\mathrm{w}} - T_{\mathrm{aw}})} \tag{2.41}$$

An engineering approach for the effect of compressibility on boundary-layer development was suggested by Eckert (1956). It assumes that the form of the boundary-layer parameters, such as δ/x, $\theta x/$, c_f, C_f and Stanton number, St, remains the same as the incompressible boundary layers, except the fluid properties, ρ, μ, c_p, k are calculated at a *reference temperature*, T^*. Eckert's reference temperature, T^*, depends on the actual and adiabatic wall temperature(s), and freestream temperature according to:

$$T^* = T_\infty + 0.50\left(T_w - T_\infty\right) + 0.22\left(T_{aw} - T_\infty\right) \tag{2.42}$$

We calculate the fluid density, ρ^*, at the reference temperature, T^*. We may use Sutherland's law of viscosity for air that relates the coefficient of viscosity at temperature, T (in K) to the reference values of μ_0 (in kg/ms) and T_0 (in K), to calculate μ^*:

$$\frac{\mu^*}{\mu_0} \cong \left(\frac{T^*}{T_0}\right)^{3/2} \frac{T_0 + 110}{T^* + 110} \tag{2.43}$$

where reference (dynamic) viscosity is $\mu_0 = 1.7894 \times 10^{-5}\ \mathrm{kg\,m^{-1}s^{-1}}$, which corresponds to $T_0 = 288.16\,\mathrm{K}$. Alternatively, we may use properties of air tabulated in gas tables, e.g. by Keenan, Chao, and Kaye (1983). Figure 2.9 shows the coefficient of viscosity of air according to gas tables' data.

We also need specific heat at constant pressure for air as a function of temperature, $c_p(T)$. We may use the tabulated values or graphs, such as Figure 2.10 where data is taken from Keenan, Chao, and Kaye (1983). Figure 2.11 shows the variation of thermal conductivity of air with temperature, $k(T)$, where again the data is taken from the gas tables of Keenan, Chao, and Kaye (1983).

Since the Prandtl number is defined as $\mu c_p/k$, using gas tables for μ, c_p, and k, we arrive at Figure 2.12 where the Prandtl number of (dry) air as a function of gas temperature is plotted. In Figure 2.12, we note that the Prandtl number of (dry) air over a wide temperature range, i.e. between e.g. 700 and 2000 K remains nearly constant at 0.71.

Therefore, reference temperature method of Eckert produces the compressible version of the boundary-layer parameters based on T^*. Boundary-layer thickness, local skin friction coefficient, laminar friction drag coefficient, and local Stanton number for a flat plate in compressible flow, for adiabatic wall, are summarized in Table 2.3.

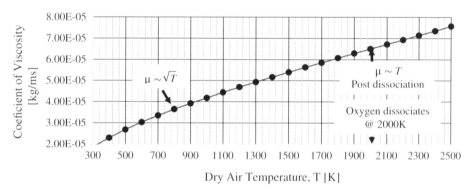

Figure 2.9 Coefficient of viscosity of dry air, $\mu(T)$, in kg/ms, at p = 1.0 atm.

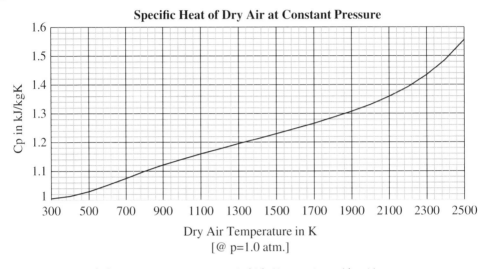

Figure 2.10 Specific heat at constant pressure, c_p in kJ/kgK, at $p = 1$ atm. (dry air).

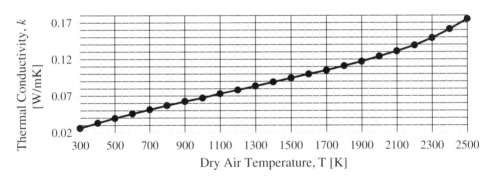

Figure 2.11 Thermal conductivity, k, of dry air in W/mK, at $p = 1$ atm.

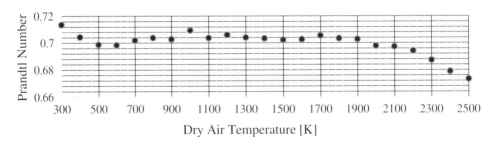

Figure 2.12 Prandtl number of dry air at $p = 1$ atm.

Table 2.3 Summary of compressible laminar boundary-layer parameters on an adiabatic wall with zero pressure gradient.

Compressible laminar boundary layer on a flat plate (adiabatic wall)	
$\delta(x) \cong \dfrac{5.0x}{\sqrt{Re_x^*}}$	Local boundary-layer thickness in compressible flow
where $Re_x^* = \dfrac{\rho^* u_e x}{\mu^*}$	Reynolds number calculated at T*
$c_f^*(x) \cong \dfrac{0.664}{\sqrt{Re_x^*}}$	Local skin friction coefficient in compressible flow
$C_f^* \cong \dfrac{1.328}{\sqrt{Re_c^*}}$	Laminar compressible friction drag coefficient
$St^*(x) = \dfrac{1}{2} c_f^* Pr^{*-2/3} = \dfrac{0.332}{\sqrt{Re_x^*}} Pr^{*-2/3}$	Stanton number calculated at T^*
where $Pr^* = \dfrac{\mu^* c_p}{k^*}$	Prandtl number calculated at T*

2.4.1.3 Case 3: Turbulent Boundary Layers

Strong viscous shear in thin laminar boundary layers inhibit the growth and breakup of the instability waves in the fluid. Through viscous diffusion that thickens the boundary layer, the shear is diminished and can no longer prevent the growth of the instability waves, which include the Tollmien-Schlichting (T-S) waves, or Görtler instability waves on concave walls or Rayleigh waves (cross-flow instability). The graphical depiction in Figure 2.13 (adapted from White (2005)) shows our understanding of a possible path to turbulence through various states of the boundary layer, in a Newtonian fluid. It shows the natural transition on a flat plate with adiabatic wall to fully turbulent flow. This depiction highlights the transition through slowly growing T-S waves, which obey linear stability theory (LST). However, the path to transition may bypass the linear stage altogether and directly enter the turbulent state through nonlinear processes such as surface roughness, screw slots and aft-facing steps, trip strip, among others (e.g. see Malik (1990)).

Turbulence, which is characterized by randomly fluctuating eddies, provides a new vehicle for momentum transport in fluids beyond the molecular viscosity, μ. Since fluctuating eddies transport momentum and kinetic energy across the stream surfaces (i.e. laterally), the fluid behaves in a heightened state of stress, known as *apparent* stress and heightened state of thermal conductivity. Since the apparent stress is caused by eddies, an *eddy viscosity*, ε, is added to the molecular kinematic viscosity, ν, to represent the total viscosity in a turbulent flow, i.e. $\rho(\nu + \varepsilon)$. The same treatment follows thermal conductivity where eddy conductivity, κ, is added to the molecular conductivity, k, to represent the total conductivity of a viscous fluid in turbulent flow, i.e. $(k + \kappa)$.

Velocity and shear profiles in laminar and turbulent boundary layers (TBLs) are shown in Figure 2.14. The presence of eddy viscosity has caused transport of higher momentum fluid to the layers closer to the wall; thus, the velocity profile exhibits lower mass deficit in TBL. Due to larger velocity gradient at the wall, TBL experiences higher

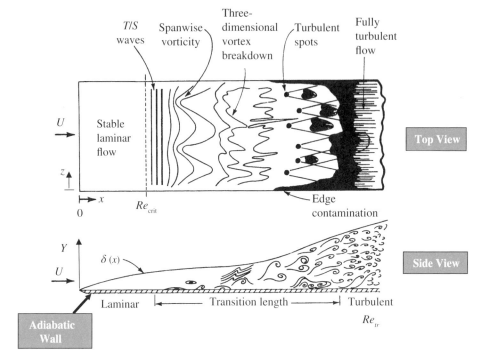

Figure 2.13 Process of natural transition from laminar to turbulent boundary layer for a Newtonian fluid on a flat plate with adiabatic wall. *Source:* Adapted from White (2005).

wall shear stress, as shown. It is indeed this heightened state of shear stress at the wall in TBLs that can withstand adverse pressure gradient in freestream and delay separation. Also note that turbulent fluctuations away from the wall create higher apparent stress within the boundary layer (than the wall). The presence of the wall suppresses turbulent fluctuations; thus, we have a *laminar sublayer* at the wall beneath the turbulent outer layers.

The time-averaged streamwise velocity at the edge, $\bar{U}_e$ shown in Figure 2.14b, in TBL, reminds us of the time-dependent nature of the turbulent flow, which is absent in the

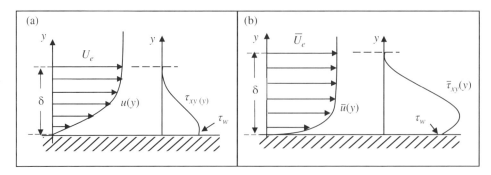

Figure 2.14 Velocity and mean shear profiles in laminar and turbulent boundary layers: (a) laminar boundary-layer profile (velocity and shear); (b) turbulent boundary-layer profile (time-averaged velocity and shear).

laminar flow case. The fluctuating velocity in turbulent flow is thus described by a mean speed, which is the time-averaged $u(t)$ and a root-mean-square of the fluctuation, or *rms* value of the random fluctuation. Denoting the fluctuating components in a 3-D flow as u', v', and w', the *rms* of their product is proportional to the eddy viscosity, ε. Equation (2.44) shows the nine terms in the turbulent stress (also known as Reynolds stress) tensor in Cartesian coordinates:

$$\overline{\overline{\tau}} = -\rho \begin{pmatrix} \overline{u'^2} & \overline{u'v'} & \overline{u'w'} \\ \overline{v'u'} & \overline{v'^2} & \overline{v'w'} \\ \overline{w'u'} & \overline{w'v'} & \overline{w'^2} \end{pmatrix} \tag{2.44}$$

The diagonal terms in the stress tensor (with the multiplier of $-\rho$) in Eq. (2.44), are turbulent normal stresses and off-diagonal terms are turbulent shear stresses. Turbulent kinetic energy, per unit volume, is related to the sum of the diagonal terms, namely,

$$TKE = \frac{1}{2}\rho\left(\overline{u'^2} + \overline{v'^2} + \overline{w'^2}\right) \tag{2.45}$$

The turbulent shear stresses are related to the gradient of mean shear flow, following the Boussinesq assumption, for example:

$$-\rho\overline{u'v'} = \rho\varepsilon\left(\frac{\partial\overline{u}}{\partial y}\right) \tag{2.46}$$

Turbulent stress tensor is symmetrical, i.e. $\tau_{ij} = \tau_{ji}$. For TBLs, the Reynolds number dependence of boundary-layer thickness, displacement thickness, and friction drag coefficient appears as $Re^{-1/5}$ as compared to the laminar boundary-layer case where we had $Re^{-1/2}$. The TBL characteristic lengths and skin friction coefficient are given (without proof) in Eqs. (2.47)–(2.49) where the dependence on $Re^{-1/5}$ appears. For further studies or review of viscous flows and turbulence, textbooks by Schlichting and Gersten (2017), White (2005), or Sherman (1990), among others, are recommended.

$$\frac{\delta(x)}{x} \cong \frac{0.37}{Re_x^{1/5}} \tag{2.47}$$

$$\frac{\delta^*(x)}{x} \cong \frac{0.046}{Re_x^{1/5}} \tag{2.48}$$

$$C_f \cong \frac{0.074}{Re_c^{1/5}} \tag{2.49}$$

The adiabatic wall temperature in TBLs is nearly the same as the stagnation temperature in the freestream, i.e.

$$T_{aw} = T_\infty + r\frac{V_\infty^2}{2c_p} \tag{2.50}$$

Except for the appearance of recovery factor. Here, the recovery factor, r, for TBLs on a flat plate is equal to the cube root of the Prandtl number, i.e.

$$r = \sqrt[3]{Pr} \tag{2.51}$$

The corresponding compressible TBL develops the same form as in (2.47)–(2.49), except the Reynolds number is evaluated at the reference temperature, T^*, and the wall is treated as an adiabatic wall. For example, the skin friction drag coefficient on a flat plate in a compressible TBL is:

$$C_f^* \cong \frac{0.074}{Re_c^{*1/5}} \tag{2.52}$$

The Stanton number for compressible boundary layers from laminar to turbulent cases is summarized in Table 2.4. All "*" quantities in the correlations are evaluated at the reference temperature, T^*.

Table 2.4 Summary of Stanton number correlations in compressible boundary layers on a flat plate for a Newtonian fluid.

Correlations for Stanton number in a compressible boundary layer	
$St_x^* \, Pr^{*2/3} = 0.332 \, Re_x^{*-1/2} \quad Re_x < 5x10^5$ Laminar BL, flat plate	(2.53)
$St_x^* \, Pr^{*2/3} = 0.0296 \, Re_x^{*-1/5} \quad 5x10^5 < Re_x < 10^7$ Turbulent BL, flat plate	(2.54)
$St_x^* \, Pr^{*2/3} = 0.185\left(\log Re_x^*\right)^{-2.584} \quad 10^7 < Re_x < 10^9$ Turbulent BL, flat plate	(2.55)

2.4.1.4 Case 4: Transition

Transition in real flows is a complex phenomenon even for a Newtonian fluid. Complexity stems from numerous contributors to the onset and the length of transition region. The phenomenon of the T-S instability wave growth and subsequent saturation describes the *natural transition* in the boundary layer.

In engineering applications, it is customary to define a *critical Reynolds number*, Re_{crit}, which corresponds to the point where the instability waves, e.g. T-S waves, become neutrally stable. It is indeed descriptive of the state of the boundary layer, which undergoes a *transition* phase between its stable and unstable states. Here, *stable* refers to the state of the flow where instability waves decay, *transition* starts after the instability waves are neutrally stable, and *unstable* state corresponds to the breakup of the saturated instability waves into turbulent spots, which subsequently grow into full turbulence. Surface roughness, curvature, vibration, freestream turbulence intensity, and wall thermal boundary condition all impact the onset and length of the transition. Therefore, the detailed process/stages of transition is very complex and is still under intense fundamental study.

Morkovin (1969), Schlichting and Gersten (2017), Smith (1956), and Faber (1995), among others, outline the theoretical foundation for transition and turbulence introducing concepts of *receptivity* and *stability theory*. An example of the pathway to transition that bypasses the classical T-S wave instability model is transition on swept wings.

In this case, *cross-flow instability*, which is an inherent inviscid instability, is the dominant mechanism for transition (see Glauser et al. 2014). We address these issues in this section, albeit briefly.

In elementary fluids, we learned that a dozen factors affect transition location and its length in real flows:

1) Surface curvature, i.e. concave, convex or 3-D surfaces with compounded curvature, due to centrifugal instability
2) Surface (relative) roughness, i.e. standard roughness or environmentally induced roughness
3) Freestream turbulence intensity, $Tu_\infty \equiv 100x\sqrt{\frac{1}{3}[\overline{u'^2} + \overline{v'^2} + \overline{w'^2}]}/U_\infty$ (for nonisotropic turbulence), or $Tu_\infty \equiv 100x\sqrt{\overline{u'^2}}/U_\infty$ for isotropic turbulence where $\overline{u'^2} = \overline{v'^2} = \overline{w'^2}$
4) Wall heat transfer, e.g. heated, cooled, or adiabatic wall, where wall cooling stabilizes the boundary layer and heating promotes earlier transition in air
5) Two- and three-dimensional pressure gradient, in part related to surface curvature and in part caused by rotation, e.g. rotor blades in turbomachinery
6) Boundary-layer bleed or mass injection at the wall, where bleed stabilizes boundary layer and delays transition
7) Gaps or steps in multielement surfaces, e.g. wing slats, flaps
8) Compressibility effects, i.e. Mach effect and wave formations
9) Hypersonic, high-temperature effects with thermal boundary condition at the wall
10) Freestream oscillation, i.e. $U_\infty(t)$, or body oscillation, $V_{body}(t)$, e.g. pitching and plunging airfoil
11) Compliant wall/skin
12) Use of surfactants

Due to the complexity of transition phenomenon, any empirical "rule" on transition location is, by necessity, case-specific, and an estimation. One such rule is known as the critical Reynolds number rule, which states that the transition location on a flat plate, for example, occurs in a specific range of Reynolds numbers. The range starts with a *lower limit* and extends to an upper limit of critical Reynolds number. For instance, transition Reynolds number on a flat plate is specified within the following (broad) range:

$$Re_{crit} = 350{,}000 \text{ to } 2.8\times10^6 \tag{2.56}$$

For example, in compressor aerodynamics, we often take 500 000 as the lower limit of the critical Reynolds number based on blade chord. We recognize the limitation of this simplistic approach. Another approach examines the ratio of displacement-to-momentum thickness in a boundary layer. This ratio is called the *shape factor*, H_{12}:

$$H_{12} \equiv \frac{\delta^*}{\theta} \tag{2.57}$$

There is a correlation between the shape factor and the critical Reynolds number. Figure 2.15a shows the approximate correlation of the shape factor and its relation to transition on a flat plate according to Wazzan et al. (1981) and Figure 2.15b shows shape factor variation on NACA 63_3-018 at $\alpha = 2°$ airfoil. We note from Figure 2.15 that the onset of transition for flat plate and an airfoil at small angle of attack corresponds to H_{12}

(a)

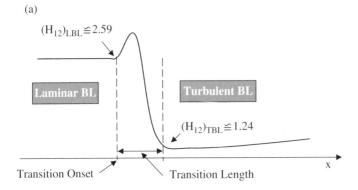

(b)

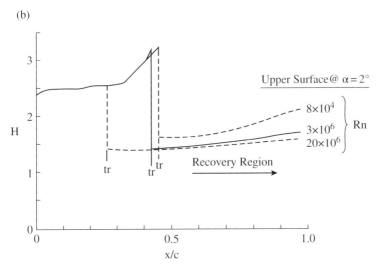

Figure 2.15 Distribution of shape factor, H_{12}, on a flat plate and on an airfoil through transition, (for an adiabatic wall): (a) flat plate; (b) shape factor variation on the NACA 63_3-018. *Source:* From McMasters and Henderson 1979.

of about 2.59. However, the location where transition ends is widely different and depends on Reynolds number as well as freestream turbulence and other factors listed earlier in this section.

Another popular and physically sound approach to establishing transition onset is based on LST. LST was used by Smith (1952) and confirmed by van Ingen (1956), and they showed that a relation exists between the amplification rate of the unstable disturbances (i.e. TS waves) with pressure gradient and freestream turbulence and the transition point. Their approach is called e^N-method, where N is the natural log of the ratio of the amplitude of the most unstable wave to the amplitude of the neutral instability wave. The empirically established values of N predominantly fall in the 8–10 range. For the application of e^N-method to a variety of airfoils, see Cebeci (1999). For a tabulated catalog of experimental flows and their correlation to the N-values, Malik (1990) may be consulted. Further treatment of transition is beyond the scope of this chapter.

2.4.2 Profile Drag of an Airfoil

As noted earlier, profile drag is the sum of the friction drag and pressure (or form) drag on airfoil. Figure 2.16 shows a definition sketch of a plain airfoil. The characteristic length scales in an airfoil are (a) the chord length, (c, b) the maximum thickness, t_{max}, maximum camber, z_{max} and the leading-edge radius, $r_{l.e.}$ The shape of the trailing edge is often sharp with either cusp or finite trailing edge angle, or rounded, or with finite TE thickness, as shown in Figure 2.17.

There is a physical boundary condition, known as Kutta condition, imposed at the sharp trailing edge of airfoils, which states that the static pressure is continuous at the trailing edge. For finite TE angle (as shown in Figure 2.17a), velocity at the TE is zero (i.e. it becomes an aft stagnation point), whereas in cusp TE the velocity on the upper and lower surfaces emerge at the TE with equal magnitude and direction. Since thin airfoil theory is based on the vortex sheet model, the Kutta condition demands that the strength of the vortex sheet to vanish at the TE, to guarantee continuous static pressure at TE. There is no counterpart of the Kutta condition at a *rounded TE* or finite TE thickness airfoil, as shown in Figure 2.17c,d. Finite TE thickness (Figure 2.17d) is used in the design of supercritical airfoils.

The airfoil forces and moments are defined in Figure 2.18 where an airfoil is in a freestream, with V_∞ at a positive angle of attack (AoA), α. Lift per unit span, L', is the force generated by the airfoil that is normal to the freestream flow direction, drag force per unit span, D', is the force that acts on the airfoil in the direction of the freestream, and the pitching moment about point O is shown positive in the clockwise direction (i.e. in the direction of increasing α). The vector sum of the lift and drag forces is shown as the airfoil resultant force.

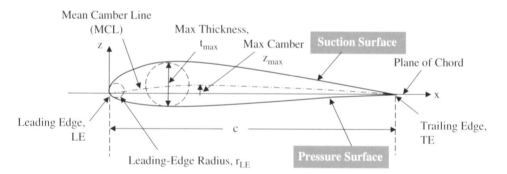

Figure 2.16 Definition sketch of a plain airfoil.

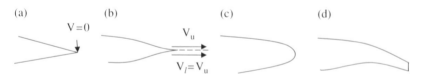

Figure 2.17 Airfoil trailing edge shapes and boundary conditions: (a) finite TE angle; (b) cusp TE; (c) rounded TE; and (d) finite TE thickness.

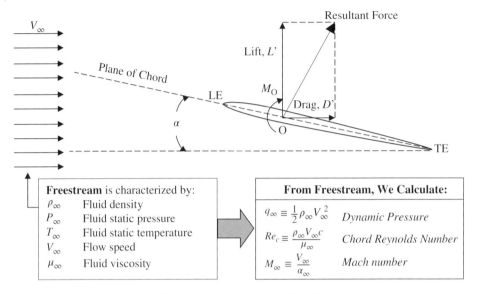

Figure 2.18 Definition sketch of forces and moments on an airfoil in angle of attack, α.

For plain airfoils, the sectional (i.e. 2-D) lift, drag, and pitching moment coefficients about O are nondimensional parameters that are defined in functional form in Eqs. (2.58)–(2.60):

$$c_l \equiv \frac{L'}{q_\infty c} = c_l(M_\infty, Re_c, \alpha, \gamma, \text{shape}) \tag{2.58}$$

$$c_d \equiv \frac{D'}{q_\infty c} = c_d(M_\infty, Re_c, \alpha, \gamma, \text{shape}) \tag{2.59}$$

$$c_{m,O} \equiv \frac{M'_O}{q_\infty c^2} = c_{m,O}(M_\infty, Re_c, \alpha, \gamma, \text{shape}) \tag{2.60}$$

Note that all prime quantities, e.g. L', indicate per-unit span, forces, and moment. Mach dependence in aerodynamic coefficients (Eqs. (2.58)–(2.60)) signifies the compressibility effect. Reynolds number based on chord signifies the probable state of the boundary layer on the suction and pressure surfaces of the airfoil, which is closely tied to the airfoil drag. Dependence of aerodynamic forces and moments on angle of attack and airfoil shape stems from their impact on pressure distribution. In addition to these, we recall that in high-speed compressible boundary layers, the wall thermal boundary condition, T_w/T_∞ and the Prandtl number entered the fray. Hence, the functions on the right-hand side of Eqs. (2.58)–(2.60) need to be amended to include T_w/T_∞ and Pr, in high-speed (e.g. hypersonic) flows.

The lift coefficient of thin, cambered airfoils in inviscid, incompressible fluid is theoretically

$$c_l = 2\pi(\alpha - \alpha_{L=0}) \tag{2.61}$$

where α is the angle of attack in radians, $\alpha_{L=0}$ is the angle-of-zero lift in radians, and 2π is the lift curve slope, i.e. $dc_l/d\alpha$ (per radian). Angle-of-zero lift is due to airfoil camber, which is a small negative angle for positively cambered airfoils. Typically, it is about $-1°$ to $-5°$ for plain airfoils. Figure 2.19 shows the angle-of-zero lift for a cambered airfoil. It is defined as the angle of the freestream with respect to chord that a cambered airfoil produces zero lift. Therefore, it takes a negative AoA to overcome the positive camber effect on lift, for a net result of zero lift.

Also, since airfoil stall has its roots in boundary-layer separation, which is a viscous-dominated phenomenon, the inviscid theory that produced Eq. (2.61) is unable to predict stall (or alternatively, $c_{l,max}$). Including the effect of fluid viscosity, thin airfoils (prior to stall) produce lift coefficient, c_l, according to:

$$c_l = a\left(\alpha - \alpha_{L=0}\right) \tag{2.62}$$

where a is the lift curve slope, per radian, which is slightly less than the theoretical value, i.e. 2π. Figure 2.20 shows the lift coefficient on a symmetrical and cambered

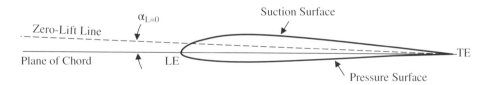

Figure 2.19 Graphical depiction of zero-lift line and $\alpha_{L=0}$.

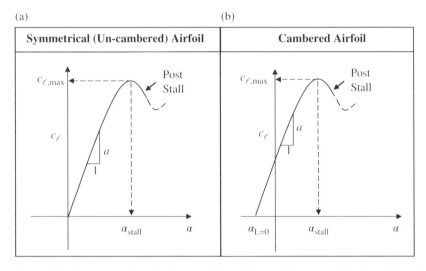

Figure 2.20 Pre- and post-stall behavior of lift coefficient with angle of attack on thin airfoils in high-Reynolds number flows (e.g. @ $Re_c = 6 \times 10^6$) (Lift curve slope "a" is $\leq 2\pi$, per radian): (a) The lift coefficient for a thin symmetrical airfoil ($\alpha_{L=0}$); (b) cambered airfoil lift coefficient.

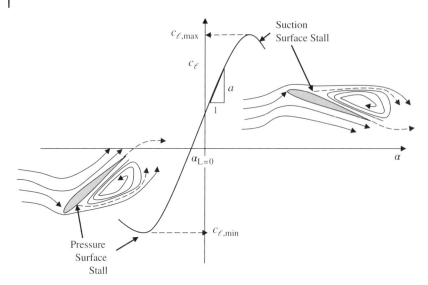

Figure 2.21 Positive and negative stall characteristics of a thin airfoil in high-Reynolds number flow (e.g. @ $Re_c = 6 \times 10^6$).

airfoil. Note that the behavior of airfoil lift with AoA is linear at first; then it becomes nonlinear as it gets closer to maximum lift and finally it exhibits profound nonlinearity in post stall.

The *plain airfoil*, i.e. without leading- and trailing-edge devices (e.g. slats and flaps), has its stall angle of attack in the range of 12°–16°, its maximum lift coefficient about 1.2–1.6 and the angle of zero-lift, as noted earlier, between –1° and –5°. In the range of negative AoA, the thin airfoil exhibits a linear behavior until a negative stall AoA where the negative stall is initiated on the pressure surface of the airfoil (see Figure 2.21). Note the $c_{l,min}$ in Figure 2.21.

In Eq. (2.60), we have expressed the functional dependence of pitching moment coefficient about point O as a function of the angle of attack of the airfoil, among other parameters. For a general point O, that expression is correct. However, in aerodynamic theory, we have proved that an *aerodynamic center* exists (on the airfoil chord) such that the pitching moment coefficient, about that point, becomes independent of the AoA. Hence the α-dependence of pitching moment coefficient vanishes about the aerodynamic center. For thin airfoils, the aerodynamic center is *theoretically* located at the quarter chord, i.e.

$$x_{a.c.} = \frac{c}{4} \tag{2.63}$$

Therefore, c_{mac} is independent of AoA. The thin airfoil calculations yield pitching moment about aerodynamic center as negative, i.e. in the counter-clockwise direction or in the direction of reduced AoA. The behavior of pitching moment coefficient about aerodynamic center of a thin airfoil in pre- and post-stall is shown qualitatively in Figure 2.22. Pitching moment at quarter chord varies between 0 (for symmetrical airfoils) to ~−0.1 for cambered airfoils at high chord Reynolds numbers, e.g. $Re_c = 6 \times 10^6$.

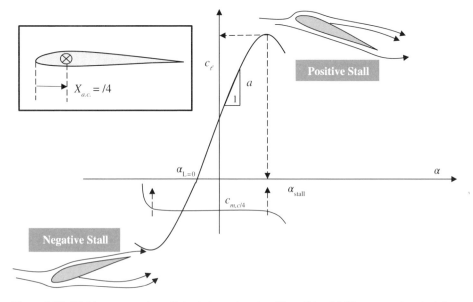

Figure 2.22 Pitching moment coefficient at quarter-chord for a thin airfoil in pre- and post-stall.

In conjunction with pitching moment, we also define a *center of pressure,* on the airfoil chord where the pitching moment, about which, vanishes. For symmetrical airfoil, center of pressure is co-located with the aerodynamic center at quarter chord.

Reynolds number impacts aerodynamic behavior of airfoils through their boundary layers, i.e. laminar, transitional, and turbulent. In general, a higher-chord Reynolds number creates higher $c_{l,max}$ and postpones stall AoA by few degrees. It does not impact the lift curve slope. The effect of Reynolds number on $c_{l,max}$ and α_{stall} is shown in Figure 2.23.

The effect of leading-edge devices (LEDs), such as slats, leading-edge droop, and Kruger flaps on airfoil lifting characteristics is shown in Figure 2.24. LEDs produce higher stalling angle-of-attack and thus higher $c_{l,max}$, as shown in Figure 2.24.

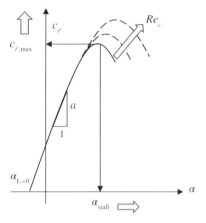

Figure 2.23 Effect of chord Reynolds number on lift coefficient in high-Reynolds number flows.

Trailing-edge devices, e.g. flaps, in effect increase the *effective camber* of the airfoil. When activated, effective camber increases and thus angle-of-zero lift shifts in the negative direction. The maximum lift coefficient thus increases in harmony with the increased effective camber. These effects (and typical values) for split flaps are shown in Figure 2.25.

The airfoil profile drag coefficient is a strong function of airfoil geometry, chord Reynolds number, Mach number, and angle of attack. The geometry and the AoA, for

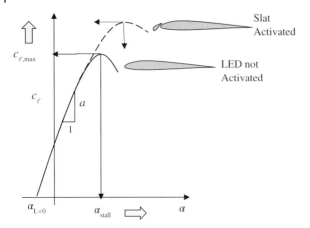

Figure 2.24 Effect of leading-edge devices (LED) such as slats, leading-edge droop and Kruger flaps on lifting characteristics of thin airfoils in high-Reynolds number flows (e.g. @ $Re_c = 6 \times 10^6$).

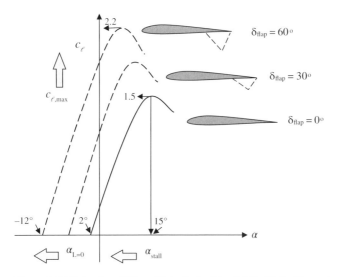

Figure 2.25 Behavior of lift coefficient of a thin airfoil with trailing-edge (split) flap at high-Reynolds number flows (e.g. @ $Re_c = 6 \times 10^6$).

any given Mach and Reynolds number, establish the pressure distribution on the airfoil. Angle of attack causes the transition to move forward, and thus the extent of TBL over the airfoil increases. TBL can withstand adverse pressure gradient and thus remain attached. The attached boundary layer at AoA on the suction and pressure surfaces of an airfoil then generates lift.

Hence, Reynolds number and pressure distribution establish the laminar zone (that always starts) at the leading edge (LE), transitional length, and the turbulent zone on the airfoil. At a small angle of attack, the pressure or form drag is small compared to friction drag. As the boundary layer thickens, the pressure drag increases and beyond stall dominates the airfoil drag. The behavior of C_f with Reynolds and Mach number is shown in Figure 2.8. The friction drag coefficient on (one side of) a flat plate is shown

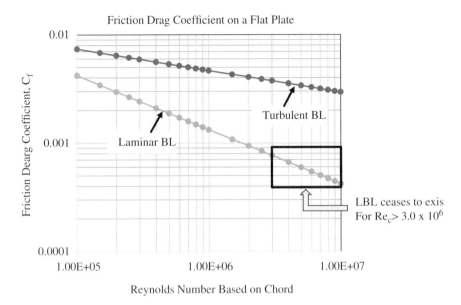

Figure 2.26 Friction drag coefficient on (one side of) a flat plate in the incompressible limit.

in Figure 2.26 (using log-log scale) in terms of chord Reynolds number from 10^5 to 10^7. The laminar branch is the Blasius solution and the turbulent branch follows Eq. (2.48). Note that the laminar branch only extends to about 2.8×10^6 and thus it does not extend to 10^7. It is shown here for comparative purposes.

The behavior of airfoil profile drag is measured in wind tunnels and is extensively reported in various airfoil data collections, e.g. Abbott and Von Doenhoff (1945, 1959). Figure 2.27 shows an airfoil drag polar (c_l vs. c_d) for NACA 64-series of airfoils at different thicknesses (12–21%) and cambered for a design lift coefficient of 0.4. The characteristic "bucket" shape in the low-drag area is due to the extensive laminar boundary layer design of the 64-series airfoils at its design lift coefficient. When the airfoil moves past the design lift coefficient, the transition point abruptly moves upstream and thus causes a bucket-shaped drag polar.

The profound impact of pressure gradient on velocity profile in a laminar boundary layer is shown in Figure 2.28. Note that *adverse pressure gradient* in the streamwise direction, where dp/dx > 0, creates a point of inflection in the velocity profile that marks the onset of laminar separation. The inflection point is labeled as I on the dp/dx > 0 graph in Figure 2.28.

laminar flow control (LFC) offers promising drag reduction and the design tools to achieve them. For an overview on LFC, Joslin (1998) is recommended.

2.5 Drag Due to Lift

2.5.1 Classical Theory

In this section we present the classical finite wing theory based on Prandtl's 1922 paper. In 3-D flows over lifting surfaces, e.g. wing, the trailing streamwise vortices in the wake induce a velocity normal to the wing, which is in the direction of negative lift.

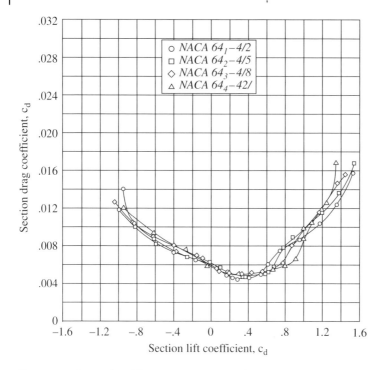

Figure 2.27 Drag polar of four NACA 64-series airfoil sections of various thickness and cambered for a design lift coefficient of 0.4, $Re_c = 9 \times 10^6$. *Source:* From Abbott and Von Doenhoff 1945.

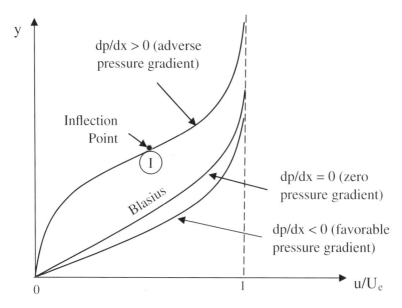

Figure 2.28 The effect of pressure gradient on velocity profile in a laminar boundary layer ($dp/dx < 0$ is called *favorable pressure gradient*).

This induced velocity is called *downwash*. Figure 2.29 shows a finite wing of span b in a uniform flow, V_∞, and the induced downwash, $w(y)$.

In a finite wing, the local chord is $c(y)$, as shown in Figure 2.30. We define the integral of the local chord along the wingspan as the wing *planform area*, S:

$$S \equiv \int_{-b/2}^{b/2} c(y)\, dy \tag{2.64}$$

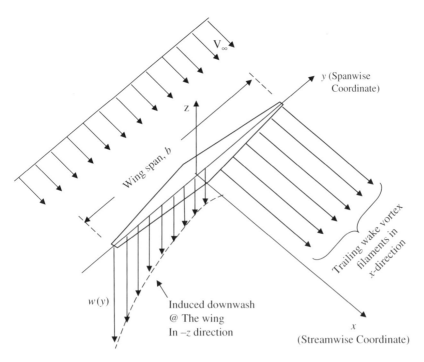

Figure 2.29 Finite wing with its trailing wake (shown for half-span) in uniform flow with the induced downwash.

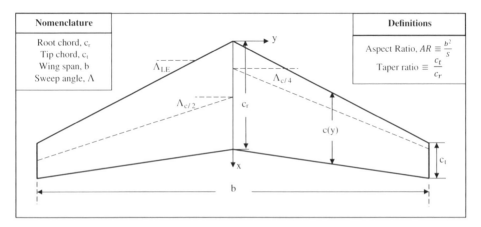

Figure 2.30 Geometric parameters of a finite wing of trapezoidal planform with sweep.

Wing aspect ratio is defined as:

$$AR \equiv \frac{b^2}{S} \tag{2.65}$$

In case of trapezoidal wing, we define a wing taper ratio as

$$\text{Taper ratio} \equiv \frac{c_t}{c_r} \tag{2.66}$$

Figure 2.31 shows an airfoil section at a spanwise station on a finite wing where the downwash creates an induced angle of attack and thus the local wind senses an effective AoA.

In Figure 2.31, we have defined four angles related to an airfoil located on a finite wing:

1) Geometric angle of attack, α, is the angle between flight direction and the local plane of chord. A wing with *geometric twist* α varies along the span, i.e. $\alpha = \alpha(y)$.
2) Angle-of-zero lift, $\alpha_{L=0}$, is a sectional characteristic, as it depends on the sectional airfoil camber. Therefore, in general $\alpha_{L=0} = \alpha_{L=0}(y)$, which is called *aerodynamic twist*.
3) Induced AoA, α_i, is the angle between the flight direction and the local wind direction. The local wind direction is affected by the *downwash*, w, on the wing.
4) Effective AoA, α_{eff}, is the angle between the local relative wind direction and the plane of chord, i.e. $\alpha_{eff} = \alpha - \alpha_i$.

Note that the local relative wind direction is *tilted* by the downwash; therefore, the local aerodynamic force $F'(y)$ that is normal to the local relative wind is *tilted* as well. The local aerodynamic force is then resolved in the lift and drag directions (i.e. normal to and in the direction of flight). The drag force created by the induced downwash is called induced drag, or drag due to lift. The sectional induced drag and lift coefficient are related through induced AoA, namely,

$$D_i'(y) = L'(y).\tan\alpha_i \cong L'(y).\alpha_i \tag{2.67a}$$

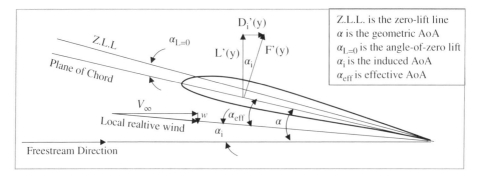

Figure 2.31 Definition sketch of four angles on a cambered airfoil on a finite wing at spanwise station, *y*.

In Eq. (2.67a) tangent of the induced AoA is replaced by the angle itself, α_i (in radians), since the induced AoA is small.

We may define a sectional induced drag coefficient, c_{di} by dividing Eq. (2.67a) by q_∞ and local chord, $c(y)$, to get

$$c_{di} = c_\ell . \alpha_i \tag{2.67b}$$

The sectional lift coefficient, $c_l(y)$ may be expressed in terms of the effective AoA, as:

$$c_\ell(y) = a_0(\alpha_e - \alpha_{L=0}) \tag{2.68}$$

Here, a_0 is the 2-D airfoil lift curve slope (theoretically, 2π for thin airfoils). The sectional lift, $L'(y)$ is defined based on the sectional lift coefficient, sectional chord length $c(y)$, and the dynamic pressure, q_∞:

$$L'(y) = q_\infty c_\ell(y) c(y) \tag{2.69a}$$

The Kutta-Joukowski theorem that relates lift to circulation is still valid at the sections, and the sectional induced drag follows the same form, namely

$$L'(y) = \rho_\infty V_\infty \Gamma(y) \tag{2.69b}$$

$$D'_i(y) = \rho_\infty w(y) \Gamma(y) \tag{2.69c}$$

where $\Gamma(y)$ is the circulation around the wing at a spanwise station, y. The wing lift force is the integral of the sectional lift force along the span, i.e.

$$L = \int_{-b/2}^{b/2} L'(y) dy = q_\infty \int_{-b/2}^{b/2} c_\ell(y) c(y) dy \tag{2.70a}$$

For constant c_l along the span, i.e. when $(\alpha_e - \alpha_{L=0})$ = constant, we integrate the chord length along the span, which results in the wing planform area, S, i.e.

$$L = q_\infty c_\ell \int_{-b/2}^{b/2} c(y) dy = q_\infty c_\ell S \tag{2.70b}$$

This equation shows that the wing lift coefficient and the sectional lift coefficient are equal to each other, namely

$$C_L \equiv \frac{L}{q_\infty S} = c_\ell \tag{2.70c}$$

Returning to the condition of constant sectional lift along the span, i.e. $(\alpha_e - \alpha_{L=0})$ = constant, we note that a wing with no geometric and aerodynamic twists along its span will generate the same lift coefficient if the induced AoA is constant along the span, i.e. α_i = constant. Prandtl showed that for a wing with *elliptic loading*, i.e. elliptic lift or circulation distribution:

$$L'(y) = L'_0 \sqrt{1 - \left(\frac{y}{b/2}\right)^2} = \rho_\infty V_\infty \Gamma_0 \sqrt{1 - \left(\frac{y}{b/2}\right)^2} \tag{2.71}$$

The induced AoA is constant along the span of a straight wing and it is inversely proportional to wing aspect ratio and directly proportional to the wing-lift coefficient, namely

$$\alpha_i = \frac{C_L}{\pi\,AR} = \text{constant} \qquad \left[\text{Elliptical loading}\right] \tag{2.72}$$

Now, if we express the wing-lift coefficient in terms of the wing-lift curve slope, a, and the geometric angle of attack, α and the $\alpha_{L=0}$, we get:

$$C_L = a\left(\alpha - \alpha_{L=0}\right) = c_\ell = a_0\left(\alpha - \alpha_i - \alpha_{L=0}\right) \tag{2.73a}$$

If we substitute for α_i from Eq. (2.72) in Eq. (2.73a) and do minor manipulation, we get the following relation between wing- and airfoil-lift curve slopes, a and a_0, for elliptically loaded wing:

$$a = \frac{a_0}{1 + \dfrac{a_0}{\pi\,AR}} \tag{2.73b}$$

The effect of wing aspect ratio on lift curve slope is shown in Figure 2.32. The wing-induced drag coefficient is related to the wing-lift coefficient through (2.67b), for an elliptically loaded wing:

$$C_{Di} = \frac{C_L^2}{\pi\,AR} \tag{2.74}$$

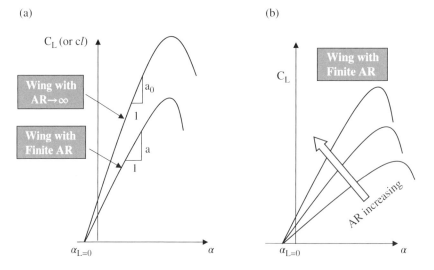

Figure 2.32 Lifting characteristic of straight wings of finite or infinite AR: (a) wing and airfoil lift characteristics; (b) effect of aspect ratio on lift.

The elliptical loading of a wing, requires the circulation in the spanwise direction to follow an elliptic profile, namely,

$$\Gamma(y) = \Gamma_0 \sqrt{1 - \left(\frac{y}{b/2}\right)^2} \tag{2.75}$$

The downwash, for an elliptically loaded wing, remains constant along the span:

$$w(y) = -\frac{\Gamma_0}{2b} = \text{constant} \tag{2.76}$$

The negative sign in Eq. (2.76) identifies downwash as a velocity component in the $-z$ direction. Using Prandtl lifting-line theory of 1922, we can prove that the wing elliptical loading leads to a minimum induced drag on a wing. There are correction factors that are introduced for nonelliptically loaded wings, for example:

$$C_{Di} = \frac{C_L^2}{\pi AR}(1+\delta) = \frac{C_L^2}{\pi eAR} \tag{2.77}$$

where δ is the nonoptimal loading correction factor for induced drag, or inverse of $(1 + \delta)$, which is called "e" in Eq. (2.77) and is known as the Oswald span efficiency factor. The lift curve slope for a wing with nonelliptic lift distribution is:

$$a = \frac{a_0}{1 + \frac{a_0}{\pi AR}(1+\tau)} \tag{2.78}$$

The correction factor τ for the lift curve slope in Eq. (2.78) can be calculated from the Fourier coefficients of the expansion of the general lift distribution function along the span. These are summarized in Table 2.5.

The impact of aspect ratio on straight wing-lift curve slope is summarized in Table 2.6.

2.5.2 Optimal Spanloading: The Case of Bell Spanload

Prandtl in his 1933 paper revised his assertion that elliptic spanloading resulted in minimum induced drag, or maximum aerodynamic efficiency. Prandtl found that the bell-shaped lift distribution along the span was the preferred spanloading criterion that created superior aerodynamic performance. Bowers et al. (2016) published a detailed aerodynamic analysis that presented a unifying wing theory for high efficiency and coordinated control that extended Prandtl's 1933 optimal spanloading concept. They concluded that *bell spanload* creates six favorable aerodynamic effects:

1) It brought the *wingtip vortex inboard*, as compared to elliptic loading that created a strong wingtip vortex.
2) The inboard vortex then created *induced upwash* near the wingtips instead of continuous downwash along the span as in elliptic loading.
3) It created the resultant force near the wingtip that is *tilted into the relative wind*.
4) It created *induced thrust* instead of induced drag near the wingtip.

Table 2.5 Summary of finite wing aerodynamic parameters of elliptic and general lift distributions for straight, high-AR wings.

Straight (unswept) wing of high aspect ratio	
Elliptic lift distribution (high AR)	**General lift distribution (high AR)**
	$L'(y)$ or $\Gamma(y)$; a general function of y
$L'(y) = L'_0 \sqrt{1 - \left(\dfrac{y}{b/2}\right)^2}$	
$L'_0 = \rho_\infty V_\infty \Gamma_0$	$L'_0 = \rho_\infty V_\infty \Gamma_0$
$w = -\dfrac{\Gamma_0}{2b} = \text{constant}$	$w(y) = -\dfrac{1}{4\pi}\displaystyle\int_{-b/2}^{b/2} \dfrac{(d\Gamma/dy)\,dy}{y_0 - y}$
$\alpha_i = \dfrac{C_L}{\pi AR} = \text{constant}$	$\alpha_i(y) = -\dfrac{w(y)}{V_\infty}$
$C_{Di} = \dfrac{C_L^2}{\pi AR}$	$C_{Di} = \dfrac{C_L^2}{\pi AR}(1+\delta) = \dfrac{C_L^2}{\pi e AR}$
$a = \dfrac{a_0}{1 + \dfrac{a_0}{\pi AR}}$	$a = \dfrac{a_0}{1 + \dfrac{a_0}{\pi AR}(1+\tau)}$

Table 2.6 Impact of wing aspect ratio on lift curve slope for unswept wings.

Aerodynamic correlations for straight wings of high- and low- AR	
High aspect ratio wing, lift curve slope	$a = \dfrac{a_0}{1 + (a_0/\pi AR)(1+\tau)}$
Low-AR wing, lift curve slope	$a = \dfrac{a_0}{\sqrt{1 + (a_0/\pi AR)^2} + a_0/(\pi AR)}$

5) It created *proverse yaw* phenomenon instead of the adverse yaw in elliptic loading.
6) It has the potential to eliminate any *auxiliary yaw device*, e.g. rudder to accomplish coordinated turn.

A scaled model of two experimental aircraft were designed and flight tested by Bowers et al. to measure the aerodynamic superiority of bell spanload as Prandtl outlined in his 1933 paper. Figure 2.33 (courtesy NASA) shows one of the experimental aircraft in flight. The inboard and outboard airfoil sections in the sub-scale model were designed by the Eppler code (1980). The bell spanload was accomplished through wingtip twist. The resulting induced flow along the wingspan, the vortex rollup, both analytically and in flight are shown in Figure 2.34 (courtesy of NASA). Bowers et al. (2016) demonstrated that the bell spanload concept showed superior aerodynamic efficiency that may be utilized to improve future aircraft design, such as all-wing and blended wing body aircraft. For additional detail, see Bowers et al. (2016).

Figure 2.33 Experimental aircraft with bell spanload design in flight. *Source:* Courtesy of NASA.

(a)   (b)

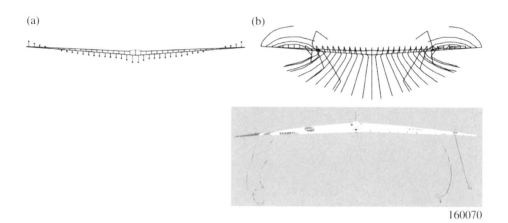

160070

Figure 2.34 Induced flow along the wing span (a), and vortex rollup (b). *Source:* Courtesy of NASA.

2.6 Waves in Supersonic Flow

Waves are essentially fluid navigation means around bodies in supersonic flow. Concepts of zone of action and silence are introduced through speed of sound.

2.6.1 Speed of Sound

Sound waves are infinitesimal pressure waves propagating in a medium. The propagation of sound waves, or acoustic waves, is reversible and adiabatic, hence isentropic. Since sound propagates through collision of fluid molecules, the speed of sound is higher in liquids than gas. The speed of sound, a, in gas follows:

$$a^2 = \left(\frac{\partial p}{\partial \rho} \right)_s \tag{2.79a}$$

For a perfect gas, it becomes:

$$a = \sqrt{\gamma RT} = \sqrt{\gamma \frac{\bar{R}}{MW} T} \tag{2.79b}$$

where the ratio of universal gas constant, $\bar{R}$, and MW is the molecular mass of the gas.

The equation for the speed of propagation of sound that we derived is for a gas at rest. Let us superimpose a uniform collective gas speed in a particular direction to the wave front; then the wave propagates as the vector sum of the two; namely, $\bar{V} + \bar{a}$. For waves propagating normal to a gas flow, we either get $(V + a)$ or $(V - a)$ as the propagation speed of sound. It is the $(V - a)$ behavior that is of interest here. In case the flow is sonic, then $(a - a = 0)$, which will not allow the sound to travel upstream and hence creates a zone of silence upstream of the disturbance. In case the flow speed is even faster than the local speed of sound, i.e. known as supersonic flow, the acoustic wave will be confined to a cone downstream of the source. These two behaviors for small disturbances are shown in Figure 2.35.

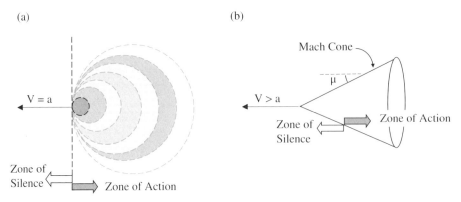

Figure 2.35 Acoustic wave propagation in sonic and supersonic flows (or the case of a moving source): (a) sonic; (b) supersonic.

The ratio of local gas speed to the speed of sound is called Mach number, M:

$$M \equiv \frac{V}{a} \tag{2.80}$$

The envelope of the waves that create the zones of action and silence is the Mach wave. It makes a local wave angle with respect to the flow, μ, which from the geometry of wave propagation, as shown in Figure 2.36, is:

$$\mu = \sin^{-1}\left(\frac{a.t}{V.t}\right) = \sin^{-1}\left(\frac{1}{M}\right) \tag{2.81}$$

The mechanism for turning in supersonic flow is through compression and expansion Mach waves.

In addition, compression Mach waves converge and may coalesce to form shock waves. This behavior is the opposite of the expansion Mach waves that diverge and never coalesce to form an expansion shock wave. The wave pattern around a 2-D-diamond airfoil is shown in Figure 2.37. The attached plane oblique shocks (OSs) at the

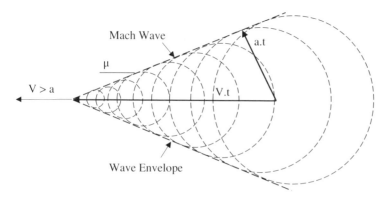

Figure 2.36 Wave front created by a *small disturbance* moving at a supersonic speed.

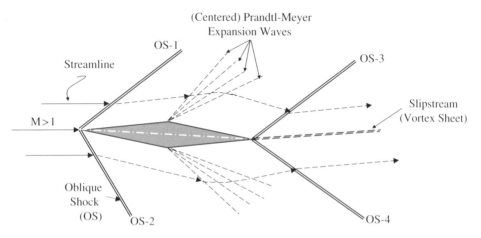

Figure 2.37 Sketch of oblique waves: four oblique shock (OS) waves and two centered expansion fans, about a diamond airfoil at an angle of attack in supersonic flow.

leading edge turn the flow parallel to the surface and the expansion waves, known as the Prandtl-Meyer waves, cause the flow to turn at the airfoil shoulders. The trailing-edge waves turn the flow parallel to each other and the slipstream that emanates from the trailing edge. The flow downstream of the tail waves, in general, is not parallel to freestream flow upstream of the airfoil. However, the flows that emerge from the upper and lower surfaces are locally parallel to each other and the static pressure is continuous across the slipstream. This is also in contrast to the subsonic flow that recovers the upstream flow direction further downstream of the trailing edge. The angle of the slip-stream is determined by the continuity of static pressure in parallel flows.

2.6.2 Normal Shock Wave

A special case of an oblique shock wave is the normal shock. This wave is normal to the flow and causes a sudden deceleration: static pressure, temperature, density, and entropy rise; Mach number and total pressure drop across the wave. The *jump conditions* across a normal shock are established through conservation principles and are derived in gas dynamic or propulsion textbooks, e.g. Anderson (2010) or Farokhi (2014). These are listed here for reference:

$$M_2^2 = \frac{2+(\gamma-1)M_1^2}{2\gamma M_1^2-(\gamma-1)} \tag{2.82}$$

$$\frac{\rho_2}{\rho_1} = \frac{(\gamma+1)M_1^2}{2+(\gamma-1)M_1^2} \tag{2.83}$$

$$\frac{p_2}{p_1} = 1+\frac{2\gamma}{\gamma+1}\left(M_1^2-1\right) \tag{2.84}$$

$$\frac{T_2}{T_1} = \left[1+\frac{2\gamma}{\gamma+1}\left(M_1^2-1\right)\right]\left[\frac{2+(\gamma-1)M_1^2}{(\gamma+1)M_1^2}\right] \tag{2.85}$$

$$\frac{p_{t2}}{p_{t1}} = \left[1+\frac{2\gamma}{\gamma+1}\left(M_1^2-1\right)\right]\left[\frac{1+\frac{\gamma-1}{2}\left(\frac{2+(\gamma-1)M_1^2}{2\gamma M_1^2-(\gamma-1)}\right)}{1+\frac{\gamma-1}{2}M_1^2}\right]^{\frac{\gamma}{\gamma-1}} \tag{2.86}$$

$$\Delta s/R = -\ell n\left(p_{t2}/p_{t1}\right) \tag{2.87}$$

2.6.3 Oblique Shock Waves

As stated earlier, compression Mach waves tend to coalesce to form an oblique shock wave. A normal shock is a special case of an oblique shock with a wave angle of 90°. Figure 2.38 shows a schematic drawing of an oblique shock flow with a representative streamline that abruptly changes direction across the shock. The shock wave angle with respect to upstream flow is called β and the flow-turning angle (again, with respect to

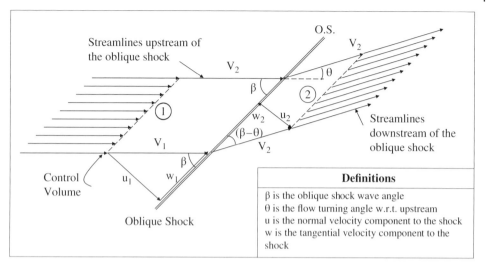

Figure 2.38 Definition sketch of velocity components normal and parallel to an oblique shock, with the wave angle and flow turning angle definitions.

the upstream flow) is θ. A representative control volume is also shown in Figure 2.38. The flow is resolved into a normal and a tangential direction to the shock wave. The velocity components are u and w that are normal and tangential to the shock wave front, respectively. Two velocity triangles are shown upstream and downstream of the shock with vertex angles β and $(\beta - \theta)$, respectively. The control volume is confined between a pair of streamlines and the entrance and exit planes are parallel to the shock. We choose the area of the entrance and exit streamtube to be one for simplicity.

Since oblique shock waves involve two angles: (i) the wave angle; β and (ii) turning angle, θ, one of the angles need to be specified for closure. Consequently, the application of conservation laws, mass, two momentum and energy, and the perfect gas law as the equation of state will produce the following relations for oblique shocks:

$$\frac{\tan \beta}{\tan(\beta - \theta)} = \frac{(\gamma + 1)M_1^2 \sin^2 \beta}{2 + (\gamma - 1)M_1^2 \sin^2 \beta} \tag{2.88}$$

$$M_{2n}^2 = \frac{2 + (\gamma - 1)M_1^2 \sin^2 \beta}{2\gamma M_1^2 \sin^2 \beta - (\gamma - 1)} \tag{2.89}$$

$$M_2 = \frac{M_{2n}}{\sin(\beta - \theta)} \tag{2.90}$$

The tangential velocity to an oblique shock, w, is conserved since the strength of a straight oblique shock is constant along the shock. Hence, $w_1 = w_2$. All the jump conditions on density, static pressure, temperature, total pressure, and entropy across an oblique shock are identical to normal shocks except the upstream Mach number, M_1, is replaced by M_{1n}, which is $M_1 \sin \beta$ in an oblique shock, e.g. $\frac{p_2}{p_1} = 1 + \frac{2\gamma}{\gamma + 1}(M_1^2 \sin^2 \beta - 1)$.

Figure 2.39 is a graph of Eq. (2.88) with Mach number as a running parameter (from NACA Report 1135, 1953). There are two solutions for the wave angle, β, for each turning angle, θ. The high wave angle is referred to as the *strong solution* and the lower wave angle is called the *weak solution*. A continuous line and a broken line distinguish the weak and strong solutions in Figure 2.39, respectively. The weak solution is preferred by nature, whereas the strong solution can only be simulated in a laboratory. Also, there is a maximum turning angle, θ_{max}, for any supersonic Mach number. For example, a Mach

(a)

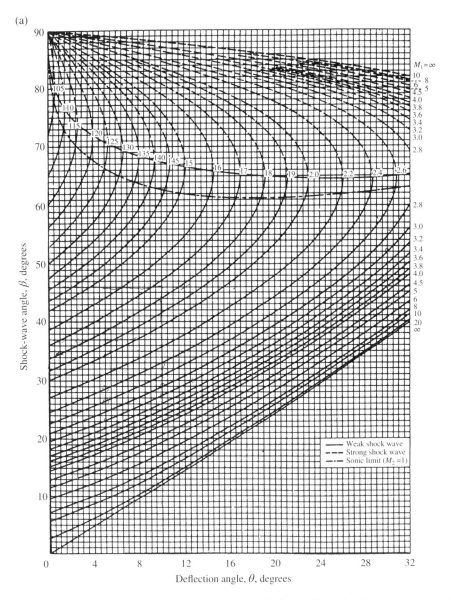

Figure 2.39 (a) Plane oblique shock chart for $\gamma = 1.4$.; (b) Plane oblique shock chart – continued. *Source:* From NACA Report 1135, 1953.

(b)

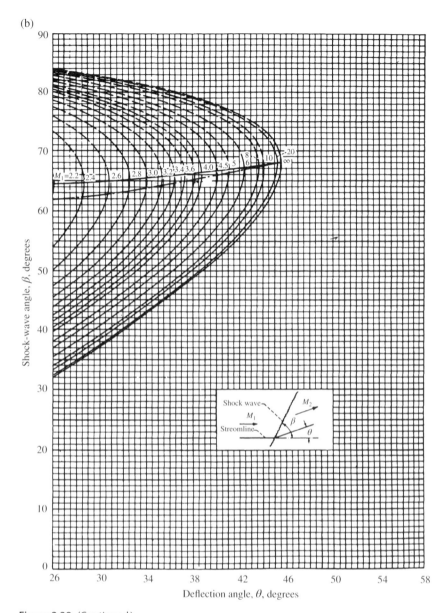

Figure 2.39 (Continued)

2 flow can only turn ~23° via an attached plane oblique shock wave. For higher turning angles than θ_{max}, a detached shock wave will form upstream of the body. This shock, due to its shape, is referred to as the *bow shock*.

2.6.4 Expansion Waves

The oblique shocks are compression waves in a supersonic flow that abruptly turn the flow and compress the gas in the process. An expansion wave causes a supersonic

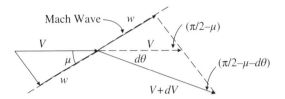

Figure 2.40 Incremental turning across a Mach wave with velocity resolved in the normal and tangential directions to the wave.

flow to turn and the static pressure to drop. The expansion waves are Mach waves that make a Mach angle with respect to the local flow. The flow through expansion waves is inherently isentropic, as the Mach waves are of infinitesimal strength and hence reversible. An expansion Mach wave is only capable of turning the flow infinitesimally and accelerates the flow incrementally. Such a flow turning and flow acceleration is shown in Figure 2.40 with exaggeration of the turning angle and the flow acceleration increment.

The local wave angle with respect to upstream flow is at the Mach angle, μ. The incremental turning angle is labeled $d\theta$. The change of velocity is shown by a magnitude of dV and a direction $d\theta$ with respect to upstream flow. Since the velocity downstream of the Mach wave has to have the same tangential component, w, as the upstream flow, the right triangles that share the same tangential velocity are labeled with V and $V + dV$ vectors in Figure 2.40.

We may apply the law of sines to the triangle with sides V and $V + dV$ to relate the turning angle and incremental speed change:

$$\frac{V+dV}{\sin\left(\dfrac{\pi}{2}-\mu\right)} = \frac{V}{\sin\left(\dfrac{\pi}{2}-\mu-d\theta\right)} \tag{2.91}$$

Let us simplify the sines and isolate the velocity terms, such as

$$1+\frac{dV}{V} = \frac{\cos\mu}{\cos(\mu+d\theta)} \cong \frac{\cos\mu}{\cos\mu - d\theta.\sin\mu} = \frac{1}{1-d\theta.\tan\mu} \approx 1 + d\theta.\tan\mu \tag{2.92}$$

We applied a small angle approximation to sine and cosine $d\theta$ and in the last approximation on the RHS, we used a binomial expansion and truncation at the linear term for a small ε:

$$\frac{1}{1\pm\varepsilon} \approx 1\mp\varepsilon \tag{2.93}$$

Now from the Mach triangle shown in Figure 2.41, we deduce that

$$\tan\mu = \frac{1}{\sqrt{M^2-1}} \tag{2.94}$$

Figure 2.41 Sketch of a Mach wave and its right triangle.

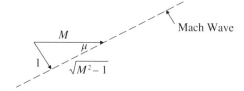

Substitute this expression in Eq. (2.92) to get a simple relationship between the flow turning and the incremental acceleration across a Mach wave in a at a supersonic flow, i.e.

$$d\theta = \sqrt{M^2 - 1} \frac{dV}{V} \tag{2.95}$$

We may change the velocity as a variable in Eq. (2.95) to a Mach number:

$$d\theta = \frac{\sqrt{M^2 - 1}}{1 + \frac{\gamma - 1}{2} M^2} \frac{dM}{M} \tag{2.96}$$

Equation (2.96) may be integrated in the limits of zero angle for $M = 1$ and $v(M)$ at a general supersonic Mach, M. The angle $v(M)$ is called the Prandtl-Meyer function:

$$\int_0^{v(M)} d\theta = v(M) = \int_1^M \frac{\sqrt{M^2 - 1}}{1 + \frac{\gamma - 1}{2} M^2} \frac{dM}{M} \tag{2.97}$$

The definite integral on the RHS of Eq. (2.97) is integrable in closed form by partial fraction and simple change of variables to cast the integrand in the form of the derivative of an inverse tangent function. The result of this integration is the Prandtl-Meyer function:

$$v(M) = \sqrt{\frac{\gamma + 1}{\gamma - 1}} \tan^{-1} \sqrt{\frac{\gamma - 1}{\gamma + 1}(M^2 - 1)} - \tan^{-1} \sqrt{M^2 - 1} \tag{2.98}$$

Figure 2.42 shows a log-linear plot of the Prandtl-Meyer function and the Mach angle as a function of Mach number for $\gamma = 1.4$. Both $v(M)$ and $\mu(M)$ are plotted in degrees.

2.7 Compressibility Effects and Critical Mach Number

The subsonic flight Mach number that leads to the first appearance of sonic flow on a body, e.g. on an airfoil, is called *critical Mach number*. We relate the highest Mach number on a body, M_{max}, to the minimum pressure point, or equivalently $c_{p,min}$, on the body. The pressure coefficient is defined as:

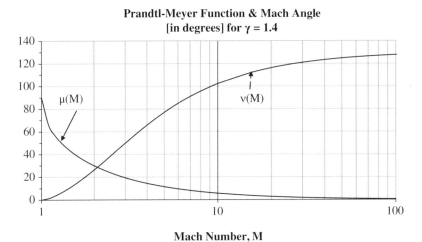

Figure 2.42 Prandtl-Meyer function and Mach angle (in degrees) for a diatomic gas In supersonic flow. *Source:* From Farokhi 2014.

$$c_p \equiv \frac{p - p_\infty}{q_\infty} = \frac{2}{\gamma M_\infty^2}\left[\frac{p}{p_\infty} - 1\right] = \frac{2}{\gamma M_\infty^2}\left\{\left[\frac{1 + \frac{\gamma-1}{2}M_\infty^2}{1 + \frac{\gamma-1}{2}M^2}\right]^{\frac{\gamma}{\gamma-1}} - 1\right\} \quad (2.99)$$

The minimum pressure on the airfoil corresponds to the highest Mach number, i.e. in this case, $M = 1$; therefore, the *critical Mach number* is established from $c_{p,\mathrm{min}}$ according to:

$$c_{p,\mathrm{min}} = \frac{2}{\gamma M_\mathrm{crit}^2}\left\{\left[\frac{1 + \frac{\gamma-1}{2}M_\mathrm{crit}^2}{\frac{\gamma+1}{2}}\right]^{\frac{\gamma}{\gamma-1}} - 1\right\} \quad (2.100)$$

There are many compressibility correction models that allow the incompressible $c_{p,\mathrm{min}}$ results to be extended into the compressible domain. The classical Prandtl-Glauert compressibility correction factor in subsonic flow is an example:

$$c_{p,\mathrm{M}} = \frac{c_{p0}}{\sqrt{1 - M^2}} \quad (M < 1) \quad (2.101)$$

where $c_{p,\mathrm{M}}$ is the compressible c_p corresponding to Mach, M, and c_{p0} is the incompressible c_p (corresponding to Mach zero). Note that Prandtl-Glauert compressibility correction as expressed in Eq. (2.101), is very simple and is independent of γ! It is also singular at Mach 1.0, which points to the failure of linearized aerodynamics at transonic Mach numbers. As the angle of attack on the airfoil increases, the $c_{p,\mathrm{min}}$ drops. A family of

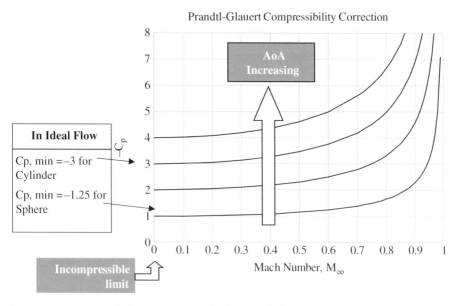

Figure 2.43 Compressibility correction applied to an airfoil pressure coefficient.

Prandtl-Glauert compressibility corrected $c_{p,\min}$ curves for an airfoil at successively higher angles of attack is shown in Figure 2.43. In this graph the incompressible values of $c_{p,\min}$ are taken as -4, -3, -2, and -1. It is customary to plot $-c_p$, since $c_{p,\min}$ is itself a negative quantity.

To establish the critical Mach number for an airfoil in angle of attack, we graph the negative $c_{p,\min}$ from Eq. (2.80) on a compressibility-corrected $c_{p,\min}$, e.g. Prandtl-Glauert, and the intersection of the two curves establish the M_{crit}. Figure 2.44 shows the graphical determination of M_{crit} for air (i.e. $\gamma = 1.4$).

The Kármán-Tsien compressibility correction is based on the nonlinear hodograph solution that uses a linear approximation to pressure-density relation, instead of the isentropic p/ρ^{γ} relation. This new gas model is known as *tangent gas*, and allows for the compressible and incompressible pressure coefficients to be related, as follows:

$$\left(C_p\right)_{M\infty} = \frac{\left(C_p\right)_{M_\infty=0}}{\sqrt{1-M_\infty^2} + \left(\frac{M_\infty^2}{1+\sqrt{1-M_\infty^2}}\right)\frac{\left(C_p\right)_{M_\infty=0}}{2}} \tag{2.102}$$

Kármán-Tsien subsonic compressibility correction is plotted against the critical pressure coefficient for the graphical solution of the critical Mach number in Figure 2.45. We note that Karman-Tsien predicts a lower critical Mach number than the Prandtl-Glauert. In essence, Prandtl-Glauert underpredicts the compressibility effects. Kármán-Tsien is indeed more accurate than Prandtl-Glauert based on extensive experimental measurements. Note that in Figures 2.44 and 2.45, we have listed $C_{p,\mathrm{crit}}$ and M_{crit} for cylinder and sphere, as reference.

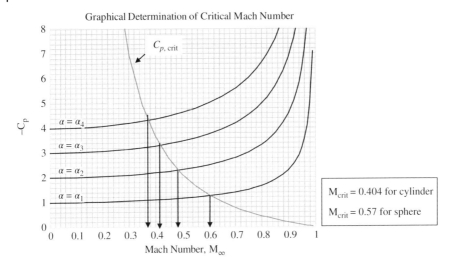

Figure 2.44 Graphical construction of critical Mach number on an airfoil at AoA.

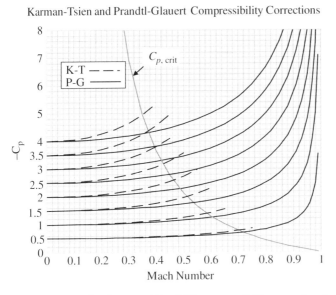

Figure 2.45 Family of Karman-Tsien (K-T) compressibility correction factors and their comparison with the Prandtl-Glauert (P-G) corrections.

The effect of compressibility on friction drag coefficient is correlated with the incompressible value using Johnson and Rubesin (1949) correlation:

$$\frac{C_f}{C_{fi}} = \frac{1}{\left(1 + 0.1305 M^2\right)^{0.12}} \tag{2.103}$$

2.8   Drag Divergence Phenomenon and Supercritical Airfoil

The phenomenon of drag divergence is closely tied to flow separation induced by a strong shock wave on an airfoil in transonic flow, i.e. when $M > M_{crit}$. Whitcomb and Clark (1965) envisioned and developed a new airfoil shape that flattened its supercritical suction pressure to weaken/eliminate the terminal shock and thus postpone drag rise associated with the drag divergence Mach number. Another prominent feature of supercritical (SC) airfoils is their larger leading-edge radius as compared to the conventional airfoils, e.g. NACA 64 series airfoils. This feature avoids the typical peaky suction pressure distribution and allows for smooth flow expansion/acceleration near the leading edge. The 8-ft Transonic Pressure Tunnel at NASA-Langley provided the testbed for Whitcomb's ideas. Of particular interest, in addition to design-point drag characteristics at transonic speeds, was the airfoil drag behavior at off-design Mach number conditions, namely at low speed for takeoff. Figure 2.46 shows one member of the supercritical airfoil family, i.e. SC(2)-0714, developed at NASA-Langley.

The reduced curvature of the suction surface of the supercritical airfoil was designed to flatten the peaky suction pressure distribution that appears on conventional airfoils and thus weaken/eliminate the terminal shock (known as *shockless* transonic airfoil). The modified pressure distribution shown in Figure 2.47 spreads the supersonic bubble on the suction surface from leading edge to clearly past 50–60% chord on the upper surface. It also shows a flat rooftop pressure distribution, or flat peak suction C_p before a weak terminal shock transitions the flow to subsonic. The dashed line passing through C_p^* separates the subsonic and supersonic regions on the airfoil. Here C_p^* is the *critical pressure coefficient*. The airfoil is operating at high lift, with normal force coefficient, $c_n = 0.871$, at Mach 0.721. The chord Reynolds number is 40×10^6. Another characteristic of SC-airfoils is their increased camber near the trailing edge. This is accomplished by convex curvature on the suction surface near TE and the *curvature reversal* on the pressure surface, i.e. concave curvature, at the trailing edge that increases C_p on the aft portions of the airfoil and thus increases the *aft loading*. Another design feature that helps minimize the boundary-layer separation near the airfoil aft region is the airfoil's relatively thick trailing edge, typically ~1–1½% chord. Note that in Phase 2 SC airfoil development, leading-edge radius is 3% chord for a 14% thick airfoil.

For early transonic airfoil development, analysis and design, Garabedian (1978) and Carlson (1976), among others, may be consulted. Lynch (1982) provides an industry's perspective on supercritical airfoil and wing design for commercial transports.

The phenomenon of drag divergence is clearly demonstrated in Figure 2.48 where an airfoil, at different AoA, is placed in airstreams with increasing Mach number. With successively increasing AoA, the critical Mach number drops, as shown in Figure 2.48. Therefore, locally supersonic flow on the airfoil starts at a reduced Mach number, which explains the appearance of drag divergence behavior at lower Mach numbers with increasing α in Figure 2.48. It is customary to relate drag divergence phenomenon to the slope of the curve C_d vs. M, i.e. $\dfrac{\partial c_d}{\partial M} = 0.10$. M_{dd} is defined in Figure 2.48.

Figure 2.46 The 14% thick, NASA SC(2)-0714 airfoil shape. *Source:* From Jenkins 1989 (SC(2) designation signifies Phase 2 SC airfoil development at NASA).

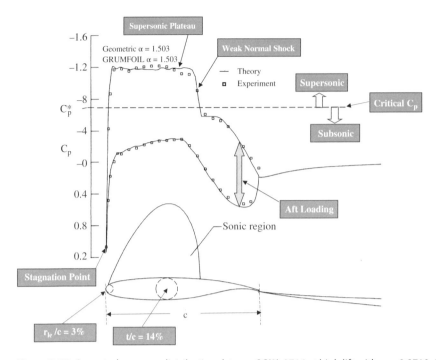

Figure 2.47 Corrected pressure distribution data on SC(2)-0714 at high lift with $c_n = 0.8710$, $M = 0.721$, and $Re_c = 40 \times 10^6$. *Source:* Adapted from Jenkins 1998.

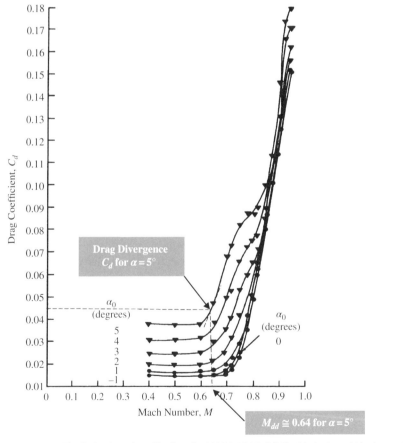

Figure 2.48 The behavior of profile drag for NACA 2315 airfoil with AoA and Mach number. *Source:* Adapted from Loftin 1985.

2.9 Wing Sweep

In pursuit of supersonic aerodynamics, the drag divergence of straight wings in transonic flight was seen as a barrier to high-speed flight that became known as the *sonic barrier*. In 1935, Busemann (1935) introduced the concept of wing sweep to alleviate drag divergence at supersonic/transonic Mach numbers in what became a historic gathering of aerodynamics pioneers at the 5th Volta Congress held in Rome, Italy. Busemann resolved the freestream velocity, V_∞, relative to a swept wing with Λ leading-edge sweep angle, into a component normal to the wing, i.e. $V_\infty\cos\Lambda$, and a component of freestream parallel to the wing $V_\infty\sin\Lambda$. Busemann postulated that the aerodynamic force and moment characteristics of a swept wing depends on the *normal component* of flow to the wing, i.e. $V_\infty\cos\Lambda$ and as long as the normal component is subsonic, the wing will not experience supersonic drag rise. On the subsonic/transonic side, R.T. Jones (1947) is credited with the application of wing sweep to subsonic aircraft to increase their critical Mach number in the United States. The increase in critical Mach number of a swept wing with constant aspect ratio and streamwise thickness ratio is shown in Figure 2.49 (from Loftin 1980). For example, a straight wing of aspect ratio 7 with 12% thick symmetrical airfoil sections will experience a critical Mach number of 0.75 (see Loftin 1980). A 40° swept wing of the same aspect ratio and streamwise thickness ratio (of 12%) will have its critical Mach number raised to 0.86, according to Figure 2.49.

Expressing the aerodynamic results due to sweep are often linked to the sweep angles at other locations besides the leading edge, namely, quarter-chord or half-chord.

Finally, introducing simple theory of sweep, we express the pressure coefficient, lift and pitching moment coefficients and the critical pressure coefficient in terms of sweep angle as:

$$C_{p,\Lambda} = C_{p,\Lambda=0}\cos^2\Lambda \tag{2.104}$$

$$C_{\ell,\Lambda} = C_{\ell,\Lambda=0}\cos^2\Lambda \tag{2.105}$$

$$C_{m_0,\Lambda} = C_{m_0,\Lambda=0}\cos^2\Lambda \tag{2.106}$$

$$C_{p,\text{crit}} = \frac{2}{\gamma M_{\text{crit}}^2}\left[\left[\frac{1+\dfrac{\gamma-1}{2}M_{\text{crit}}^2\cos^2\Lambda}{\dfrac{\gamma+1}{2}}\right]^{\frac{\gamma}{\gamma-1}}-1\right] \tag{2.107}$$

And the compressibility-corrected C_p as:

$$\left(C_p\right)_{M_\infty} = \frac{\left(C_p\right)_{M_\infty=0}}{\beta} = \frac{\left(C_p\right)_{M_\infty=0}}{\sqrt{1-M_\infty^2\left[\cos^2\Lambda-\left(C_p\right)_{M_\infty=0}\right]}} \tag{2.108}$$

The lift curve slope of a swept wing (with $\Lambda_{c/2}$), and aspect ratio, AR, is related to the airfoil lift curve slope, a_0, as

$$a = \frac{a_0\cos\Lambda}{\sqrt{1+\left(a_0\cos\Lambda/\left(\pi AR\right)\right)^2}+a_0\cos\Lambda/\left(\pi AR\right)} \tag{2.109}$$

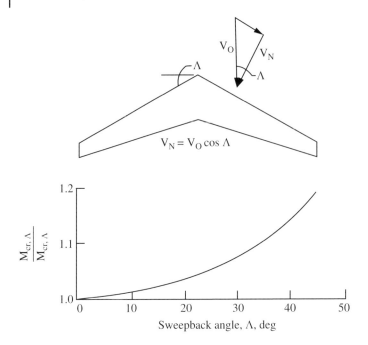

Figure 2.49 Increase in critical Mach number for a swept wing with constant aspect ratio and streamwise thickness ratio. *Source:* From Loftin 1980.

The drag divergence Mach number for a swept wing (of $\Lambda_{c/4}$) is related to straight wing drag divergence Mach number by:

$$(M_{dd})_\Lambda = \frac{(M_{dd})_{\Lambda=0}}{\sqrt{\cos\Lambda_{c/4}}} \tag{2.110}$$

The critical and drag divergence Mach numbers of a swept wing are related to each other following Shevell and Kroo (1992):

$$M_{dd} = M_{crit}\left[1.02 + 0.08(1 - \cos\Lambda_{c/4})\right] \tag{2.111}$$

In Eq. (2.111), M_{dd} is defined as the Mach number where $\dfrac{\partial c_d}{\partial M} = 0.05$.

For more detailed discussion, Vos and Farokhi (2015) may be consulted.

The impact of wing sweep on lift-drag ratio is shown in Figure 2.50a at three different flight conditions. At Mach 0.6 at 30 kft altitude, optimum sweep is 20°. At Mach 2.2, 60 kft altitude, 75° marks the optimum sweep angle and finally at sea level, Mach 1.2, the optimum sweep is at the extreme, 110°. The "hinge" point in Figure 2.50a represents the wing rotation about a fixed point (i.e. hinge or pivot) as the mechanism for creating variable sweep. Three variable-sweep wing positions are shown in Figure 2.50a. An operational fighter aircraft, F-14, that uses variable sweep is also shown in Figure 2.50b, for reference.

(a)

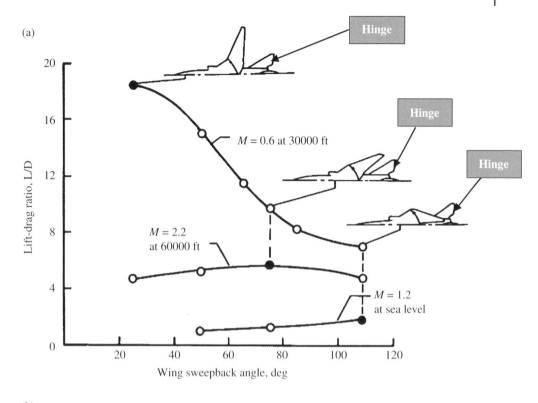

(b)

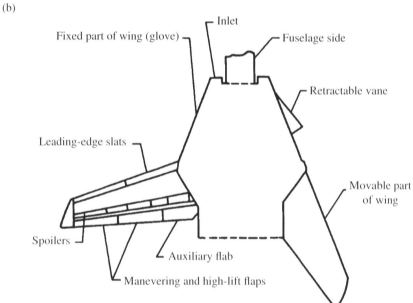

Figure 2.50 Aerodynamic behavior of variable-sweep aircraft: (a) impact of variable sweep on lift-drag ratio shows the optimum sweep angle at different phases of flight. *Source:* Adapted from Polhamus and Toll 1981; (b) Approximate wing platform shape of F-14A variable sweep jet fighter. *Source:* From Loftin 1985.

2.10 Delta Wing Aerodynamics

The lift generated on a delta wing is the sum of the *potential lift* and the *vortex lift*. As the angle of attack, α, increases the contribution of vortex lift increases. Figure 2.51 shows four delta wing planforms. The stall phenomenon in delta wings is related to a *breakdown* in vortex lift, which is known as "vortex breakdown (VB)." Figure 2.52 shows the top, aft- and side-views of a delta wing with the leading edge vortex and the vortex core shown schematically. The suction associated with the vortex core is responsible for the *vortex lift*.

As noted, delta wings derive their lift from two sources: (i) lift due to potential flow and (ii) vortex lift. With increasing AoA, the share of the vortex lift increases as it extends to the point of vortex breakdown. In the incompressible limit, following Polhamus suction analogy (1966, 1971), the wing-lift coefficient is:

$$C_L = C_{L,\,P} + C_{L,V} = K_P \sin\alpha\,\cos^2\alpha + K_V \cos\alpha\,\sin^2\alpha \qquad (2.112)$$

where K_P and K_V are the proportionality constants for potential and vortex lift and they are correlated with the wing aspect ratio. Figure 2.53 show the proportionality constants K_P and K_V as a function of delta wing aspect ratio.

The details of LE-vortex separation and reattachment is shown in Figure 2.54. Wing-lift coefficient and lift-to-drag ratio are shown in Figure 2.55 for a 60° delta wing with sharp leading edge. We note that the lift curve slope for a delta wing is significantly lower, e.g. half, of the unswept wing. The stall angle of attack, in contrast, is significantly higher than the straight wing, e.g. 35° vs. 15°.

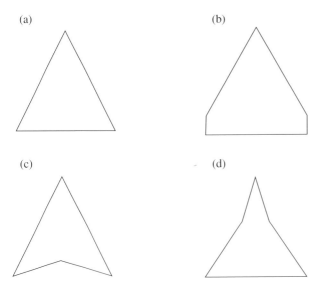

(a)　　　　　　　　　　(b)

(c)　　　　　　　　　　(d)

Figure 2.51 Four delta-wing planforms: (a) simple delta wing; (b) cropped delta wing; (c) notched delta wing; (d) double delta wing.

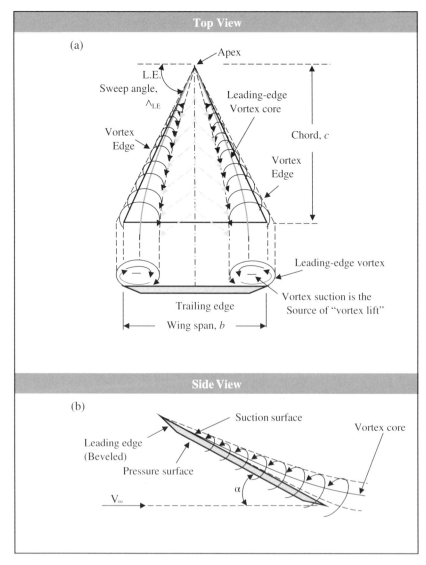

Figure 2.52 Three-views of a delta wing with its leading-edge vortex formation: (a) top and aft views of a delta wing at AoA with the leading-edge vortex; (b) side-view of the delta wing.

2.10.1 Vortex Breakdown

Delta wing stall is closely tied to the phenomenon of *vortex breakdown* (VB). The loss of vortex lift occurs when the leading-edge vortex formation on the suction surface is faced with adverse (streamwise) pressure gradient in high-α environment and thus the vortex core experiences instabilities that grow and cause breakdown.

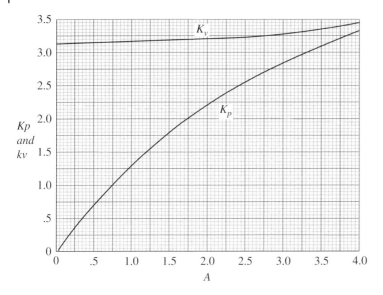

Figure 2.53 Variation of K_P and K_V for delta wings at $M \cong 0$. *Source:* From Polhamus 1968.

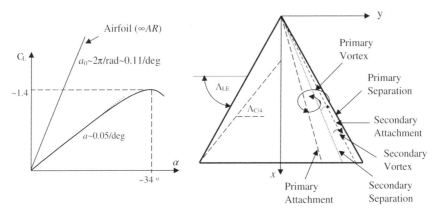

Figure 2.54 Lift curve slope of a delta wing and the details of LE-vortex separation reattachment, and secondary separation and vortex formation on the suction surface.

Three types of vortex breakdown are observed (see Sarpkaya 1971) in water tunnels, wind tunnels and in flight: (i) symmetrical or bubble breakdown; (ii) helical or spiral breakdown; and (iii) double-helix breakdown. The first two types of vortex breakdown are more dominant, i.e. more frequently observed (e.g. see the classic image from Lambourne and Bryer 1962, and Figure 2.56), where both types of vortex breakdown are observed simultaneously on a delta wing in water tunnel. Between the two dominant vortex breakdowns, the bubble type represents the most severe instability with significant loss of vortex lift. Finally, we note that vortex breakdown is found in a variety of high-intensity vortical flows, from free swirling jets to confined swirling flows, and

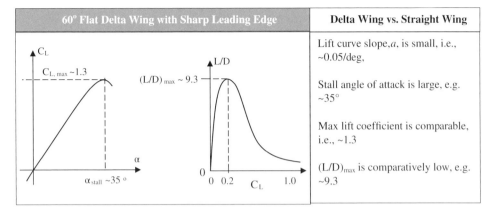

60° Flat Delta Wing with Sharp Leading Edge	Delta Wing vs. Straight Wing
	Lift curve slope,a, is small, i.e., ~0.05/deg,
	Stall angle of attack is large, e.g. ~35°
	Max lift coefficient is comparable, i.e., ~1.3
	$(L/D)_{max}$ is comparatively low, e.g. ~9.3

Figure 2.55 Trends of lift and lift-to-drag variation on a flat delta wing with sharp leading edge.

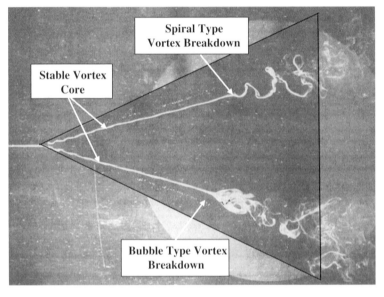

Figure 2.56 Dye injection in the vortex core of a delta wing in water tunnel (speed: 2 in. s^{-1}).
Source: Adapted from Lambourne and Bryer 1962.

swept wings, leading-edge extensions, strakes and canards. For a comprehensive review of vortex breakdown, the paper by Lucca-Negro, O. and O'Doherty, T (2001) is recommended (see Figure 2.57).

2.11 Area-Rule in Transonic Aircraft

In wing–body combinations, dramatic zero-lift drag rise in transonic flight proved to be a daunting obstacle to reach supersonic speeds. In the 1950s, Richard Whitcomb of NASA conducted systematic studies on scaled wing–body models in a wind tunnel and concluded:

(a)

(b)

(c)

(d)

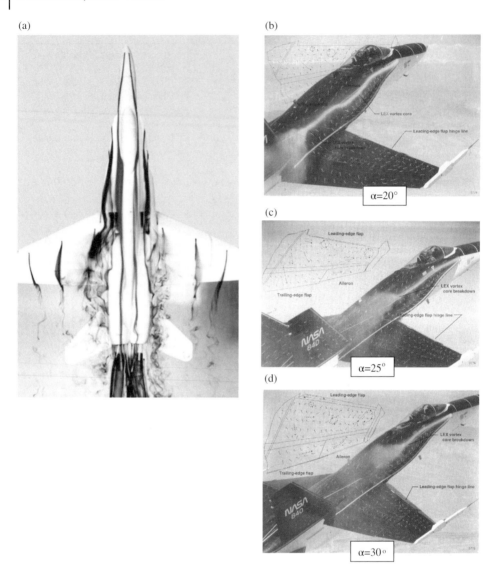

Figure 2.57 Colored dye flow visualization in a water tunnel on a 1/48th scale F-18 model at NASA Dryden Flow Visualization Facility is shown in (a) where helical or spiral type of breakdown is observed; the group of pictures taken in flight in parts (b)–(d) show successive increase in AoA from 20° to 30° and in-flight flow visualization of vortex breakdown on F-18 at high-α. The vortex breakdown of bubble type is observed in these flights, and we note that the breakdown location progressively moves upstream with increasing α. *Source:* From Fisher, Del Frate, and Richwine 1990: (a) Water tunnel dye injection and flow visualization; (b) in-flight; (c) in-flight; (d) in-flight.

> Near the speed of sound, the zero-lift drag rise of a low aspect ratio thin wing–body combination is primarily dependent on the axial development of the cross-sectional areas normal to the airstream. Therefore, it follows that drag rise for any such configuration is approximately the same as that for any other with the same development of cross-sectional areas.
>
> On the basis of the concept proposed, it would be expected that indenting the body of a wing–body combination, so that the combination has the same axial distribution of cross-sectional area as the original body alone, would result in a large reduction or elimination of the drag rise associated with the wing.
>
> *Whitcomb 1956*

Whitcomb proved this hypothesis through detailed measurements. Figure 2.58 from Whitcomb (1956) report show the drag rise behavior with transonic Mach number of three bodies.

In part (a):

1) A slender axisymmetric body
2) The same slender axisymmetric body with low-AR unswept wing
3) The same as 2, but with area-rule applied to the wing–body intersection

In part (b):

4) A slender axisymmetric body
5) The same slender axisymmetric body with low-AR delta wing
6) The same as 2, but with area-rule applied to the wing–body intersection

In part (c):

7) A slender axisymmetric body
8) The same slender axisymmetric body with low-AR swept wing
9) The same as 2, but with area-rule applied to the wing–body intersection

Transonic area rule requires the cross-sectional area of the vehicle normal to the flight direction, i.e. $S(x)$, to be constant, or $dS/dx = 0$ in the wing–body intersection area. This is called area-rule, or as Whitcomb called it: *indenting the body.*

Zero-lift drag coefficient is defined as:

$$C_{D_0} = C_D - C_{D_l} \tag{2.113}$$

Figure 2.59a,b show the F-102 aircraft before and F-102a after the area-rule implementation in the wing-fuselage section.

2.12 Optimum Shape for Slender Body of Revolution of Length ℓ in Supersonic Flow

Consider a pointed slender axisymmetric body along the x-axis of length ℓ, in a supersonic flow at zero angle of attack. Figure 2.60 is the definition sketch of the slender axisymmetric body in uniform supersonic flow along x-direction.

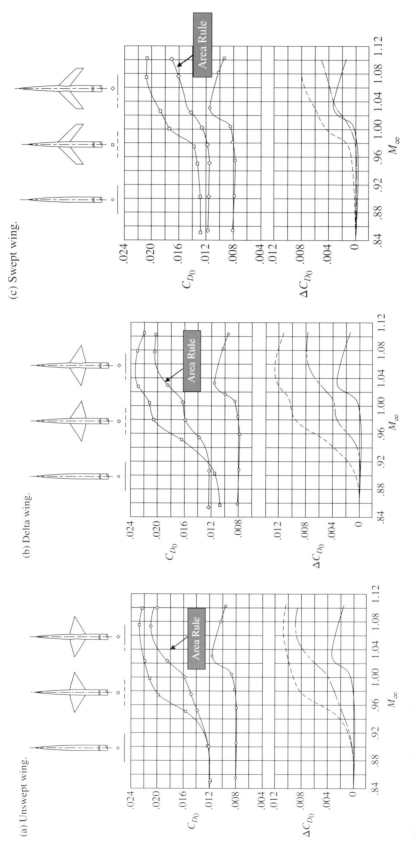

Figure 2.58 Application of transonic area-rule to three wing–body combinations and its impact on transonic drag rise. *Source:* Adapted from Whitcomb 1956: (a) zero-lift drag rise behavior of an axisymmetric body with unswept wing with and without area-rule; (b) zero-lift drag rise behavior of a slender axisymmetric body with low-AR delta wing with and without area-rule; (c) zero-lift drag rise behavior of a slender axisymmetric body with low aspect ratio swept wing with and without area-rule.

(a) Unswept wing.

(b) Delta wing.

(c) Swept wing.

(a) (b)

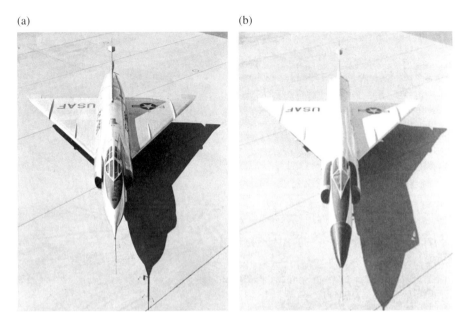

Figure 2.59 Supersonic fighter, F-102, that flew supersonically only after area-rule design modification (NASA Photos): (a) original F-102, a supersonic fighter, without area-rule; (b) F-102a with area-rule.

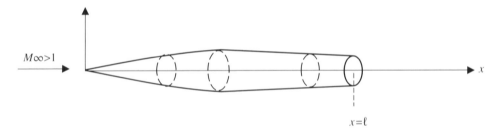

Figure 2.60 A pointed slender axisymmetric body of length, *l*, in supersonic flow.

The assumption of *slender body* makes this a small perturbation problem where linearized aerodynamics govern such flows in supersonic flow. Classical method of distributing sources and sinks along the *x*-axis (i.e. line source) simulates inviscid, non-lifting flows. The unknown source strength, $f(x)/V_\infty$, is then related to the slope of the area distribution of the axisymmetric body. First, we introduce the Glauert variable, θ:

$$x = \frac{l}{2}(1+\cos\theta) \tag{2.114}$$

which places the body nose at $\theta = \pi$ and the TE at $\theta = 0$. Also, by way of Fourier sine series, we express the source strength distribution, $f(\theta)$, on a generalized body of length ℓ as:

$$f(\theta) = \ell \sum_{n=1}^{\infty} A_n \sin(n\theta) \tag{2.115}$$

where $f(\theta) = 0$ at both ends. By factoring the length scale in Eq. (2.115), we make the Fourier coefficients, A_n, nondimensional:

$$A_n = \frac{2}{\pi l} \int_0^\pi f(\theta) \sin n\theta d\theta \qquad (2.116)$$

The derivative of the cross-sectional area, $S'(x)$, is $f(x)$, or the cross-sectional area, $S(x)$ is the integral of $f(x)$ according to:

$$S(x) = \int_0^x f(\xi) d\xi \qquad (2.117)$$

In terms of the new variable, θ,

$$S(\theta) = \int_\pi^\theta f(\theta) \frac{d\xi}{d\theta} d\theta = -\frac{l^2}{2} \int_\pi^\theta \sum_{n=1}^\infty A_n \sin(n\theta) \sin\theta d\theta \qquad (2.118)$$

By integrating the cross-sectional area $S(\theta)$ along the x-axis (or θ-axis), we get the volume of the axisymmetric body, V:

$$V = \frac{\pi l^3}{8} \left(A_1 - \frac{1}{2} A_2 \right) \qquad (2.119)$$

Therefore, the wave drag of a pointed slender axisymmetric body based on Fourier coefficients is:

$$D_{wave} = \frac{\pi \rho_\infty V_\infty^2 l^2}{8} \sum_{n=1}^\infty n A_n^2 \qquad (2.120)$$

2.12.1 Sears-Haack Body

For a pointed body at the nose and base, we have $S(0) = S(\pi) = 0$. This forces the $A_1 = 0$ and for a body of given volume, V, A_2 becomes:

$$A_2 = -\frac{16V}{\pi l^3} \qquad (2.121)$$

Since all higher harmonics in D_{wave} are squared in amplitude, then the minimum of this function occurs when they all vanish, i.e.

$$D_{wave} = \frac{64 V^2}{\pi l^4} \rho_\infty V_\infty^2 \qquad (2.122)$$

and the slender axisymmetric body with $S(\theta)$ cross-sectional area that describes the minimum wave drag in supersonic flow, is known as the Sears-Haack body, with

$$S(\theta) = \frac{4V}{\pi l} \left(\sin\theta - \frac{1}{3} \sin 3\theta \right) \qquad (2.123)$$

Figure 2.61 shows the radius and area distribution of Sears-Haack body.

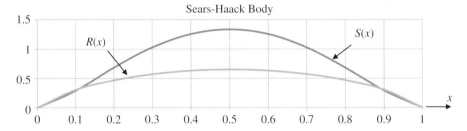

Figure 2.61 The shape of Sears-Haack body and its cross-sectional area distribution.

2.12.2 Von Karman Ogive of Length 𝓵 and Base Area, S(𝓵), for Minimum Axisymmetric Nose Wave Drag

The slender body of revolution that is pointed at the nose, has a length, ℓ, and possesses a base area, $S(\ell)$ attains its minimum wave drag if the slope of the body vanishes at the trailing edge. This is called the von Karman ogive. Figure 2.62 shows the von Karman ogive.

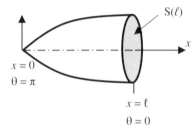

Figure 2.62 Von Karman Ogive with base area, S(l).

Since $S(\ell)$ is finite, A_1 is not zero in the Fourier series, rather

$$A_1 = -\frac{4S(\ell)}{\pi l^2} \tag{2.124}$$

with its wave drag expressed as:

$$D_{wave} = \frac{2S(\ell)^2}{\pi \ell^2} \rho_\infty V_\infty^2 \tag{2.125}$$

The area distribution, $S(\theta)$, for the von Karman ogive, for a given base area, is

$$S(\theta) = \frac{S(\ell)}{\pi}\left(\pi - \theta + \frac{1}{2}\sin 2\theta\right) \tag{2.126}$$

Figure 2.63 shows von Karman ogive with minimum wave drag for a base area of 0.1, body length, 1.0.

2.13 High-Lift Devices: Multi-Element Airfoils

Plain airfoil's lift generation is limited by the circulation that it creates without flow separation. As a result, to create higher $c_{l,\,max}$, a host of leading- and trailing-edge devices are employed that essentially boost the airfoil circulation. Some increase effective camber and some energize the suction surface boundary layer to withstand higher angles of attack. These high-lift devices are leading-edge droop (not shown), slats, and

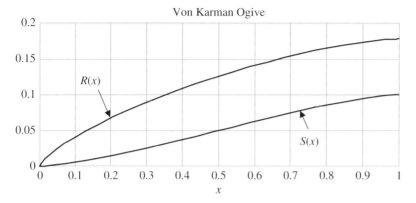

Figure 2.63 Von Karman ogive for $\ell = 1.0$ and a base area, $S(\ell) = 0.1$.

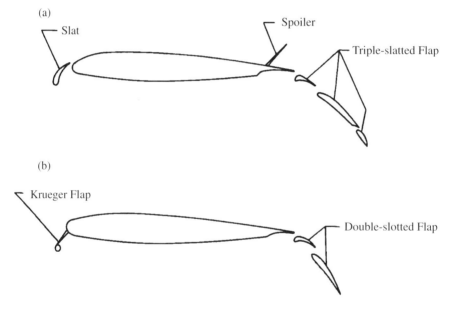

Figure 2.64 Typical LE and TE devices on aircraft wing sections. *Source:* From Loftin 1985: (a) airfoil with triple-slotted flap, slat, and spoiler; (b) airfoil with double-slotted flap and Krueger flap.

Krueger flaps as LE devices and a variety of flaps as TE device. Figure 2.64 shows some of the leading and trailing edge devices, as well as spoilers (shown in part (a)) where they can be deployed in landing.

The combination of leading-edge droop nose and single Fowler flap on trailing edge for a natural laminar airfoil was studied by Jirásek and Amoignon (2009). Khodadoust and Washburn (2007) and Khodadoust and Shmilovich (2007) studied active control of flow separation on high-lift systems. Figure 2.65 shows the effect of various LE and TE devices on sectional $c_{l,\max}$. A plain airfoil's $c_{l,\max}$ is enhanced from ~1.45 to 2.4 when a plain flap or a LE slat is deflected. The $c_{l,\max}$ enhancement increases to 2.6 when a split flap is deployed. Single-slotted flap takes $c_{l,\max}$ to 2.95 and the double-slotted flap

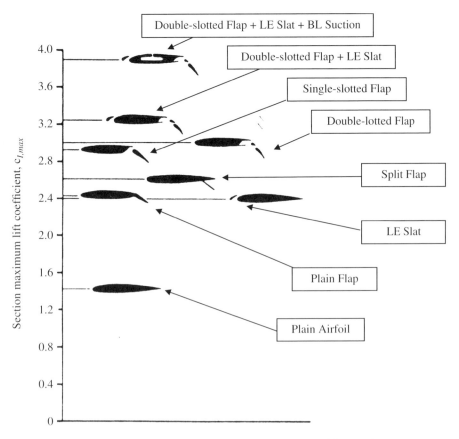

Figure 2.65 Effect of some high-lift devices on $C_{l,max}$. *Source:* Adapted from Loftin 1985.

increases $c_{l,max}$ to 3.0. The combination of the LE slat and the double-slotted flap pushes $c_{l,max}$ to 3.2. Finally, in Figure 2.65, we note that the addition of upper surface-boundary-layer suction to an airfoil with LE slat and double-slotted flap has the potential of increasing $c_{l,max}$ to close to 4.0.

The effect of high lift devices on an aircraft in approach and landing and their interaction with the exhaust jet are shown in Figures 2.66 and 2.67. In Figure 2.66 four lift curves are shown. The first is the clean wing with the slat deployed. The second is deployment of the trailing-edge device: single slotted flap. The third is the vane/main double slotted flap where the interaction with jet is shown in Figure 2.67. Finally, the fourth is the deployment of the main aft double-slotted flap. For detail reference, Rudolph (1996) should be consulted.

Figure 2.68 shows the $C_{L,max}$ on the swept wings of modern transport aircraft with moderate aspect ratio as a function of high-lift system complexity (data from Dillner, May, and McMasters 1984). The wing sweep angle is measured at quarter chord and the $C_{L,max}$ is based on statistical analysis performed by the authors. Triple-slotted flap with leading-edge slat generates the highest $C_{L,max}$, as shown in Figure 2.68. Although $C_{L,max}$ in the range of 2–3 is suitable for conventional takeoff and landing aircraft, it falls short for the STOL (short takeoff and landing) aircraft application. For a modern survey of

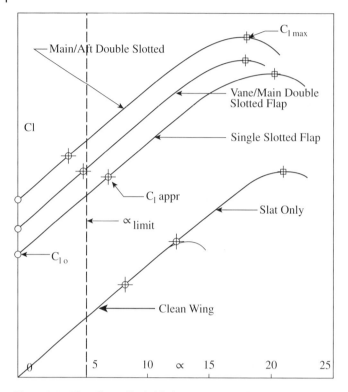

Figure 2.66 The effect of high-lift devices on aircraft C_L vs. AoA in approach and landing. *Source:* From Rudolph 1996.

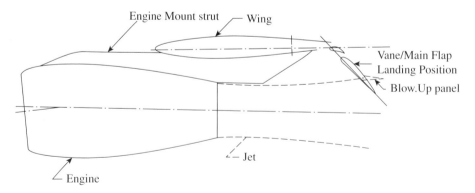

Figure 2.67 Interaction of the trailing-edge flap system with the jet in landing. *Source:* From Rudolph 1996.

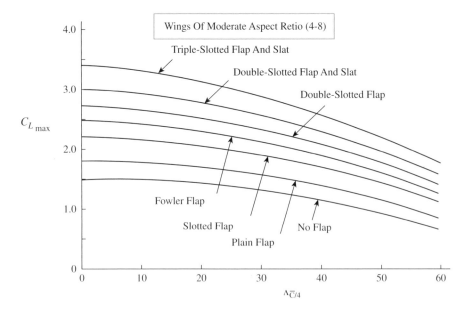

Figure 2.68 The effect of wing sweep angle on $C_{L,max}$ of wings of moderate aspect ratio with high-lift devices of various system complexity. *Source:* Data from Dillner, May, and McMasters 1984.

aerodynamic design of multielement high-lift systems, see van Dam (2002). For applications to airplane performance, Lan and Roskam (2018) and Roskam (2018) should be consulted.

2.14 Powered Lift and STOL Aircraft

STOL aircraft require significantly higher $C_{L,max}$ than conventional aircraft, e.g. $C_{L,max}$ of 10. Circulation control (CC) airfoils use Coanda effect for a rounded trailing edge airfoil and a jet issued from a slot that forces the aft stagnation point to the pressure surface. This method is very powerful and capable of creating $C_{l,max}$, of 5.0 when coupled with LE droop, at zero incidence angle. The drawback of circulation-controlled airfoil is its need for a pressurized source of air. A new operational parameter, jet momentum coefficient, emerges in the design of CC airfoils, defined as

$$C_j \equiv \frac{\dot{m}_j V_j}{q_\infty S} \tag{2.127}$$

In modern studies, CC airfoils have both leading as well as trailing-edge blowing slots, on the suction and pressure surfaces of the airfoil with the mission of flow control in all phases of flight, including cruise drag efficiency. The combination of boundary-layer control (BLC) and CC through tangential jet blowing is its powerful means of lift generation. Extreme short takeoff and landing (ESTOL) aircraft is an advanced application area where pneumatic CC wing and other powered lift concepts, e.g. internally blown flap (IBF), external blown flaps (EBFs), and upper surface blowing (USB), enter as suitable

Figure 2.69 NASA Extreme Short Takeoff and Landing (ESTOL) aircraft concept (see Jones and Joslin 2004).

candidates with a goal of $C_{L,\,max}$ of 10 and cruise L/D of 16 at cruise Mach number of 0.8. Figure 2.69 shows NASA's concept of ESTOL aircraft (see Jones and Joslin 2004). Review paper by Englar (2005) on powered lift concepts is also recommended.

2.15   Laminar Flow Control, LFC

LFC is a powerful flow control strategy that delays boundary-layer transition. It is thus attractive, especially to long-range commercial transports, since it offers major improvement in cruise L/D and consequently block fuel (BF) reduction. The natural laminar flow (NLF) is accomplished through surface geometry tailoring and requires no power. Holmes and Obara (1992) provide an extensive review of the NLF technology and flight experiments, including test and flow visualization techniques in flight. Hybrid laminar flow control (HLFC), by contrast, requires power to provide suction of the boundary layer and promote its laminar state over extended regions of the wing, nacelle, fuselage, and empennage. In this arena, NASA has conducted fundamental and pioneering work since 1930. An extensive overview of the LFC spanning the years of 1930–1990s, is provided by Joslin in 1998. The strip near the leading edge of a wing represents the most effective suction port for LFC over the wing. Low-level suction beyond the strip near the leading edge and partial chords station, e.g. 40%, has proven effective in transition delay. LFC was extended to supersonic flow. Studies of supersonic laminar flow control (SLFC) was applied to NASA High-Speed Civil Transport (HSCT) program by Boeing (see Parikh and Nagel 1990) and McDonnel-Douglas Aircraft (see Powell, Agarwal, and Lacey 1989). Both studies concluded major performance gains in cruise L/D, reduction in takeoff gross weight (TOGW), and operating empty weight (OEW) and FB reduction of ~12%. Concepts of LFC and its applications are shown in Figure 2.70 (from Collier 1993).

The benefits of the HLFC on a supersonic transport are shown in Figure 2.71 (from Parikh and Nagel 1990). Figure 2.72 shows the laminar flow winglet on a Boeing 737

(a)

Natural laminar flow (NLF)

Laminar flow control (LFC)

Hybrid laminar flow control (HLFC)

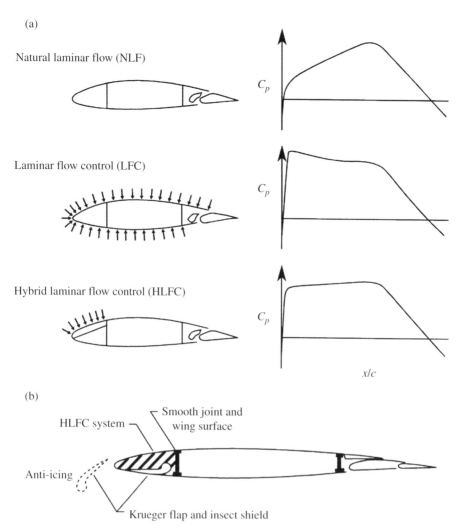

(b)

Figure 2.70 Laminar flow control concepts. *Source:* From Collier 1993: (a) NLF, LFC, HLFC concepts for wing; (b) practical application of HLFC wing.

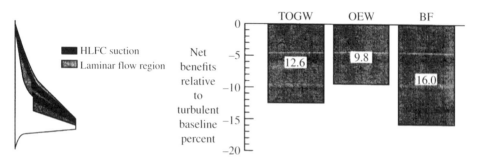

Figure 2.71 Benefits of HLFC on advanced supersonic transport at Mach 2.4, 6500 n.mi, range for 247 passengers. *Source:* From Parikh and Nagel 1990.

Figure 2.72 Boeing 737-800 ecoDemonstrator with laminar flow winglet for 737 MAX. *Source:* Courtesy of Boeing.

Figure 2.73 The first production aircraft with HLFC. *Source:* Courtesy of Boeing.

MAX. Figure 2.73 shows first production aircraft using HLFC, i.e. Boeing 787-9 that uses suction strips on the leading edges of the fin and tailplane.

Use of fluidic control on the fin/rudder to improve its L/D and thus reduce its size is dictated by one engine inoperative (OEI) directional control requirements (technology demonstrated on Boeing 757 in NASA $40' \times 80'$ tunnel). This technology is shown in Figure 2.74.

2.16 Aerodynamic Figures of Merit

Breguet range equation in unaccelerated level flight is derived in Chapter 1, as well as in basic aerodynamics/flight mechanics books, e.g. Anderson (2005):

Figure 2.74 NASA 40′×80′ tunnel with tufts on Boeing 757 fin and rudder. *Source: Courtesy NASA.*

$$R = \eta_0 \frac{Q_R}{g_0} \frac{L}{D} \ln \frac{W_i}{W_f} \tag{2.128}$$

where η_0 is the engine overall efficiency, Q_R is the fuel heating value, L/D is the aircraft lift-drag ratio, W_i is the initial weight of the aircraft, and W_f is the final weight of the aircraft (note that W_f in this expression is not fuel weight). L/D is called the aircraft aerodynamic efficiency.

The overall efficiency in a jet engine is defined as:

$$\eta_o = \frac{F_n V_0}{\dot{m}_f Q_R} \tag{2.129}$$

By substituting Eq. (2.129) into (2.128) and rearranging, we get:

$$R = \left(M_0 \frac{L}{D} \right) \frac{a_0/g_0}{TSFC} \ln \frac{W_i}{W_f} \tag{2.130}$$

The coefficient $(M_0 L/D)$ is called the *range factor or cruise efficiency* parameter, TSFC is the thrust-specific fuel consumption and a_0 is the speed of sound at flight altitude. Here we note that the aircraft range is maximized when $(M_0 L/D)$ is a maximum, when we maintain the specific fuel consumption (TSFC) constant. Anderson (2005) shows

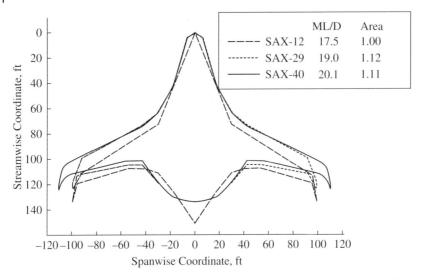

Figure 2.75 Conceptual blended wing–body design for low-noise and high-cruise efficiency. *Source:* From Hileman et al. 2010.

that the range factor is a maximum when $C_L^{1/2}/C_D$ is maximum. Therefore, we conclude that range of a jet aircraft is maximized when $C_L^{1/2}/C_D$ is maximum. In a blended wing-body (BWB) configuration designed for low noise and aerodynamic efficiency, Hileman et al. (2010) showed that ML/D varied between 17.5 and 20.1 for the three conceptual designs, SAX-12, SAX-29, and SAX-40, as shown in Figure 2.75.

Now, in terms of endurance, E, we use Eq. (2.130) to derive endurance in unaccelerated level flight:

$$E = \frac{R}{V_0} = \frac{\eta_0}{V_0} \frac{Q_R}{g_0} \frac{L}{D} \ln \frac{W_i}{W_f} = \left(\frac{L}{D}\right) \frac{1/g_0}{TSFC} \ln \frac{W_i}{W_f} \qquad (2.131)$$

Here, we note that the endurance of a jet aircraft is maximized when L/D attains its maximum value. We may graphically establish $(L/D)_{max}$ from aircraft drag polar by drawing a tangent to the C_L vs. C_D curve. The slope of the tangent is the maximum lift-drag ratio. Figure 2.76 shows a typical drag polar and the graphical depiction of maximum L/D, as well as C_{D0}, $C_{D,min}$, and $C_{L,max}$. Figure 2.77 shows the subsonic and supersonic aircraft drag Polars. C_l' is the lift coefficient at minimum drag.

In supersonic flow, L/D as well as $(L/D)_{max}$ drops. For reference, the drag polar of a 10% thick diamond airfoil in Mach $\sqrt{2}$ stream is shown in Figure 2.78. The angle of attack that maximized L/D is 6° and the value of the theoretical $(L/D)_{max}$ is only 4.72, as shown in Figure 2.79.

Stalling speed, V_S, of the aircraft demands $C_{L,max}$ to satisfy:

$$C_{L,max} \frac{1}{2} \rho_\infty V_S^2 S = W \qquad (2.132)$$

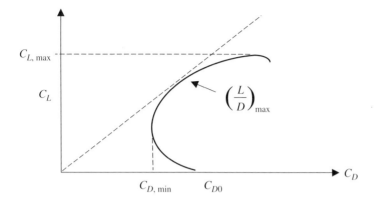

Figure 2.76 Aircraft drag polar with graphical construction of $(L/D)_{max}$.

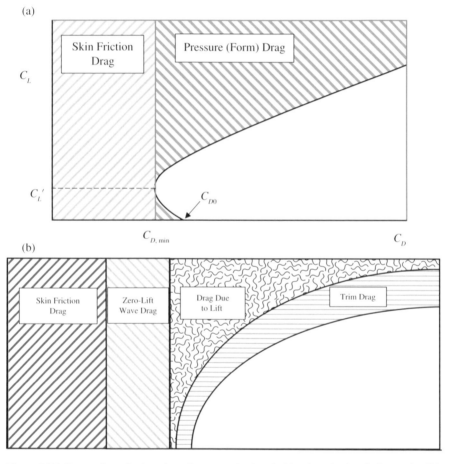

Figure 2.77 Drag polars of subsonic and supersonic aircraft: (a) subsonic aircraft drag polar; (b) supersonic aircraft drag polar.

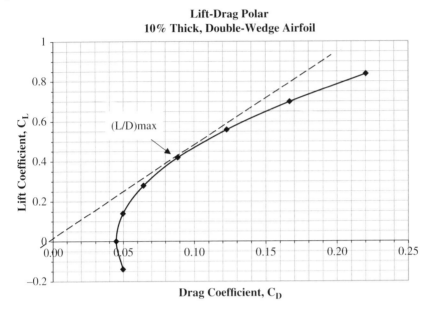

Figure 2.78 The drag polar of a diamond airfoil in AoA.

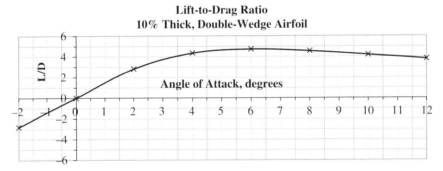

Figure 2.79 Lift-drag ratio of a 10% thick diamond airfoil in Mach $\sqrt{2}$ flow as a function of AoA.

By isolating stalling speed, we get:

$$V_S = \sqrt{\frac{2W}{\rho_\infty S C_{L,\max}}} = \sqrt{\left(\frac{W}{S}\right)\frac{2}{\rho_\infty C_{L,\max}}} \tag{2.133}$$

The ratio of aircraft weight to its planform area, W/S, is called the *wing loading,* and we note that Eq. (2.133) connects three important aircraft flight parameters, namely stalling speed, wing loading, and $C_{L,\max}$:

$$V_S = \sqrt{\frac{2}{\rho_\infty}} \left(\frac{W}{S}\right)^{1/2} \left(C_{L,\max}\right)^{-1/2} \tag{2.134}$$

The approach speed, V_A is governed by airworthiness requirement, which is ≥1.23 times the stalling speed, i.e. 1.23 V_S; therefore, the minimum approach speed is:

$$V_A = 1.23 \left[\sqrt{\left(\frac{W}{S} \right) \frac{2}{\rho_\infty C_{L,\max}}} \right] \qquad (2.135)$$

Hence, the approach lift coefficient, $C_{L,A}$ is $1/(1.23)^2$ times $C_{L,\max}$, i.e.

$$C_{L_A} = \frac{C_{L_{\max}}}{1.51} \qquad (2.136)$$

Therefore, the aerodynamic parameter of importance to takeoff/landing is $C_{L,\max}$. Typical C_L versus angle of attack for a conventional aircraft is shown in Figure 2.80 (from Heffley et al. 1977). Another interest in higher $C_{L,\max}$ stems from its effect on reducing the stall speed and thus reducing aircraft noise in takeoff and landing. Concept of *super lift* is presented by Yang and Zha (2017) that addresses $C_{L,\max}$.

For reference, we note here that aircraft L/D also impacts the *climb gradient*, which is stipulated by airworthiness requirements for OEI in 2-, 3-, and 4-engine aircraft, along with thrust-to-weight ratio, T/W:

$$\tan \gamma \cong \frac{T}{W} - \frac{1}{L/D} \qquad (2.137)$$

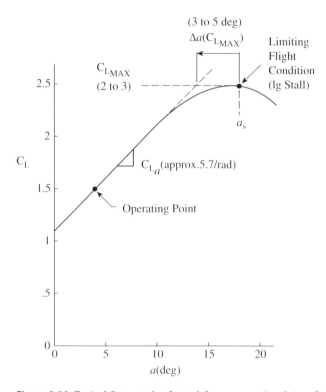

Figure 2.80 Typical C_L vs. angle of attack for a conventional aircraft. *Source:* From Heffley et al. 1977.

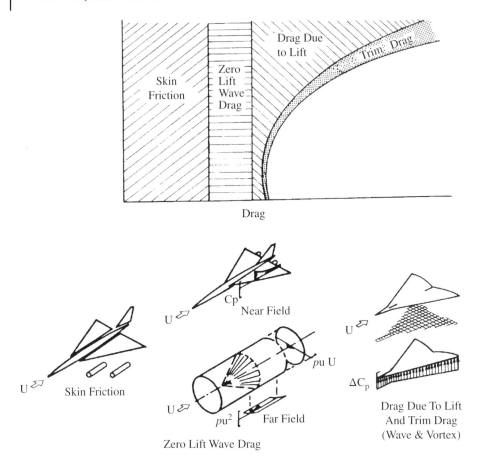

Figure 2.81 Supersonic drag buildup. *Source:* From Jobe 1985.

For airworthiness requirements for climb gradients in OEI, see van Dam (2002).

To summarize, the buildup of aircraft drag at supersonic speed is calculated based on (compressible) skin friction, zero-lift wave drag, induced drag or drag due to lift, and trim drag. These elements are shown in Figure 2.81 drag polar (from Jobe 1985).

2.17 Advanced Aircraft Designs and Technologies for Leaner, Greener Aviation

Environmentally Responsible Aviation (ERA) Project was born at NASA to achieve three ambitious goals simultaneously by the year 2025 for commercial aircraft flying at Mach 0.85, for 7000 mi, carrying between 50 000 and 100 000 pounds of payload:

1) 50% less fuel burn (compared to engines that entered service in 1998)
2) 50% reduction in harmful emissions
3) 83% reduction in the size of the geographic areas affected by objectionable airport noise

Figure 2.82 Blended wing–body aircraft for reduced noise and higher cruise efficiency. *Source:* Courtesy of NASA/Boeing.

Three aircraft companies in the United States – Boeing, Lockheed Martin, and Northrop Grumman – answered the call (under contract from NASA in 2010). In this section, we briefly examine their concept planes as well as other promising concepts for greener, leaner aircraft. As these concepts and technologies are "work-in-progress," the three challenging goals of NASA ERA Project are not simultaneously met by the proposed designs.

In the tradition of X-planes, X-48 from Boeing's advanced vehicle concepts is a BWB design with P&W geared turbofan engines on top and with two vertical tails that help shield communities from noise. The BWB design in X-48 uses HLFC that produces a high lift-to-drag-ratio vehicle that offers passenger capacity in the wing and thus reduced drag over the conventional tube and wing aircraft with the same load factor (i.e. number of passengers). In the BWB design, *fuselage* provides lift, thereby improving cruise efficiency. Figure 2.82 shows the X-48 aircraft from Boeing (courtesy of NASA/ Boeing). Qin et al. (2004) conducted a comprehensive study of blended wing body aircraft aerodynamics and demonstrated that progressive improvements can be achieved through detailed design, e.g. spanwise loading. In a side-by-side comparison of a 480-passenger (tube and wing) aircraft (A380-700) and a BWB aircraft, Liebeck (2004) concluded that BWB saves 19% in OEW, produces 18% saving in *maximum takeoff weight* (MTOW), 19% saving in *total sea-level static thrust* and 32% savings in *fuel burn per seat* for an 8700 nautical-mile flight.

Flying wing (FW) concept that dates back to the Jack Northrop era (1930s and 1940s) is revisited by Northrop Grumman. Figure 2.83 shows the flying wing, commercial aircraft version, with four high bypass Rolls-Royce engines embedded on the upper surface for noise shielding. The tailless flying wing commercial airliner employs LFC and advanced composite structure. This concept airplane is expected to provide the highest

Figure 2.83 Northrop Grumman's concept aircraft designed for commercial aviation market. *Source:* Courtesy NASA/Northrop Grumman.

Figure 2.84 Box wing airliner concept plane for reduced drag. *Source:* Courtesy of NASA/Lockheed Martin.

aerodynamic efficiency for large passenger aircraft in subsonic-transonic range. The tailless flying wing design with advanced technology airframe and propulsion system is a strong candidate for the future of commercial aviation.

Another aerodynamic concept revisited is the box wing aircraft that offers the potential for higher lift-to-drag ratio (by an expected 16%) and is proposed as a concept airliner by Lockheed-Martin. Figure 2.84 shows the box wing concept plane (courtesy NASA/Lockheed Martin) that employs HLFC, advanced lightweight composite materials and ultra-high bypass (UHB) turbofan engine by Rolls-Royce Liberty Works, known as an ultra-fan engine. Induced drag is reduced in a box wing since it lacks wing-tips. In addition, lifting surfaces are extended aft continuously, i.e. beyond the front portion of the wing box, that create higher L/D as compared to a conventional wing of the same span.

To explore lifting fuselage concept, a team of MIT, Aurora Flight Sciences, and Pratt and Whitney designed a wide fuselage (the D8, or "double bubble") concept plane with low swept wings and aft-propulsion system for boundary-layer ingestion (BLI) and shielded by the empennage. The airplane is designed for the domestic market, flying at Mach 0.74, carrying 180 passengers 3000 nautical miles. Aurora Flight Sciences predicts (see Aurora Flight Sciences website: http://www.aurora.aero) that D8 could reduce fuel burn by 71%, reduce noise by 60 EPNdB, and reduce NO_x by 87% in landing and takeoff cycles as compared to the best-in-class 737-800 narrow-body aircraft. Figure 2.85 shows the double-bubble D8 for high lift-to-drag ratio, low emissions, and low noise (courtesy of NASA/MIT and Aurora Flight Sciences). D8 has the technology readiness level to enter service in 2030–2035 time frame.

Concept of high aspect ratio is supported by truss-braced design in the Volt hybrid propulsion aircraft designed by Boeing. Figure 2.86 shows the Volt that was designed under NASA Subsonic Ultra Green Aircraft Research, SUGAR, program at Boeing (courtesy NASA/Boeing).

The hybrid wing-body design, as proposed by MIT, is shown in Figure 2.87 (courtesy NASA/MIT). This is a tailless configuration with BLI electric propulsors in the aft.

Finally a low-boom X-Plane (X-59A) developed by Lockheed Martin Skunk Works shown in Figure 2.88 ends the aerodynamics chapter with a promise of "Greener-Quieter-Safer" and faster flight. X-59A is designed for Mach 1.4 cruise at 55 000 ft altitude. Its sonic boom loudness on the ground has a goal of 75 PLdB as compared to 105–110 PLdB for Concorde (see Warwick, G 2018).

The common thread in the advanced design and technology aircraft is the *integrated system design* approach that produces synergy.

Improved aerodynamic efficiency is envisioned through:

1) LFC, or hybrid LFC
2) Lifting "fuselage," e.g. BWB, FW, D8
3) Tailless configuration, e.g. Flying Wing
4) Higher effective aspect ratio, e.g. Volt
5) BLI by aft installation of propulsors
6) Propulsion system integration
7) Folding wing in ground operation, taxiways, and airport gates

Improved fuel burn will be accomplished through:

1) UHB turbofan engines with advanced core
2) Hybrid gas-electric propulsion
3) Propulsion system integration
4) Extensive use of lightweight advanced composites
5) Folding wings produce high AR in flight and fold to accommodate ground operation
6) Use of fluidic devices to improve lift, reduce control surface size
7) Use of electric operations at the gate and taxi

Figure 2.85 Wide fuselage for enhanced lift, boundary-layer ingestion (BLI) propulsion and low swept wings for drag reduction are combined in MIT concept plane. *Source:* Courtesy of NASA/MIT.

Reduced noise footprint will be achieved through:

1) Propulsion system integration
2) UHB turbofan engines with reduced fan pressure ratio and jet speed
3) Upper surface installation, i.e. wing shielding
4) Empennage shielding
5) Flight profile management and ground operations
6) Noise mitigating design for the landing gear and high-lift devices in LTO cycle
7) Low-boom technology

Figure 2.86 The Volt is a hybrid gas turbine and battery technology plane designed by Boeing with high aspect ratio truss-braced wing. *Source:* Courtesy of NASA/Boeing.

Figure 2.87 Hybrid wing body, H series, future aircraft concept from MIT. *Source:* Courtesy of NASA/MIT.

Figure 2.88 X-59A Low Boom X-Plane: Greener-Quieter-Safer. *Source:* Courtesy of NASA.

Reduced harmful emissions will be accomplished through:

1) Alternative jet fuels with lower life-cycle emissions
2) Advanced (staged) combustor design
3) Hybrid, e.g. gas-electric propulsion or multifuel hybrid propulsion
4) Lower-altitude cruise, e.g. 25 kft, to mitigate persistent contrail
5) Propulsion system integration
6) Advanced air traffic management (ATM) and ground operations
7) Electric propulsion with lower lifecycle emissions

2.18 Summary

This chapter reviewed aircraft aerodynamics both at the fundamental and applied levels. Pressure and shear distribution on bodies were examined at different flight Mach numbers. Elementary treatment of laminar and TBLs was presented for both incompressible as well as compressible flows. Reynolds number and transition were identified as the main contributors to friction drag estimation. Elementary transition models were discussed. The chapter first presented finite wing using classical Prandtl wing theory, followed by the modern view on bell spanloading and its impact on aerodynamic efficiency and *proverse yaw* phenomenon. Mach number is related to pressure distribution, compressibility effects, wave formations, and wave drag. Concept of critical Mach number and drag divergence is presented and tied to modern supercritical airfoil development. The role of area-rule on zero-lift drag in transonic flow was discussed. The optimal shape of Sears-Haack body and von Kármán ogive are derived in supersonic flow. Wing aerodynamics, including the effects of aspect ratio, sweep, planform shape, e.g. delta wing, and stall characteristics are presented. High-lift devices in the context of landing and takeoff performance and powered lift for STOL/ESTOL application were introduced. LFC was briefly discussed, and its potential benefits in aircraft performance were outlined. Finally, the chapter presented some modern aircraft design and technologies that lead to greener, leaner aviation. It discussed aircraft concepts, mainly commissioned by NASA through ERA Project, for low noise, low emissions, and low fuel burn. Finally, a low-boom flight demonstrator, X-59A, developed by Lockheed Martin's Skunk Works under Quiet Supersonic Technology (QueSST) Program, was shown with the schedule to fly in 2021.

References

Abbott, I.H. and Von Doenhoff, A.E. (1945). Summary of Airfoil Data. *NACA Report No. 824*, Washington, DC.

Abbott, I.H. and Von Doenhoff, A.E. (1959). Theory of Wing Sections. New York: Dover Publications.

Alexander, M.G. (1991). Subsonic Wind Tunnel Testing Handbook. Version 1.1, *Technical Report, Flight Dynamics Directorate*, WL-TR-91-3073.

Anderson, J.D. (2012). Modern Compressible Flow, 3e. New York: McGraw-Hill & Co.

Anderson, J.D. (2016). Fundamentals of Aerodynamics, 6e. New York: McGraw-Hill.

Anderson, J.D. Jr. (1989). Hypersonic and High Temperature Gas Dynamics. New York: McGraw-Hill.

Anderson, J.D. Jr. (2005). Introduction to Flight, 6e. New York: McGraw-Hill.

Bertin, J.J. and Cummings, R.M. (2013). Aerodynamics for Engineers, 6e. Upper Saddle River, NJ: Prentice Hall.

Bowers, A.H., Murillo, O.J., Jensen, R., Eslinger, B., and Gelzer, C. (2016). On Wings of the Minimum Induced Drag: Spanload Implications for Aircraft and Birds. *NASA/ TP-2016-219072*, Washington, D.C.

Busemann, A. (1935). Aerodynamischer Auftrieb bei Überschallgeschwindigkeit. In: Proceedings of the 5th Volta Congress held in Rome, Italy, 328–360. Volta Congress.

Carlson, L.A. (1976). Transonic airfoil analysis and design using cartesian coordinates. Journal of Aircraft 13 (5): 349–356.

Carlson, H.W. and Harris, R.V. Jr. (1969). A unified system of supersonic aerodynamic analysis. In: Analytic Methods in Aircraft Aerodynamics, NASA SP 228, 639–658. Washington, D.C.: NASA.

Cebeci, T. (1999). Engineering Approach to the Calculation of Aerodynamic Flows. Berlin: Springer Verlag.

Collier, F.S, Jr. 1993. An Overview of Recent Subsonic Laminar Flow Control Flight Experiments. *AIAA Paper Number: AIAA-93-2987*.

Covert, E.E. (ed.) (1985). Thrust and drag: its prediction and verification. In: AIAA Progress in Astronautics and Aeronautics, vol. 98, 1–346. Washington, D.C.: AIAA.

Dillner, B., May, F.W., and McMasters, J.H. (1984). Aerodynamic issues in the design of high-lift systems for transport aircraft. AGARD Fluid Dynamics Panel Symposium, Brussels, Belgium, May.

Eckert, E.R.G. (1956). Engineering relations for heat transfer and friction for high-velocity laminar and turbulent boundary layer flow over surfaces with constant pressure and temperature. Transactions of the ASME 78 (6): 1273.

Eckert, E.R.G. (1960). Survey of Boundary Layer Heat Transfer at High Velocities and High Temperatures. *WADC Tech. Report*, April, 59–264.

Englar, R.J. (2005). Overview of circulation control pneumatic aerodynamics: Blown force and moment augmentation and modification as applied primarily to fixed-wing aircraft. In: Paper #2 presented at the NASA/ONR Circulation Control Workshop, Hampton, VA, March, 2004. Washington, DC: NASA CP 2005–213509.Also published in Workshop Proceedings,

Eppler, R., and Somers, D.M. (1980). A Computer Program for the Design and Analysis of Low-Speed Airfoils. *NASA TM-80210*, Washington, D.C.

Faber, T.E. (1995). Fluid Dynamics for Physicists. Cambridge, UK: Cambridge University Press.

Farokhi, S. (2014). Aircraft Propulsion, 2e. Chichester, UK: Wiley.

Fisher, D.F., Del Frate, J.H., and Richwine, D.M. (1990). In-Flight Flow Visualization Characteristics of the NASA F-18 High Alpha Research Vehicle at High Angles of Attack. *NASA TM-4193*, Washington, D.C.

Garabedian, P.R. (1978). Transonic airfoil codes. In: Paper in the Proceedings of the Conference: Advanced Technology Airfoil Research, vol. 1, 45–54. Washington, D.C.: NASA *CP-2045*.

Glauser, M.N., Saric, W.S., Chapman, K.L., and Reibert, M.S. (2014). Swept-wing boundary-layer transition and turbulent flow physics from multipoint measurements. AIAA Journal 52 (2): 338–347.

Hayes, W.D. and Probstein, R.F. (1959). Hypersonic Flow Theory. New York: Academic Press.

Heffley, R.K., Stapleford, R.L., and Rumold, R.C. (1977). Airworthiness Criteria Development for Powered Lift Aircraft. Washington, D.C.: *NASA CR-2791*.

Heppenheimer, T.A. (2007). Facing the Heat Barrier: A History of Hypersonics. Washington, DC: NASA SP-2007-4232.

Hileman, J.I., Sapkovszky, Z.S., Drela, M. et al. (2010). Airframe design for silent fuel-efficient aircraft. AIAA Journal of Aircraft 47 (3): 956–969.

Hoerner, S.F. (1965). Fluid-Dynamic Drag: Practical Information on Aerodynamic Drag and Hydrodynamic Resistance. Hoerner Fluid Dynamics Publisher.

Hoerner, S.F. and Borst, H.V. (1985). Fluid-Dynamic Lift: Information on Lift and Its Derivatives in Air and Water, 2e. Hoerner Fluid Dynamics Publisher.

Holmes, B.J. and Obara, C.J. (1992). Flight research on natural laminar flow applications. In: Natural Laminar Flow and Laminar Flow Control (eds. R.W. Barnwell and M.Y. Hosseini). New York: Springer Verlag.

Jenkins, R.V. (1989). NASA SC(2)-0714 Airfoil Data Corrected for Sidewall Boundary Layer Effects in the Langley 0.3-Meter Transonic Cryogenic Tunnel. *NASA TP-2890*, Washington, D.C.

Jirásek, A. and Amoignon, O. (2009). Design of a high-lift system with a leading edge droop nose. AIAA Paper Number 2009-3614. Paper presented at AIAA Applied Aerodynamics Conference. San Antonio, Texas.

Jobe, C.E. (1985). Prediction and verification of aerodynamic drag, part I: prediction, Chapter IV. In: Thrust and Drag: Its Prediction and Verification, Progress in

Astronautics and Aeronautics, vol. 98 (ed. E.E. Covert), 173–206. Washington, D.C.: AIAA.

Johnson, H.A. and Rubesin, M.W. (1949). Aerodynamic heating and convective heat transfer – summary of literature survey. Transactions of ASME 71: 447–456.

Jones, R.T. (1947). Wing Planforms for High-Speed Flight. *NACA Report 863*, Washington, D.C. (Supersedes NACA TN 1033).

Jones, G.S., and Joslin, R.D. (Editors). (2004). Proceedings of the 2004 NASA/ONR Circulation Control Workshop. *NASA/CP-2005-213509*, Washington, D.C.

Joslin, R.D. (1998). Overview of Laminar Flow Control. Washington, D.C.: *NASA TP-1998-208705*.

Keenan, J.H., Chao, J., and Kaye, J. (1983). Gas Tables: Thermodynamic Properties of Air Products of Combustion and Component Gases Compressible Flow Functions, 2e. New York: Wiley.

Khodadoust, A. and Shmilovich, A. (2007). High Reynolds number simulations of distributed active flow control for a high-lift system. AIAA paper number 2007-4423. Paper presented at the AIAA Applied Aerodynamics Conference, Miami, FL.

Khodadoust, A. and Washburn, A. (2007). Active control of flow separation on a high-lift system with a slotted flap at high Reynolds number. AIAA paper number 2007-4424. Paper presented at the AIAA Applied Aerodynamics Conference, Miami, FL.

Kuethe, A.M. and Chow, C.Y. (2009). Foundations of Aerodynamics: Bases of Aerodynamic Design, 5e. New York: Wiley.

Lambourne, N.C. and Bryer, D.W. (1962). The Bursting of Leading-Edge Vortices – Some Observations and Discussion of the Phenomenon. London: Aeronautical Research Council, R&M No. 3282.

Lan, E.C. and Roskam, J. (2018). Airplane Aerodynamics and Performance. Lawrence, Kansas: DAR Corporation.

Liebeck, R.H. (2004). Design of the blended wing body subsonic transport. AIAA Journal of Aircraft 41 (1): 10–25.

Loftin, L.K. (1985). Quest for Performance: The Evolution of Modern Aircraft. *NASA SP-468*, Washington, D.C.

Loftin, K.L. Jr.1980. Subsonic Aircraft: Evolution and the Matching of Size to Performance. *NASA RP 1060*, Washington, D.C.

Lucca-Negro, O. and O'Doherty, T. (2001). Vortex breakdown: a review. Progress in Energy and Combustion Science 27 (4): 431–481.

Lynch, F. (1982). Commercial transports – aerodynamic design for cruise efficiency. In: Transonic Aerodynamics, AIAA Progress in Astronautics and Aeronautics, vol. 81 (ed. D. Dixon), 81–144. Washington, D.C.: AIAA.

Malik, M.R. (1990). Stability theory for laminar flow control design. In: Viscous Drag Reduction in Boundary Layers, Progress in Astronautics and Aeronautics, vol. 123 (eds. D.M. Bushnell and J.N. Hefner), 3–46. Washington, D.C.: AIAA.

McMasters, J.H. and Henderson, M.L. (1979). Low-speed single-element airfoil synthesis. Science and technology of low speed and motorless flight, Part I. NASA CP-2085.

Morkovin, M.V. (1969). On the many faces of transition. In: Viscous Drag Reduction. Berlin: Springer-Verlag.

Parikh, P.G. and Nagel, A.L. (1990). Application of Laminar Flow Control to Supersonic Transport Configurations. *NASA CR-181917*, Washington, D.C.

Peñaranda, F. E. and Freda, M. S. (Editors), (1985a). Aeronautical Facilities Catalogue, Vol. 1: Wind Tunnels, *NASA RP-1132*, Washington, D.C.

Peñaranda, F. E. and Freda, M. S. (Editors), (1985b). Aeronautical Facilities Catalogue, Vol. 2: Airbreathing Propulsion and Flight Simulators, NASA RP-1133, Washington, D.C.

Polhamus, E.C. (1966). A Concept of Vortex Lift of Sharp-Edge Delta Wings Based on a Leading-Edge Suction Analogy. NASA TN D-3767, Washington, D.C.

Polhamus, E.C. (1968). Application of leading-edge suction analogy of vortex lift to the drag due to the lift of sharp-edge delta wings. *NASA TN D-4739*, Washington, D.C.

Polhamus, E.C. (1971). Predictions of vortex lift characteristics by a leading-edge suction analogy. AIAA Journal of Aircraft 8 (4): 193–199.

Polhamus, E.C. and Toll, T.A. (1981). Research related to variable sweep aircraft development. *NASA TM-83121*, Washington, D.C.

Powell, A.G., Agarwal, S., and Lacey, T.R. (1989). Feasibility and benefits of laminar flow control on supersonic cruise airplanes. *NASA CR-181817*, Washington, D.C.

Prandtl, L. (1904). Über Flüssigkeitsbewegung bei sehr kleiner Reibung. In: Proceedings of the 3rd International Mathematics Congress, 484–491. Heidelberg-Germany: International Congress of Applied Mechanics.

Prandtl, L. (1922). Applications of Modern Hydrodynamics to Aeronautics. *NACA Report No. 116*, Washington, D.C.

Prandtl, L. (1933). Uber tragflügel kleinsten induzierten widerstandes. In: Zeitschrift für Flugtechnik und Motorluftschiffahrt, vol. 1, No. 6. Munich, Germany.

Qin, N., Vavalle, A., Le Moigne, A. et al. (2004). Aerodynamic considerations of blended wing body aircraft. Progress in Aerospace Sciences 40: 321–343.

Rae, W.H. and Pope, A. (1984). Low-Speed Wind Tunnel Testing. New York: Wiley.

Roskam, J. (2018). Airplane Flight Dynamics & Automatic Flight Controls. Lawrence, Kansas: DAR Corporation.

Rudolph, P.K.C. (1996). High-Lift Systems on Commercial Subsonic Airliners. *NASA CR-4746*, Washington, D.C.

Sarpkaya, T. (1971). Vortex breakdown in swirling conical flows. AIAA Journal 9 (9): 1792–1799.

Schlichting, H. and Gersten, K. (2017). Boundary-Layer Theory, 9e. Berlin: Springer Verlag.

Sherman, F.S. (1990). Viscous Flow. New York: McGraw-Hill.

Shevell, R.S. and Kroo, I. (1992). Compressibility drag: 3D effects and sweep. In: Aircraft Design: Synthesis and Analysis. Palo Alto, CA: Department of Aeronautics and Astronautics, Stanford University.

Smith, A.M.O. (1952). Design of DESA-2 Airfoil. El Segundo, CA, Report ES17117, AD143008: Douglas Aircraft Co.

Smith, A.M.O. (1956). Transition, pressure gradient and stability theory. In: Proceedings of the 9th International Congress of Applied Mechanics, vol. 4, 234–244. International Congress of Applied Mechanics.

Van Dam, C.P. (2002). The aerodynamic design of multi-element high-lift systems for transport airplanes. Progress in Aerospace Sciences 38: 101–144.

Van Driest, E.R. (1951). Turbulent boundary layer in compressible fluids. Journal of Aeronautical Sciences 18: 145–160.

Van Ingen, J.L. (1956). A Suggested Semi-Empirical Method for the Calculation of the Boundary-Layer Region. Delft, the Netherlands, *Report No. VTH71, VTH74*.

Vos, R. and Farokhi, S. (2015). Introduction to Transonic Aerodynamics. Heidelberg, Germany: Springer Dordrecht.

Warwick, G. (2018). NASA Wind Tunnel Tests Mature Low-Boom X-Plane Design. Aviation Week and Space Technology, Featured Content (October 30).

Wazzan, A.R., Gazley, C. Jr., and Smith, A. (1981). H-Rx method for predicting transition. Journal of Aircraft 19 (6): 810–812.

Whitcomb, R.T. (1956). A Study of the Zero-Lift Drag-Rise Characteristics of Wing-Body Combinations near the Speed of Sound. *NACA Report 1273*, Washington, D.C.

Whitcomb, R.T. and Clark, L.R. (1965). An Airfoil Shape for Efficient Flight at Supercritical Mach Numbers. *NASA TMX-1109*, Washington, D.C., July.

White, F.M. (2005). Viscous Fluid Flow, 3e. New York: McGraw-Hill.

Yang, Y. and Zha, G. (2017). Super-Lift Coefficient of Active Flow Control Airfoil: What Is the Limit? AIAA Paper No. 2017-1693, presented at the SciTech Forum, 55th AIAA Aerospace Sciences Meeting, Grapevine, Texas (January 2017).

3

Understanding Aviation's Impact on the Environment

3.1 Introduction

Sustainability in the context of aviation is based on three inter-related concepts: impact of aviation on environment (i.e. climate, air quality, and noise), the renewability of the energy sources, and the societal factors (e.g. economics) that increase our needs on air transportation. We begin with the first element of sustainability in this chapter. Aviation's impact on environment starts at the ground level with *air quality* near airports and rises through Earth's atmosphere with *greenhouse gas emissions*. Aircraft engine emissions and noise impact air quality and health at the community level.

In addition to aircraft engines, auxiliary power units (APUs), the ground support equipment, ground access vehicles, construction activities, electric power plants, and maintenance operations are also deemed as emission sources that contribute to aviation-related environmental pollution at the airports.

At the cruise altitude, greenhouse gases (GHG) in the engine emissions threaten to cause climate change through global warming. The impact of engine emissions on the climate in the form of nitrogen oxides (NO_x) threatens to deplete ozone in the protective ozone layer in Earth's atmosphere and generate harmful ozone at the ground/community level. The formation and spreading of persistent contrails in high-flying jets is the source of *aviation-induced cloudiness* (AIC), akin to naturally occurring cirrus clouds, which cover about 30% of the Earth. AIC alters radiative forcing (RF) in the upper atmosphere and thus another concern for climate change is born.

In this chapter, we address aviation-related sources and the magnitude of these *anthropogenic*, i.e. caused by human activities and emissions at the local and the global scale. Aviation noise generated in takeoff, climb, flyover, approach, and landing is presented in the second half of this chapter. Aviation noise related to supersonic flight, i.e. sonic boom, is also addressed. The impact of noise on human's physiological and psychological health is presented. The technology related to noise mitigation at the source is introduced. Finally, hopeful visions and strategies for *sustainable aviation* expressed by National Aeronautics and Space Administration (NASA), Federal Aviation Administration (FAA), Europe's vision, and International Civil Aviation Organization (ICAO) are presented. Aviation goals by 2050 and beyond are outlined.

Future Propulsion Systems and Energy Sources in Sustainable Aviation, First Edition. Saeed Farokhi.
© 2020 John Wiley & Sons Ltd. Published 2020 by John Wiley & Sons Ltd.
Companion website: www.wiley.com/go/farokhi/power

3.2 Combustion Emissions

Combustion of fossil fuel in air results in water vapor and carbon dioxide formations that belong to the category of GHG. Incomplete combustion results in carbon monoxide and unburned hydrocarbon in the exhaust jet. Nitrogen oxides, labeled as NO_x, are major concerns due to their impact on air quality. NO_x also plays a catalytic role in ozone production or depletion in lower and upper atmosphere, respectively. Sulfur content of a fuel is dictated by the aviation fuel production (or the refinery) process, and its oxides, labeled as SO_x (mainly SO_2), contribute to local air-quality concerns. Finally, particulate matter (PM) in the exhaust, known as soot and volatile gases, are combustion emissions that have condensed to solid (particle) phase, which adversely impact air quality. The US Environmental Protection Agency (EPA) has been charged with setting emission standards for aircraft engines operated in the United States.

There are two areas of concern with air pollution. The first deals with the engine emissions near airports in the landing–takeoff cycle (LTO), below 3000 ft. The second area of concern examines the effect of engine emissions at climb–cruise–descent (CCD) phase, above 3000 ft altitude. Figure 3.1 shows the type and nature of the emissions related to combustion of fossil fuels in aircraft engines, which is known as *tank-to-wake* (TTW) in *lifecycle assessment* (LCA) terminology of GHG emissions' impact on climate. The other element of LCA accounts for the *well-to-tank* (WTT) portion that contribute to GHG emissions.

3.2.1 Greenhouse Gases

Solar radiation powers the climate system on Earth. Greenhouse gases in Earth's atmosphere absorb and reemit infrared radiation and contribute to Earth's warmer climate. The average temperature on Earth's surface is 298 K, whereas without the GHG in the atmosphere it would be 255 K (i.e. 43 °C colder). The greenhouses gases in the atmosphere comprise:

1) Carbon dioxide, CO_2
2) Water vapor, H_2O

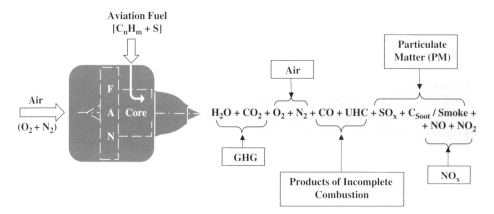

Figure 3.1 *Tank-to-wake* products of combustion of fossil fuel in an aircraft turbofan engine.

3) Methane, CH_4
4) Nitrous oxide, N_2O
5) Hydrofluorocarbons, HFCs
6) Perfluorocarbons, PFCs
7) Sulfur hexafluoride, SF_6

Aviation impacts the first four categories whereas the last three are highly regulated and their impact is minimized. Carbon dioxide, CO_2, and water vapor, H_2O, dominate the list and constitute the products of complete combustion of fossil fuels in air breathing jet engines. The large deposition of these products in the atmosphere by a growing fleet of commercial aircraft is believed to contribute to global warming through greenhouse effect. Once created, the GHG linger in the atmosphere, i.e. they have a lifespan of several decades, even centuries in the atmosphere before their absorption by the oceans, forests, plants and soil. In addition to dynamics of the GHG in the atmosphere, there is surface reflectance of sunlight from ice sheets, soil, forest, and ocean, known as the *albedo effect*, that ranges from 90% to 10%. Melting ice sheets and deforestation contribute to global warming through albedo effect. An example in the arctic is the Greenland ice sheet albedo that shows 5% reduction in reflectivity from 2000 to 2010 that has resulted in an accelerated melting rate of 6600 mi^2 of ice sheets per year.

Figure 3.2 shows the September sea ice extent and its age in the arctic in 1984 and again in 2016 (from NASA Science Visualization Studio, data: Tschudi et al. 2016). The significant reduction in the sea ice areal extent is evident, and indeed the scale of the problem is measured in millions of square miles (see Fetterer et al. 2016).

The adverse impact of the anthropogenic GHG in the atmosphere is universally acknowledged, but still a few doubt the conclusion. The *undeniable* fact of human impact on atmosphere is, however, based on scientific measurements. Figure 3.3 (adapted from Zumdahl and Zumdahl 2000) shows atmospheric CO_2 concentration since 1000 CE. It also marks the beginning of the Industrial Revolution (1784) in Europe. It shows the rapid rise in carbon dioxide concentration in the past 100 years, which coincides with increasing consumption of fossil fuels for electric power generation, home heating and transportation. At the present time, there is no emission regulation for carbon dioxide and water vapor and thus no FAA engine certification requirements. The unit of concentration in Figure 3.3 is parts per million (ppm) in molar concentration of CO_2, i.e. equal to volume fraction of CO_2 in the atmosphere. Note the rise of the atmospheric CO_2 concentration is extended from 2000 to 2013, as shown in Figure 3.3, with data based on the mean values of seasonal variations. The same dramatic rise in CO_2 concentration continues to develop, as seen in Figure 3.4 based on recent data from National Oceanic and Atmospheric Administration (NOAA), www.esrl.noaa.gov/gmd/ccgg/trends.

From the data in EPA GHG emissions report (2016), Figure 3.5 shows the percent content in GHG emissions in the United States in 2014 (see EPA Report 1990–2014). The biggest offender is carbon dioxide, with 82% share of total GHG emissions. This chart excludes water vapor as a GHG. Total emissions is reported as 6870 million metric tons of CO_2 equivalent. Figure 3.6 shows the carbon dioxide emission *by source*. Transportation sector (i.e. road transport, rail, aviation, and shipping) contributes 31% to the carbon dioxide emission in the United States. Aviation is currently responsible for 12% of CO_2 emission from all transport sources. The aviation contribution within the transport sector in the European Union is 12.4% (see Expert Group on Future Transport Fuels 2015). In another study, Lee et al. (2010) estimated the impact of

(a)

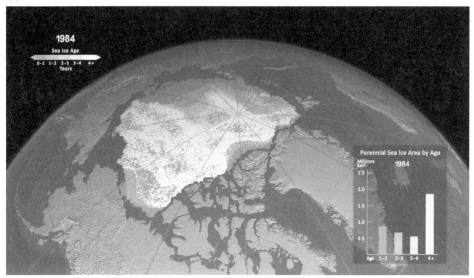

(b)

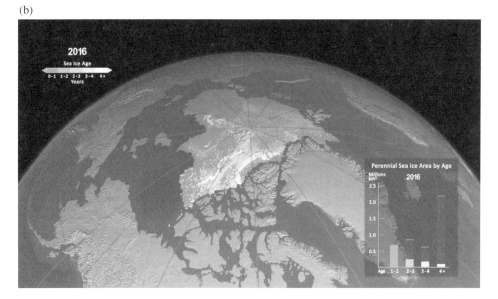

Figure 3.2 September ice sheet extent and age in the arctic in 1984 and 2016. *Source:* From Tschudi et al. 2016. (a) September 1984 sea ice extent and age; (b) September 2016 sea ice extent and age.

aviation on climate to be around 5% of all anthropogenic effects in 2005. However, aviation is the fastest-growing sector within transportation and is thus a prime candidate for a revolutionary new approach to its propulsion system, airframe design, system integration, and eco-friendly operational measures, e.g. flight altitude and cruise Mach number and airport operations.

For reference, we note that the largest US carbon dioxide pollution is generated by coal power plants. Figure 3.7 from National Renewable Energy Laboratory (NREL)

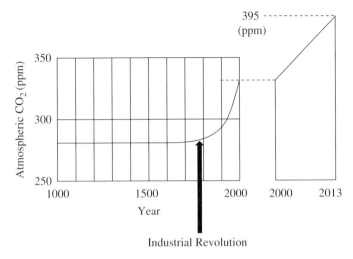

Figure 3.3 Concentration of carbon dioxide in the atmosphere since 1000 based on ice core data and direct readings since 1958. *Source:* Adapted from Zumdahl and Zumdahl 2000.

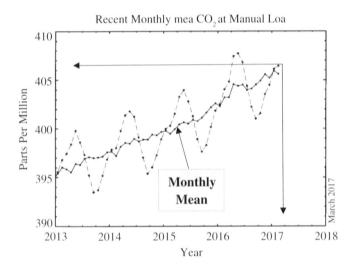

Figure 3.4 Recent monthly mean carbon dioxide measured at Mauna Loa Observatory, Hawaii (dashed line with diamonds show the monthly mean centered on the middle of each month). *Source:* Data from NOAA, www.esrl.noaa.gov/gmd/ccgg/trends.

Biofuel Atlas (see Biofuel Atlas: https://maps.nrel.gov/bioenergyatlas) shows the number and location of coal power plants in the continental United States.

3.2.2 Carbon Monoxide, CO, and Unburned Hydrocarbons, UHC

Carbon monoxide (a poisonous gas) is formed as a result of an incomplete combustion of hydrocarbon fuels. It is a source of combustion inefficiency as well as a pollutant in the engine emissions. In combustion terms, CO is a fuel with a heating value of $2267\, \text{cal}\, \text{g}^{-1}$, which is approximately 24% of the Jet-A fuel specific energy, which leaves

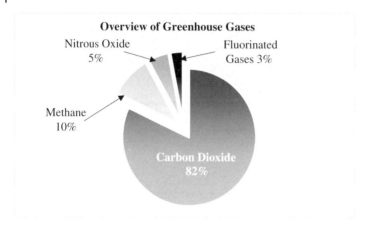

Figure 3.5 The inventory of GHG emissions in the United States in 2014. *Source:* Data from EPA 2016 Report.

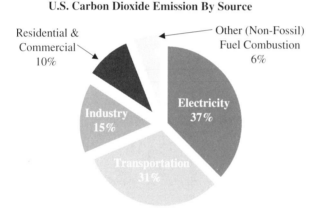

Figure 3.6 Sources of CO_2 emissions in the United States in 2014. *Source:* Data from EPA 2016 Report.

the engine in the exhaust gases without releasing its chemical energy. It is mainly generated when the engine operates in low power, e.g. idle. A fuel-lean operation at low engine power setting causes a slow reaction rate in the combustor and thus carbon monoxide formation. At a low power setting, combustor inlet temperature and pressure are reduced, which directly impact the reaction rate in the combustor. Also, partial dissociation of carbon dioxide in the hot primary zone is responsible for the carbon monoxide formation. In case of unburned hydrocarbons, UHCs, a fraction of fuel may remain unburned due to poor atomization, vaporization, and residence time or getting trapped in the coolant film in the combustor liner. The indices of carbon monoxide and unburned hydrocarbons (in $g\,kg^{-1}$-fuel) both drop with the engine power setting. An accumulation of aircraft engine data, which is reported as emission index (EI) in Figure 3.8, suggests the sensitivity of CO and UHC formations to engine power setting.

Figure 3.7 Coal power plants in the continental US. *Source:* Data from Biofuel Atlas, Department of Energy, NREL.

It is entirely plausible to relate the combustion *inefficiency*, i.e. $1 - \eta_b$, particularly in the idle setting, to the emission index of CO and UHC, as they are both *fuels* that are present in the exhaust. Blazowski (1985) correlates the combustion inefficiency, $1 - \eta_b$, to emission index EI (at idle) as:

$$1 - \eta_b = \left[0.232 (EI)_{co} + (EI)_{UHC} \right] x 10^{-3} \tag{3.1}$$

Note that the emission index EI is the mass ratio in grams of pollutants generated per kilogram of fuel consumed. As an example, let us take emission indices of CO and UHC to be ~100 and ~40 g kg^{-1} of fuel, respectively (from Figure 3.8), to represent the engine idle power setting with combustor inlet temperature ~400 K. Substitution of these emission indices in Eq. (3.1) produces combustor inefficiency (i.e. $1 - \eta_b$) of ~6.3%. However, EPA standards on emissions applied to Eq. (3.1) demands a combustor efficiency of 99% at idle setting. Low-power operations at idle and taxi as well as ground (support) equipment at airports contribute to carbon monoxide and UHC emissions with harmful effects on health. International Civil Aviation Organization (ICAO)

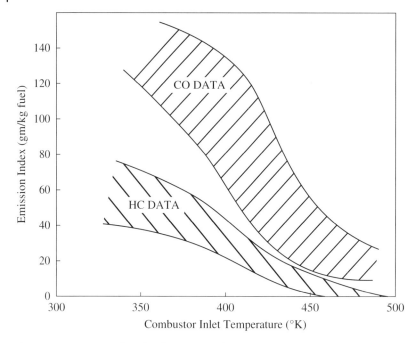

Figure 3.8 Emission index of carbon monoxide and unburned hydrocarbons (labeled HC) with engine idle setting. *Source:* From Henderson and Blazowski 1989.

Committee on Aviation Environmental Protection (CAEP) addresses operational opportunities, from airport operations to air traffic management (ATM), to minimize fuel use and reduce emissions in a circular (Cir 303- AN/176) published in 2003.

3.2.3   Oxides of Nitrogen, NO_x

Nitric oxide, NO, is formed in the high-temperature region of the primary zone at temperatures above 1800 K. The chemical reaction responsible for its production is

$$N_2 + O \rightarrow NO + N \tag{3.2}$$

Since the reaction of nitrogen is with the oxygen atom, O, in order to create nitric oxide, the regions within the flame where dissociated oxygen exists (in equilibrium) shall produce nitric oxide. This explains the high temperature requirement for NO production (in near stoichiometric mixtures). Therefore, nitric oxide is produced at high engine power settings, in direct contrast to the carbon monoxide and UHC pollutants that are generated at the low power (idle) settings. Further oxidation of NO into NO_2 takes place at lower temperatures of the intermediate or dilution zones of the combustor. Both types of the oxides of nitrogen are referred to as NO_x in the context of engine emissions. Lipfert (1972) demonstrates the temperature sensitivity of NO_x formation through an excellent correlation with numerous gas turbine engine data. Lipfert's results are shown Figure 3.9.

We note that engine types, combustor types and fuels had apparently no effect on the NO_x correlation presented by Lipfert (see Figure 3.9). All engines however correlated with the combustor inlet temperature, T_{t3}. Combustor inlet temperature is

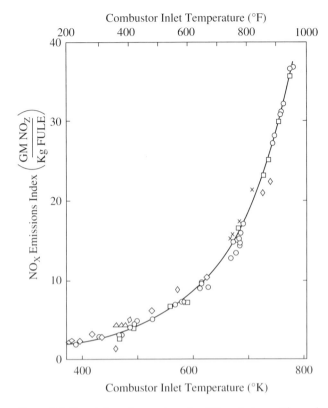

Figure 3.9 Correlation of current engine NO_x emissions with combustor inlet temperature. *Source:* From Lipfert 1972.

directly related to the compressor pressure ratio, π_c, thus we expect a similar dependence on the cycle pressure ratio, as shown in Figure 3.10a (from Henderson and Blazowski 1989). Flight Mach number impacts stagnation temperature and thus combustor inlet total temperature increases with flight Mach number. The correlation between NO_x production and flight Mach number is shown in Figure 3.10b, in supersonic flight.

3.2.4 Impact of *NO* on Ozone in Lower and Upper Atmosphere

Ozone, O_3, is toxic and a highly oxidizing agent, which is harmful to eyes, lungs, and other tissues. NO_x production in high-temperature combustion impacts the ozone concentration in the lower as well as upper atmosphere. This section discusses the chemical processes that lead to ozone creation in the lower atmosphere and to a depletion of ozone in the stratosphere.

3.2.4.1 Lower Atmosphere

Combustion at high temperatures (when local combustion temperature exceeds ~1800 K) breaks the molecular bond of N_2 and O_2 and causes NO to be formed, i.e.

$$N_{2(g)} + O_{2(g)} \rightarrow 2NO_{(g)} \tag{3.3}$$

(a)

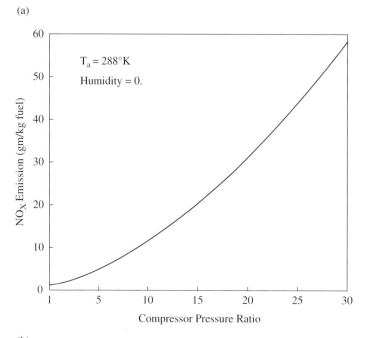

(b)

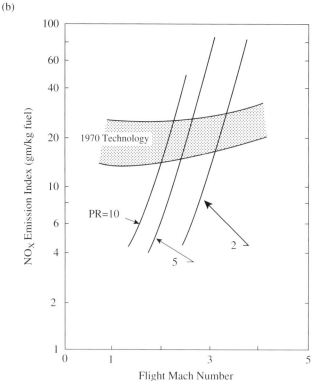

Figure 3.10 Impact of cycle pressure ratio (PR) and flight Mach number on NO_x emission: (a) the effect of compressor pressure ratio on NO_x emission in dry air; (b) the effect of flight Mach number on NO_x emission. *Source:* From Henderson and Blazowski 1989.

Nitric oxide reacts with the oxygen in the atmosphere to form nitrogen dioxide, NO_2, according to:

$$NO_{(g)} + \frac{1}{2}O_{2(g)} \rightarrow NO_{2(g)} \tag{3.4}$$

Nitrogen dioxide, NO_2, absorbs light and decomposes to:

$$NO_{2(g)} + hv\,(radiation) \rightarrow NO_{(g)} + O_{(g)} \tag{3.5}$$

The product hv in Eq. (3.5) is radiation energy with h as Plank's constant and v as the frequency of the electromagnetic wave. Oxygen atom is a highly reactive substance and, among other reactions, readily reacts with oxygen molecule to create ozone, i.e.

$$O_{2(g)} + O_{(g)} \rightarrow O_{3(g)} \tag{3.6}$$

If we sum all the reactions (3.4)–(3.6), we get the following net result, namely

$$net\,\frac{3}{2}O_{2(g)} \rightarrow O_{3(g)} \tag{3.7}$$

Since NO facilitated the creation of ozone without being consumed in the process, it serves as a *catalyst*. In the lower atmosphere, nitric oxide, NO, helps generate harmful ozone in a catalyst role. The impact of NO on ozone concentration changes as we enter the ozone layer in the stratosphere.

3.2.4.2 Upper Atmosphere

The role of ozone in the upper atmosphere (50–115 kft) is to protect Earth against harmful radiation from the sun, namely to absorb the highly energetic rays known as ultraviolet (UV) radiation, with wavelength of 100–4000 Å.

$$O_{3(g)} + hv\,(ultraviolet - radiation) \rightarrow O_{2(g)} + O_{(g)} \tag{3.8}$$

The oxygen atom is highly reactive and readily combines with O_2 to form ozone:

$$O_{2(g)} + O_{(g)} \rightarrow O_{3(g)} \tag{3.9}$$

The photochemical cycle described by (3.8) and (3.9) reactions results in no net change of the ozone concentration in stratosphere. Hence, ozone serves a stable protective role in the upper atmosphere. Flight of commercial supersonic aircraft with large quantities of NO emissions ($\sim$30 g kg^{-1} fuel) from their high-temperature combustors, over extended cruise periods and with a large fleet, promises to deplete the ozone concentration levels. Again, NO appears to be a catalyst in destroying ozone in the upper atmosphere according to the following chain reactions:

$$NO_{(g)} + O_{3(g)} \rightarrow NO_{2(g)} + O_{2(g)} \tag{3.10}$$

$$NO_{2(g)} + hv\,(radiation) \rightarrow NO_{(g)} + O_{(g)} \tag{3.11}$$

$$NO_{2(g)} + O_{(g)} \rightarrow NO_{(g)} + O_{2(g)} \tag{3.12}$$

By combining reactions (3.11) and (3.12) with *twice* the reaction in (3.10) (to get the right number of moles to cancel), we get the net photochemical reaction in the stratosphere as:

$$\text{net } 2O_{3(g)} \rightarrow 3O_{2(g)} \tag{3.13}$$

Hence, ozone is depleted through catalytic intervention of NO. The high-energy UV-radiation absorption in stratosphere by ozone is by and large the source of warmth in this layer of Earth's atmosphere. Since a rising temperature in the stratosphere with altitude is (dynamically) stable with respect to vertical disturbances, the ozone concentration levels should remain stable (i.e. constant) over time. A chart of ozone concentration with altitude that also depicts the best cruise altitude for the cruise Mach number of commercial aircraft is shown in Figure 3.11 (adapted from Kerrebrock 1992).

Now let us briefly summarize NO_x formation, combustor design parameters that impact its generation and the ozone layer.

NO_x Definition:

$$NO, NO_2$$

Combustor Reactions Leading to NO_x Formation:

$$N_2 + O \leftrightarrow NO + O$$

$$N + O_2 \leftrightarrow NO + O$$

$$N + OH \leftrightarrow NO + H$$

NO_x concentration:

$$NO_x \propto \sqrt{p_{t3}}\, e^{-2400/T_{t3}}\, t_p$$

where p_{t3} is combustion pressure, T_{t3}

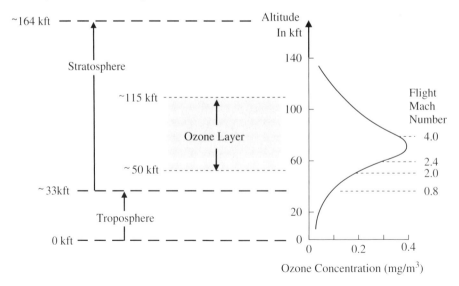

Figure 3.11 Profile of ozone concentration with altitude with a suitable cruise flight Mach number versus altitude superimposed. *Source:* Adapted from Kerrebrock 1992.

Compressor discharge temperature (K) and t_p is the residence time in the primary zone (see Kerrebrock 1992)

Role of NO_x in ozone depletion:

$$O_3 + h\nu \rightarrow O_2 + O$$

where $h\nu$ is the ultraviolet (UV)

Radiation from the Sun

$$O + O_2 \rightarrow O_3$$

$$O + O \rightarrow O_2$$

$$O + O_3 \rightarrow 2O_2$$

The last two reactions tend to limit the ozone concentration in the stratosphere:

$$NO + O_3 \rightarrow NO_2 + O_2$$

$$NO_2 + h\nu \rightarrow NO + O$$

$$NO_2 + O \rightarrow NO + O_2$$

NO is maintained, O_3 destroyed

Other ozone-depleting chemicals are chlorofluorocarbons (CFC) found in refrigerants, which since the discovery of ozone hole over Antarctica in 1985, prompted Montreal Protocol in 1987 to phase out these chemicals. Recent discovery that ozone hole over Antarctica is beginning to heal (i.e. shrink) is welcome news. As CFCs continue to reduce in the atmosphere, by 2070, the ozone hole can recover to the level of 1980s (see Strahan and Douglass 2018).

3.2.5 Impact of NO_x Emissions on Surface Air Quality

Aircraft operation below 3000 ft, i.e. in landing and takeoff, (LTO) contributes to NO_x generation and thus contributes to ozone levels in and near major airports. EPA defines the ozone standards for the duration and concentration levels. The following excerpt from the EPA on ozone designation process identifies the health risks associated with high ozone levels:

> Breathing air containing ozone can reduce lung function and increase respiratory symptoms, thereby aggravating asthma or other respiratory conditions. Ozone exposure also has been associated with increased susceptibility to:
>
> - respiratory infections,
> - medication use by asthmatics,
> - doctor and emergency department visits, and
> - hospital admissions for individuals with respiratory disease.

Ozone exposure may contribute to premature death, especially in people with heart and lung disease. High ozone levels can also harm sensitive vegetation and forested ecosystems. Along with states and tribes, we are responsible for reducing ozone air pollution. Current and upcoming federal standards and safeguards, including pollution reduction rules for power plants, vehicles and fuels, assure steady progress to reduce ozone-forming pollution and protect public health in communities across the country.

Based on the MIT report (2004), 41 of the 50 largest airports are in ozone nonattainment or maintenance areas. Although the aviation contribution to area NO_x inventory in serious and extreme status nonattainment areas, ranges from 0.7 to 6.9%, it is expected to grow. In addition to NO_x, the EPA Green Book addresses the standards in other combustion emissions (pollutants), such as particulate matter (PM), sulfur dioxide, carbon monoxide, and UHC (www.epa.gov/green-book/ozone-designation-and-classification-information). We address low-emission combustion technologies and design features in modern aircraft gas turbine engines in Section 3.4.

Aviation impact of NO_x on local air quality is estimated at less than 3% (in the national inventory) by 2020 (see FAA, Aviation and Emissions: A Primer 2005). The broader impact of aviation on environment include the ground support equipment, ground access vehicles, construction activities, electric power generating plants and maintenance operations that also contribute to aviation environmental pollution at the airports. These sources mainly contribute to carbon monoxide and volatile organic compound (VOC) emissions. US General Accounting Office (GAO) Report GAO-03-252 (2003) examines the broader impact of aviation on environment.

3.2.6 Soot/Smoke and Particulate Matter (PM)

The fuel-rich regions of the primary zone in the combustor produce carbon particulates known as soot. The soot particles are then partially consumed by high-temperature gases, i.e. in the intermediate and to a lesser extent in the dilution zones of the combustor. Visible exhaust smoke corresponds to a threshold of soot concentration in exhaust gases. Soot particles are primarily carbon (96% by weight), hydrogen, and some oxygen. An improved atomization of the fuel (using air-blast atomizers) and enhanced mixing intensity in the primary zone reduces soot formation. The parameter that quantifies soot in the exhaust gases is called the *smoke number*, SN. Its measurement is based on passing a given volume of the exhaust gas through a filter for a specific time. Then compare the optical reflectance of the stained and the clean filter by using a photoelectric reflectometer. The procedure for aircraft gas turbine engines exhaust smoke measurements is detailed in a Society of Automotive Engineers (SAE) document – ARP 1179 (1970). The smoke number is defined as:

$$SN \equiv 100\left(1 - R/R_0\right) \tag{3.14}$$

where R is the absolute reflectance of the stained filter and R_0 is the absolute reflectance of the clean filter material.

Particulate matter (PM) in aircraft exhaust emissions are classified based on their size, namely $PM_{2.5}$ and PM_{10}, which is based on the aerodynamic diameter of these solid particles in microns (μm). For example, PM_{10} includes all particulate matter less than $10\,\mu m$ in diameter, similarly, $PM_{2.5}$. The sources for the *primary* PM are found in the combustion of hydrocarbon fuels in aircraft gas turbines, and through *precursor gases* in the exhaust plume, such as sulfuric acid (from SO_x), UHC, and vaporized lubrication oil (see Whitefield et al. 2008). The National Ambient Air Quality Standards (NAAQS) regulate PM (and ozone, among other emissions) and areas that do not meet NAAQS are designated as *nonattainment areas* (see EPA, National Ambient Air Quality

Standards (NAAQS), http://www.epa.gov/air/criteria.html, 2013). Particulate matter is found to be a health hazard and 18 of the 50 largest airports operate in $PM_{2.5}$ nonattainment areas (as of 2006, reported by US Federal Aviation Administration).

3.2.7 Contrails, Cirrus Clouds, and Impact on Climate

Water vapor at low temperature first condenses and then freezes, forming ice particles. These ice particles then cling to each other and other particulate matter in the exhaust plume, such as soot and oxides of sulfur. Solid particles are always present in the products of combustion in the jet exhaust plume, forming a condensation trail. Since the condition for the contrail formation is low temperature, this is predominantly a high-altitude phenomenon, usually above 8 km (i.e. 26 kft). The main factors affecting contrail formation and persistency, i.e. lifespan, are:

1) Atmospheric temperature, humidity, and pressure
2) Water vapor content in the exhaust plume
3) Cross wind

Depending on the atmospheric conditions, contrail's lifespan may be short or long. Short lifespan, of a few minutes, is due to low humidity, which makes the ice particles rapidly liquefy and subsequently evaporate (where pressure is below vapor pressure). In contrast, long-lived contrails are called *persistent contrails* and stretch over several miles. The further subdivision of shear-induced *spreading* and *nonspreading* divides the persistent category. Naturally occurring cirrus clouds, which cover about 30% of the Earth, are formed at high altitudes, typically above 20 kft. Humidity in the air, i.e. water vapor, at low temperatures in high altitudes condenses, freezes and forms ice particles. In light of naturally occurring cirrus clouds, contrails are the source of AIC that has the potential to impact radiative forcing and thus climate change. In 2007, an Intergovernmental Panel on Climate Change (IPCC) report (IPCC 2007) contributed 2–4% of total radiative forcing to AIC. A recent study by the UK Royal Commission on Environmental Protection (RCEP), 2002 (see https://royalsociety.org/~/media/Royal_Society_Content/policy/publications/2002/9964.pdf), finds that the net effect of contrail and induced cirrus cloud is three to four times the radiative forcing of CO_2. Based on this estimation, anthropogenic radiative forcing of the climate change will be 3–15% by 2050. The uncertainties in radiative forcing are discussed by Lee et al. (2009). Sausen et al. (2005) predict AIC alters radiative forcing on a global scale by $30\,mW\,m^{-2}$ with an uncertainty range of $10–80\,mW\,m^{-2}$. This prediction is consistent with the upper value of radiative forcing predicted by Minnis et al. (2004). Decreasing cruise altitude where higher ambient temperature prevails reduces contrails, while it increases drag (due to higher density air). The solution seems to be flying at a *lower Mach number*, at a *lower altitude*, to maintain the same drag levels, and thus, in effect, reduce adverse contrail impact on climate.

3.3 Engine Emission Standards

The emission standards of the US EPA (1973, 1978 and 1982) and the ICAO (1981) define the limits on CO, UHC, NO_x and smoke production of aircraft engines near airports, known as the landing and takeoff (LTO) cycle. EPA and ICAO continually revise emission standards. Table 3.1 shows the engine emission standards regulated by EPA and ICAO.

Table 3.1 EPA and ICAO Engine Emission Standards in LTO cycle.

Pollutant	EPA standards	ICAO (1981) standards
CO	4.3 g/(kg.thrust.hr)	118 g/kN F_{00}
UHC	0.8 g/(kg.thrust.hr)	19.6 g/kN F_{00}
NO_x	3.0 g/(kg.thrust.hr)	$40 + 2(\pi_{00})$ g/kN F_{00}
Smoke	19–20	83.6 $(F_{00})^{-0.274}$ use F_{00} in kN

In the ICAO standards for the landing and takeoff (LTO) cycle, π_{00}, is the overall pressure ratio (OPR) and F_{00} is the sea-level static maximum rated engine thrust (in kN). Current ICAO standards for emissions certification of aircraft engines are contained in ICAO Annex 16, Volume II (2008). For aircraft and aircraft engine pollution control, see Environmental Protection Agency reports, 1973, 1978 and 1982.

3.4 Low-Emission Combustors

To combat the problems of CO and UHC production at the low engine power settings (e.g. idle), the concept of staged combustion is introduced. This concept breaks down a conventional combustor primary zone into a pair of separately controlled combustor stages, which may be stacked (as in parallel) or in series. The stages are referred to as a *pilot* and a *main stage*, with their separate fuel-injection systems. The pilot stage serves as the burner for the low-power setting at peak idle combustion efficiency. The main stage is for max power climb and cruise condition however operating in a *leaner* mixture ratio to control NO_x emissions. Complete combustion is achieved through separate fuel scheduling as well as an efficient (air-blast atomizer) fuel atomization, vaporization, and mixing in smaller volume combustors. The impact of percent engine load, i.e. from idle to max-power, at sea level and 11 km altitude on carbon monoxide and NO_x formation is shown in Figure 3.12 (from Penner 1999). As noted, CO is generated at low power (e.g. idle) and NO_x is the creature of the high power setting in the combustor.

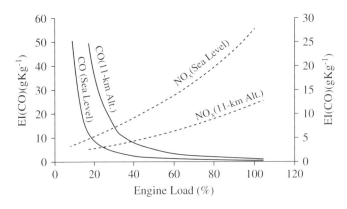

Figure 3.12 Emission characteristics of fan engine versus load at sea level and 11 km altitude. *Source:* From Penner 1999.

Table 3.2 Engine emissions with staged combustors.

	Pollutant, g/(kg.thrust.hr.cycle)			
	CO	UHC	NO$_x$	Smoke
1979 EPA standards	4.3	0.8	3.0	20
Production combustor CF6-50	10.8	4.3	7.7	13
Double-annular combustor	6.3	0.3	5.6	25
Production combustor JT9D-7	10.4	4.8	6.5	4
Vorbix combustor	3.2	0.2	2.7	30

The result of application of the double annular and Vorbix combustors to CF6-50 and JT9D-7 engines respectively is shown in Table 3.2 (Data from Lefebvre 1983).

The staged combustion concept is very effective in overall engine emission reduction in particular with CO and UHC whereas NO$_x$ continues to be a challenge. To combat NO$_x$, we should lower the flame temperature, which requires the combustor to operate at a lower equivalence ratio, of say $\phi \sim 0.6$. The problems of combustion stability and flameout accompany lean fuel–air mixtures. To achieve a stable lean mixture ratio to support a continuous combustion a pre-mixing, pre-vaporization approach has to be implemented. The combustors that are shown in Figure 3.13 employ this concept. Also the positive impact of employing air-blast atomizers is incorporated in both combustors. However, Table 3.2 indicates higher levels of smoke are generated by the double-annular and the Vorbix low-emission combustors. This reminds us of the conflict between single parameter versus system optimization problem, in which different parts/elements of the system often have opposing requirements for their optimization.

Water injection in the combustor effectively reduces the flame temperature and has demonstrated significant NO$_x$ reduction. Figure 3.13 (from Blazowski and Henderson 1974) shows the effect of water injection in the combustor on NO$_x$ emission reduction. We note that the ratio of water to fuel flow is ~ 1 for a nearly 50% NO$_x$ reduction level. This fact eliminates the potential of using water injection at the cruise altitude to lower the NO$_x$ emissions from an aircraft gas turbine engine.

Ultra-low NO$_x$ combustors with premixed prevaporized gases are shown to operate at low equivalence ratios (below 0.6) when the combustor inlet temperature is increased. The application of this concept to automotive gas turbines has produced a very promising ultra-low NO$_x$ emission level. The burner inlet conditions of 800 K and 5.5 atm. pressure in the automotive gas turbines resemble the altitude operation (or cruise operation) of an aircraft gas turbine engine. These ultra-low NO$_x$ combustor results are shown in Figure 3.14 (from Anderson 1974).

From Anderson's results in Figure 3.14 we note that a reduced residence time, which is good for NO$_x$ production, leads to a higher level of CO and UHC emissions with an attendant combustion efficiency loss. The equivalence ratio has a large impact on NO$_x$ emissions, as seen in Figure 3.14. The NO$_x$ emission index of $0.3 \, \text{g kg}^{-1}$ fuel seems to be the

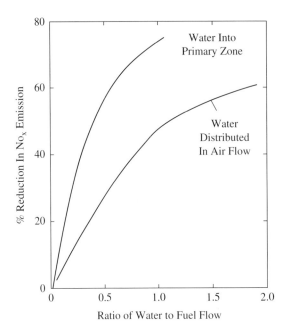

Figure 3.13 Water injection in the combustor reduces NO_x emission. *Source:* From Blazowski and Henderson 1974.

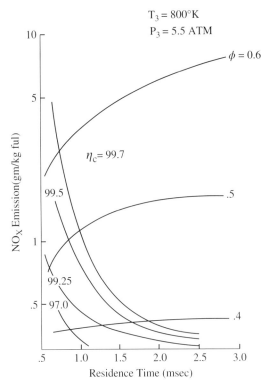

Figure 3.14 Effect of residence time (in ms) and ϕ on NO_x emission levels in ultra-low NO_x Pre-mixed/prevaporized burners. *Source:* From Anderson 1974.

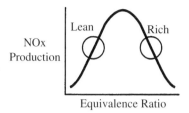

 Lean Premixed/Prevaporized   Rich Burn/Quick-Quench/Lean-Burn

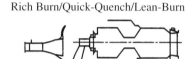

Figure 3.15 Approaches to a low-NO$_x$ combustor development. *Source:* From Merkur 1996.

absolute minimum possible, as indicated by Anderson. To achieve these ultra-low levels of the oxides of nitrogen in the exhaust emission, an increased burner inlet temperature is required. Incorporating a solid catalytic converter in an aircraft gas turbine engine burner may enable lean combustion and ultra-low NO$_x$ emission but at the expense of more complexity, weight, cost, and durability. The premixed/prevaporized burners also exhibit problems with auto-ignition and flashback for advanced aircraft engines with high-pressure ratios. Another possibility for low NO$_x$ combustor development takes advantage of *rich burn*, which results in a reduced flame temperature, and *quick quench*, which allows for the combustion to be completed while the temperature of the combustion gases remain very near the combustor exit temperature. This approach is called *rich burn-quick quench-lean burn* and has shown promising results. These approaches are summarized in Figure 3.15, from Merkur (1996). Despite these advances, we recognize that the problem of gas turbine engine emission is complex and challenging. We need to improve our understanding of the reacting flow physics in the combustor (with multi-phase, multi-time-scale simulation) and in particular in the area of auto-ignition and flashback in lean combustible mixtures on *the boundary* of lean extinction limit. We therefore need fundamental flow research on reacting mixtures that shed light on these *fuzzy interfaces* (or boundaries) in fuel–air combustion.

It is instructive to compare the 1992 subsonic engine technology and the High-Speed Civil Transport (HSCT) program goals of 2005 (from Merkur 1996); see Table 3.3. We note an order of magnitude reduction in NO$_x$ emission index and a combustor that avoids the near stoichiometric (i.e. $\phi \approx 1$) operation for the 2005 HSCT engine, as compared to the 1992 subsonic engine. The HSCT program is inactive at the present time.

Table 3.4 provides a summary of the modern combustor operational requirements and performance parameters (from Penner 1999).

3.5 Aviation Fuels

Aircraft flight envelope, i.e. altitude-Mach number, establishes the operational temperature range of the aircraft and hence its desirable fuel properties. Some of the aviation fuel properties and combustion characteristics of interest are listed in Table 3.5 for reference.

Table 3.3 Comparison between the subsonic engine of 1992 and the design goals for the high-speed civil transport of 2005.

	1992 Subsonic engine	2005 HSCT engine
Equivalence ratio	$1.0 \rightarrow 1.2$	< 0.7 or > 1.5
Gas temperature, °F	3700	$3400 \rightarrow 3750$
Liner temperature, °F	<1800	$2200 \rightarrow 2600$
Liner material	Sheet or cast superalloys	Ceramic matrix composites (CMC)
Cooling methods	Transpiration / Film	Convection
Environment	Oxidizing	Oxidizing or reducing
NO_x	$35 \rightarrow 45$ g/kg f	<5 g/kg f

Source: From Merkur 1996.

Table 3.4 Performance and operational requirements of a modern aircraft engine combustor.

Item	Requirement	Value	Max/Min
1	Combustion efficiency, η_b		
	- At takeoff thrust (%)	99.9	(Min)
	- Idle thrust (%)	99.0	(Min)
2	Low-pressure light-off capability (MPa)	0.03	(Max)
3	Lean blowout fuel/air ratio (at low engine power conditions)	0.005	(Max)
4	Ground light-off fuel/air ratio (with cold air, cold fuel), f	0.010	(Max)
5	Total pressure drop-compressor exit to turbine inlet (%), $1 - \pi_b$	5.0	(Max)
6	Exit gas temperature distribution		
	- Pattern factor	0.25	(Max)
	- Profile factor	0.11	(Max)
7	Combustion dynamics (dynamic pressure range/inlet air pressure (%))	3	(Max)
8	Liner temperature (K)	1120	(Max)
9	Cyclic life to first repair (cycles)	5000	(Min)

Table 3.5 Some fuel properties of interest to aviation.

Specific gravity or density	Dynamic viscosity	Vapor pressure	Surface tension
Spontaneous or auto- ignition temperature	Lower heating value	Volatility	Thermal stability
Initial and end boiling points	Heat capacity	Flashpoint	Flammability limits
Freezing point	Handling qualities, toxicity	Storability	Low fire risk

We omitted the most important parameter, i.e. *price and availability*, from Table 3.5, since neither can be considered a *fuel property*.

To investigate the effect of temperature on fuel thermal stability, volatility, flashpoint, and spontaneous ignition, we examine the maximum skin temperature of an aircraft in cruise and compare it to the thermal characteristics of some typical aircraft fuels. The skin temperature of an aircraft is calculated through energy balance that accounts for aerodynamic heating, radiation cooling, regenerative wall cooling, and thermal boundary conduction. The maximum skin temperature is, however, very near the stagnation temperature of flight. As an example, Figure 3.16 shows a rapid skin temperature rise with flight Mach number, assuming the cruise takes place in the constant-temperature layer (i.e. tropopause, between 11 and 20 km, or 36–66 kft).

At the elevated temperatures for high-speed flight, the fuel should exhibit thermal stability, which is partially addressed through its boiling point characteristics. It should be noted however that high-speed flight takes place at much higher altitudes than tropopause, e.g. 100 kft for SR-71 at Mach 3$^+$.

The initial boiling point temperature begins when fuel vaporization starts. The end point is the temperature where all the fuel is vaporized. The range of the initial and end boiling points, known as the distillation curve, for several fuels are shown in Figure 3.17 (from Blazowski 1985). The military fuels have a "JP" designation and Jet A is the most common commercial aircraft fuel. The original gas turbine fuel was kerosene, which is used as a basis of comparison and blending with other hydrocarbon jet engine fuels. The JP-4 was used primarily by the US Air Force (USAF) and was a highly volatile fuel. A less volatile fuel, the JP-5, was a gasoline-kerosene blended fuel that was used by the USN. The lower volatility of the JP-5 made it more suitable for long-term storage in the ship tanks as well as the blended nature of the fuel allowed a wider availability on board a US ship. The distillation curves of these two fuels may be seen in Figure 3.17, with JP-5 showing lower volatility and higher thermal stability. The low initial boiling point of the

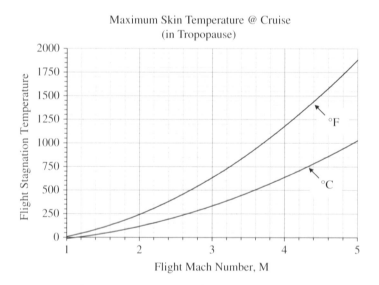

Figure 3.16 Skin temperature rise with flight Mach number ($\gamma = 1.4$, $T_{amb} \cong -70\,°F$ or $-56\,°C$).

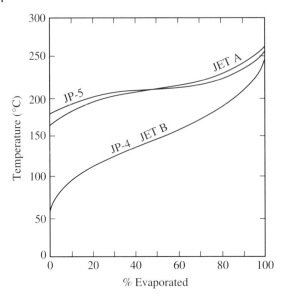

Figure 3.17 Distillation curve for common jet fuels – a measure of fuel volatility. *Source:* From Blazowski 1985.

JP-4 contributes to its volatility. We note that the fuels shown in Figure 3.17 are all evaporated at 100% when the fuels are heated to 250 °C. At high temperatures, the fuels thermally decompose to form gum (coke deposits) and clog fuel filters, fuel injector, and the fuel pump. This limit corresponds to a flight Mach number of ∼2.6.

The obvious choice for higher speed flights is the use of cryogenic fuels such as liquefied natural gas (LNG), which is mainly liquid methane, propane, or liquefied hydrogen, LH_2. Thermal stability, flight Mach number limit, and the relative fuel cost (tentative), relative to Jet A, are shown in Figure 3.18 (data from Strack 1987). The technology maturity and large-scale production efficiency of cryogenic fuels such as LH_2 or LNG will shrink the gap between the conventional jet fuel and LH_2 or LNG. The trend in cost reduction of cryogenic fuel is shown in Figure 3.18.

The specific gravity is the measure of relative liquid fuel density (relative to water at 4 °C), which is lowest for gasoline among petroleum fuels. The flammability limits, flashpoint temperature, and spontaneous ignition temperature of various fuels are key decision points in fuel selection and are shown in in Figure 3.19 (from Lefebvre 1983).

All hydrocarbon jet fuels have about the same lower heating value (nearly 18 600 BTU/lbm or 43.3 MJ kg^{-1}) regardless of their blends. The parameter that contributes to the heat of combustion of a hydrocarbon fuel is the hydrogen content of the fuel. For example, the percent hydrogen content (this is the fraction of the mass of the fuel molecule contributed by the hydrogen atoms) of kerosene (∼$C_{12}H_{26}$) is ∼15.3%, JP-4 (∼$CH_{2.02}$) is at 14.5%, propane (C_3H_8) has 18.8%, methane (CH_4) has 25%, and of course pure hydrogen is at 100% hydrogen content. On a per-atom-basis, hydrogen-to-carbon ratio in coal is about 0.8–1.0, for jet fuel is ∼2, and for methane is 4. It is interesting to examine the heating values of some hydrocarbon fuels and compare them to their hydrogen content. Table 3.6 shows the impact of hydrogen content (based on mass) on

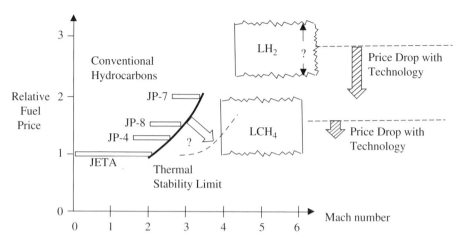

Figure 3.18 Thermal stability limit and relative price of aviation fuels (relative to Jet A). *Source:* Data from Strack 1987 (see NASA-CP-3049); with anticipated price drop with technology readiness level.

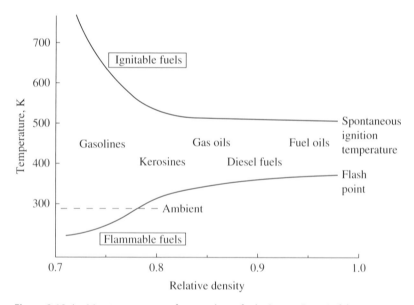

Figure 3.19 Ignition temperatures for petroleum fuels. *Source:* From Lefebvre 1983.

the heat of combustion (lower heating value). Note that the higher hydrogen content in the fuel, e.g. methane, yields a higher heat of reaction, as expected.

Ragozin (1962) presents a correlation between the fuel lower heating value (LHV) and the percent mass content of carbon, hydrogen, oxygen, sulfur, and water in the fuel. This useful and *approximate* correlation is:

$$LHV = 0.339C + 1.03H - 0.109(O - S) - 0.025W \left[\text{in MJ}/\text{kg}\right] \tag{3.15}$$

Table 3.6 Hydrogen content (based on mass) and lower heating value of common fuels.

Fuel	JP-4	Propane	Methane	Hydrogen
Hydrogen content	14.5%	18.8%	25%	100%
Lower heating value				
kcal/g	10.39	11.07	11.95	28.65
MJ/kg	43.47	46.32	49.98	119.88
BTU/lbm	18 703	19 933	21 506	51 581

The LHV in Eq. (3.15) is in $MJ\,kg^{-1}$. All parameters on the right-hand side are percent mass of carbon, hydrogen, oxygen, sulfur and water in the fuel, respectively. The coefficient of hydrogen term is the largest and hence contributes the most to an increase in the fuel heating value. Also, it is interesting to note that oxygen in the fuel reduces its heating value, which is representative of all alcohol fuels (a typical alcohol fuel, ethanol, C_2H_5OH has a lower heating value of ~62% of a hydrocarbon fuel, or methanol, CH_3OH, has only ~46% of heating value). Due to their low energy density, the use of alcohol as an alternative jet fuel is not deemed feasible. As expected, water dissolved in the fuel also lowers the heating value of the fuel. This parameter, i.e. dissolved water content in the fuel, is controlled through proper fuel handling and storage.

Fuel viscosity is an important property that determines the pressure drop in the fuel lines and the fuel pump requirements, as well as atomization of the fuel by the fuel injection system. The lower the fuel viscosity, the smaller the fuel droplets, hence faster vaporization rate and the ignition time scale. Poor atomization causes a higher fraction of UHC as well as higher CO and soot formation. On the other hand, a low viscosity fuel exhibits poor lubricity, which causes an increased fuel pump wear and reduced life. Fuel viscosity is a function of temperature and increases for a liquid fuel when temperature decreases and vice versa. At low temperatures corresponding to high altitudes, a subsonic aircraft may need an anti-freeze fuel additive or a fuel heating system integrated with its fuel tank at the propulsion system design phase.

A liquid fuel in a tank always contains a certain quantity of fuel vapor, which exerts a pressure on the liquid. This is known as the *vapor pressure* of the liquid at the given temperature. For example, JP-4, a highly volatile fuel, has a very high vapor pressure of ~0.18 atm. at 38 °C (or 100 °F), compared to JP-5 fuel with a corresponding low vapor pressure of 0.003 atm. Although a high vapor pressure is good in a combustor and leads to better vaporization rates, but it is undesirable for its lower flash point temperature. JP-5 is indeed desirable for the US Navy since it has high flashpoint, a particularly useful property for carrier safety. The excessive rate of vapor production in a fuel tank or a fuel line, especially at higher temperatures of supersonic flight, could lead to an unacceptable flash point fire risk. Figure 3.19 (from Lefebvre 1983) shows the heavier fuels have a higher flash point (therefore safer) that light fuels, such as gasoline (used in reciprocating engines) or kerosene (in jet engines). JP-8 is used by the USAF.

Aviation fuel is a blend of different hydrocarbons with different molecular structures and different reaction tendencies. We address two such compounds, olefins and aromatics, which are found in aviation fuels. Olefins are unsaturated hydrocarbons (C_nH_{2n}) that are present in aviation fuel produced in the refinery process. Aromatics (C_nH_{2n-6}) are ring compounds such as benzene (C_6H_6) that are also present in the aircraft fuel. In case

Table 3.7 Important jet fuel properties.

Property	JP-4 Spec. req.	JP-4 Typical value	Jet A (JP-8) Spec. req.	Jet A (JP-8) Typical value	JP-5 Spec. req.	JP-5 Typical falue
Vapor pressure at 38 °C (100 °F), atm	0.13–0.2	0.18	–	0.007	–	0.003
Initial boiling point (°C)	–	60	–	169	–	182
End point (°C)	–	246	288	265	288	260
Flash point (°C)	–	−25	> 49	52	> 63	65
Aromatic content (% Vol)	<25	12	< 20	16	< 25	16
Olefinic content (% Vol)	< 5	1	–	1	–	1
Saturates content (% Vol)	–	87	–	83	–	83
Net heat of combustion (cal/g)	> 10 222	10 388	> 10 222	10 333	> 10 166	10 277
Specific gravity	0.751–0.802	0.758	0.755–0.830	0.810	0.788–0.845	0.818
Approximate US yearly consumption (10^9 gal)		3.4		13.1		0.7

Source: From Blazowski 1985.

of olefins, the gum formation tendency of these compounds makes them undesirable. The aromatics have lower hydrogen content than gasoline or kerosene, which reduces the heat of reaction of the fuel mixture. Aromatic content also increases the soot formation tendencies in the combustor, which again makes them undesirable. The volume fraction of fuel containing aromatic and olefinic compounds is included in the table of fuel properties. Table 3.7 from Blazowski (1985) presents important jet fuel properties for three widely used aircraft fuels, JP-4, JP-5, and JP-8, including their aromatic and olefinic contents (in % volume).

It is instructive to examine the petroleum products in crude oil that are mainly comprised of gasoline, diesel fuel and jet fuel (or kerosene), as shown in Figure 3.20.

We note that a barrel of oil produces about 53% (by volume) gasoline and jet fuel that are the main ingredients of the aviation fuel. We may add to our shopping list 27% for the diesel fuel, which is the fuel of choice in small engines, e.g. unmanned aircraft systems (UAS) and ground support vehicle, to our products. Now, the attractiveness of fossil fuel becomes even more apparent, i.e. with a yield of >80% conversion into aviation fuel.

3.6 Interim Summary on Combustion Emission Impact on the Environment

We identified carbon monoxide (CO) and UHC as products of incomplete combustion. Through advanced combustor design, we have reduced their levels dramatically. Therefore, carbon monoxide and UHC are primary emissions of less concern (see *For Greener Skies: Reducing Environmental Impacts of Aviation* 2002). Nitrogen oxides (NO_x)

Petroleum Products in Crude Oil

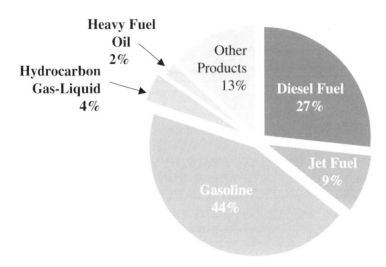

Figure 3.20 Petroleum products extracted from crude oil (in percent volume).

are produced in the high-temperature region of the primary zone in a combustion chamber. Interestingly, they are not related to burning of fossil fuels. NO_x is indeed a high-temperature phenomenon related to the dissociation of oxygen and nitrogen that forms NO_x. Ultra-low NO_x combustor designs have reduced NO_x emissions significantly. The history of ICAO NO_x regulations and NASA Combustor Programs are summarized in Figure 3.21 for reference (adapted from Chang 2012). The progressive R&D programs at FAA and NASA has resulted in 80% reduction since 1996 (i.e. relative to ICAO regulation of 1996).

Modern GE, Rolls-Royce, and Pratt & Whitney engines use ultra-low-NO_x technology combustors with >50% reduction in NO_x below CAEP 6 levels (see Table 3.8)

Oxides of sulfur are produced in aviation fuel combustion in small amounts, i.e. below regulatory levels. However, they promote particulate matter and aerosol formations. Water vapor and carbon dioxide are dominant greenhouse gas emissions from aircraft engines that are the products of complete combustion of fossil fuels in air. Water vapor, which already exists in air due to natural surface evaporation from oceans and lakes, is a greenhouse gas that affects radiative forcing of the atmosphere. The aviation share in water vapor deposition in the atmosphere is minute. However, the persistent contrail formation at high altitude (above 25 000 ft) and its impact on climate as cirrus cloud is the subject of research on AIC and radiative forcing. The concern about anthropogenic CO_2 is clearly born by its exponential rise in concentration in the atmosphere since the industrial revolution, which is still growing at a yet faster rate. To answer the question of lifetime of anthropogenic CO_2 emission, we examine the time constants of the carbon dioxide dynamics in the environment. For example, it takes 5 years for a change in carbon dioxide concentration on a local scale to reach equilibrium, and it has a lifetime of about 200 years on a global scale. Here we note that the current aviation's contribution to global carbon dioxide emissions into the atmosphere is small compared to the emissions from the energy sector and land use, i.e. biomass burning. The energy sector includes power plants that produce electricity (by burning coal and other fossil fuels), industrial

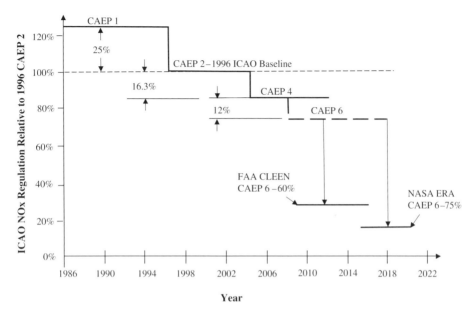

Figure 3.21 History of ICAO NO_x regulations for jet engines (with thrust > 30 kN and OPR 30) and NASA combustor programs. *Source:* Adapted from Chang 2012. CAEP, Committee on Aviation Environmental Protection; CLEEN, Continuous Lower Emissions, Energy and Noise; ERA, Environmentally Responsible Aviation (see Collier 2012).

Table 3.8 Modern aeroengines and their NO_x emission.

GEnx-1B	55% below CAEP 6
RR Trent 1000	~50% below CAEP 6
PW 810	~50% CAEP 6

processes, transport, fossil-fuel processing, and energy use in buildings. Despite its small current share, aviation CO_2 emission is rising rapidly with the popularity of air travel on a global scale. For an illuminating discussion on energy sources, emissions and possible solutions (e.g. sustainable, renewable sources of energy), read MacKay (2009).

3.7 Aviation Impact on Carbon Dioxide Emission: *Quantified*

First, we estimate the mass of carbon dioxide generated by a kilogram of aircraft fuel. For example Jet A fuel has an approximate chemical formula as $CH_{1.92}$ (or $C_{12}H_{23}$). Chemical reaction of this fuel that completely consumes carbon in the fuel and produce carbon dioxide may be written as:

$$C_{12}H_{23} + n\left(O_2 + 3.76N_2\right) \rightarrow 12CO_2 + mH_2O + \ldots \tag{3.16}$$

where n corresponds to the number of moles of $(O_2 + 3.76 N_2)$, which is the dry air composition, and m is the number of moles of water vapor in the products of combustion. In aircraft gas turbine engines, the fuel-to-air ratio (on the mass-basis) is 2–3%, which establishes n.

In Eq. (3.16), we note that 1 mol of fuel creates 12 mol of carbon dioxide. We use this proportion to arrive at their mass ratio:

$$\frac{mass_{CO_2}}{mass_{fuel}} = \frac{12(12+32)}{12(12)+23(1)} \cong 3.16 \tag{3.17}$$

From this simple consideration, we conclude that 1 kg of Jet A produces 3.16 kg of CO_2. Based on the chemical formulation of kerosene, $C_{12}H_{26}$, LNG, here assumed to be entirely methane, CH_4, and approximate chemical formulations of JP-4, Table 3.9 lists the mass of CO_2 produced per unit mass of the fuel consumed.

Interestingly, switching from Jet-A to a low-carbon fuel, such as methane (or LNG) only reduces the carbon dioxide production (per mass basis) by only 13%, i.e. we go from 3.16 kg CO_2 kg^{-1} fuel for Jet-A to 2.75 kg CO_2 kg^{-1} fuel for LNG. The other jet fuels listed in Table 3.8 (kerosene, JP-4) produce nearly the same ratio as the Jet A (in Eq. (3.17)).

Fuel burn in aviation is a function of many parameters, e.g. distance flown, altitude of flight, head- or tail wind, weather condition, passenger and cargo load factors, among others. Therefore, any representation of fuel burn versus distance, for a class of aircraft, has to be based on *averages* over large numbers. A representative calculation of fuel burn for a short-haul commercial transport, e.g. B737-400, is shown in Table 3.10 (adapted from European Environment Agency report and Jardine 2008). The fuel burn is divided into two categories of LTO (landing/takeoff) below 3000 ft. and CCD (climb/cruise/descent) above 3000 ft, over distances flown (in nautical miles and kilometer). Since LTO fuel burn is independent of the distance flown, its fraction to the total fuel burn drops as distance flown increases. For example, the ratio of the fuel consumed in LTO phase to the total fuel consumed varies between 6.8% for the 2000 nm flight to 51.5% for 125 nm flight (in a short-haul aircraft, e.g. B737-400).

Taking the data from Table 3.10 and using Eq. (3.17) (for Jet A fuel), we construct carbon dioxide emissions in LTO and CCD phases of flight as a function of distance flown (see Table 3.11). The graph of the total carbon dioxide emission (in kg) for a typical short-haul commercial transport (e.g. 737-400) is shown in Figure 3.22. Short-haul flights are defined as those with less than 3500 km distance. Medium- to long-haul flights are defined as those with >3500 km distance.

Table 3.9 Carbon dioxide production by various hydrocarbon fuels.

Fuel	Chemical formula	Mass of CO_2/Mass of fuel
Methane (or LNG)	CH_4	2.75
Kerosene	$C_{12}H_{26}$	3.11
Jet A	$CH_{1.92}$	3.16
JP-4	$CH_{2.02}$	3.14

Table 3.10 Fuel burn data for B737-400.

B737 400		Standard flight distances (nm) (1 nm = 1.852 km)							
		125	250	500	750	1000	1500	2000
Emission index for carbon monoxide (EICO) (g/kg fuel)	Taxi out	30.11	30.11	30.11	30.11	30.11	30.11	30.11
	Take off	0.90	0.90	0.90	0.90	0.90	0.90	0.90
	Climb out	0.90	0.90	0.90	0.90	0.90	0.90	0.90
	Climb/cruise/descent	3.11	2.78	2.04	1.75	1.56	1.37	1.29
	Approach landing	3.40	3.40	3.40	3.40	3.40	3.40	3.40
	Taxi in	30.11	30.11	30.11	30.11	30.11	30.11	30.11

Source: Adapted from European Environment Agency Report (2008); https://www.eea.europa.eu.

Table 3.11 Estimated CO_2 Emission for B737-400.

B737–400	CO_2 (kg)	CO_2 (kg)	CO_2 (kg)
Distance (nm)	LTO	CCD	total
125	2608.264	2457.532	5065.796
250	2608.264	4558.616	7166.88
500	2608.264	8808.184	11 416.448
750	2608.264	13 066.284	15 674.548
1000	2608.264	17 307.952	19 916.216
1500	2608.264	26 424.868	29 033.132
2000	2608.264	35 841.352	38 449.616

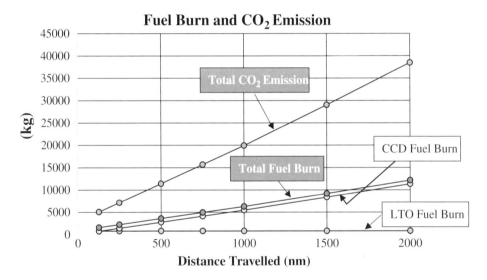

Figure 3.22 Fuel burn and CO_2 emissions in a short-haul commercial transport (e.g. B737–400).

In order to calculate carbon dioxide emission *per passenger per kilometer flown*, we need to know the number of passengers flown (e.g. assuming 180 seats on B737, or 400 seats on B747, or 295 seats on A340). Assuming a load factor of 0.80–0.90, i.e., 80–90% of seats occupied by passengers, and using the number of seats per aircraft, we may estimate the carbon dioxide emission per passenger per kilometer flown. Tables of the fuel burn and emissions (of NO_x and SO_x) per aircraft type and stage length (125–8180 nm) are available for all major types of aircraft. This is published by the European Environment Agency (EEA) pollutant emission inventory guidebook (2016).

What is missing from our calculations so far is the *lifecycle assessment* that accounts for the *full environmental impact* of aviation, which is not limited to aircraft flight emissions alone (i.e. *tank-to-wake*), rather on *well-to-tank*, as well. These include:

1) Aviation fuel supply chain (production, refinery, storage, transportation)
2) Influences of aviation-induced climate change on agriculture, forestry, eco system, energy production, and social effects

In the first category, we enlist emissions from (fossil-fuel) refineries, fuel storage, transportation, airport ground support vehicles, and other sources of emission in the supply and support chain. In the second category, we include changes in radiative forcing in the upper atmosphere due to GHG emissions, climate change that causes sea-level rise, extreme events and changes in temperature, precipitation and soil moisture. David MacKay (2009) suggests using a factor of two to three times the flight emissions to estimate carbon footprint (i.e. to capture the full impact of aviation emissions). This agrees with Jardine (2008) who suggested the range of two to four times the flight emissions. Recent studies (Rogelj et al. 2015) suggest a 1.9 factor.

Our goal here is not to get the "exact" multiplier. Rather, our aim is to alert the reader to the factors that are involved and the extent that aviation emission impacts the environment on a global scale (see RCEP 2002). Essentially, here we addressed only the flight portion of aviation on GHG emissions. As noted, the flight portion, i.e. *tank-to-wake,* is only a fraction of the total impact, and the contribution of *well-to-tank* on GHG emissions is even more significant. An example of full LCA simulation is GREET (greenhouse gases, regulated emissions, and energy use in transportation) simulation tool addresses LCA, which is a valuable tool to account for the full impact of GHG emissions (see Argonne National Laboratory, GREET Tool http://www.transportation.anl.gov/modeling_simulation/GREET/). GREET1 release in 2011 addresses LCA for alternative jet fuels in aviation (Wang 2011).

Figure 3.23 shows the emission *per seat* for three aircraft, in *equivalent* kilogram of CO_{2e}, which accounts for the overall impact of aircraft CO_2 emissions that include the lifecycle GHG emissions attributable to the aircraft fuel that includes its broader environmental impact on eco system. To put some of these numbers in perspective, a 7000 km (or 4350 mi) trip creates 1 metric ton of CO_2 emission per seat. This is roughly the distance between New York City and Rome. Now, if we multiply this number by 2 for round trip and the number of family members in a party for summer vacation, say 4, we get 8 metric tons of carbon dioxide emission that is dumped in the atmosphere. This is the impact of one family, one trip/year!

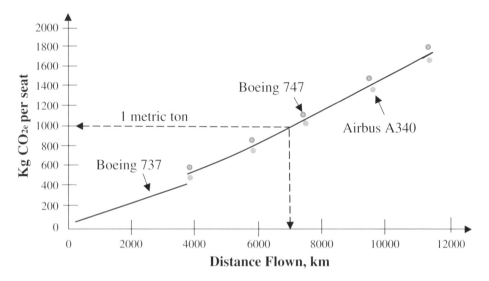

Figure 3.23 CO_{2e} emissions *per seat* as a function of flight distance. *Source:* Adapted from European Environment Agency and Jardine 2008.

Now, what happens when millions of families take multiple trips (flights)/year? To put these numbers in perspective, it is estimated that global air traffic will grow from 2.5B passengers in 2011 to 16B passengers in 2050 (based on an average of 4.9% growth/year assumption). For reference, in near-term, the International Air Transport Association (IATA) projects 5.3% passenger growth/year, which is higher than the long-term average. Boeing estimates, on average, 5% growth in commercial aviation in the next 20 years (see Bradley and Droney 2011). The growth of flying public from 2.5B passengers in 2011 to 16B in 2050 introduces a factor of 6.4 in our calculations of human impact on climate through aviation. According to projections of the 1999 report *Aviation and the Global Atmosphere*, issued by the IPCC (see IPCC 1999), the impact on carbon dioxide emissions in the year 2050, will be up to 10 times greater than in 1992. However, the ultimate goal of sustainable aviation is to *decouple* aviation emissions from the growth in global air traffic.

Before we leave the subject of combustion emissions of fossil fuels in aviation, Table 3.12 is produced to serve as a summary of the health and environmental effects of air pollutants; and a reminder that we must *pay heed!* Also, Table 3.13 is an EPA (2013) product on NAAQS, which establish criteria on six categories of air pollutants. These tables outline the standards and impact of pollutants on health and welfare of human beings.

3.8 Noise

3.8.1 Introduction

Environmental impact of aviation is measured in emissions and noise. The communities in the vicinity of airports bear the brunt of aircraft noise in takeoff, climb, flyover, approach, and landing. The exposure to loud noise is harmful to human physiological and psychological health and welfare. A brief primer on noise is first presented to bring the reader up to speed on the topic, before the aviation impact and measures to alleviate it are discussed. For more detailed exposition on aeroacoustics, see Goldstein 1974, Smith 1986 or Kerrebrock 1992, among others.

3.8.1.1 General Discussion
Annoying sound is called noise. Therefore, human *subjectivity* enters the picture. The frequency range that we can hear is called the *audible range* and lies between 20 and 20 000 Hz (cps). In general, any pressure fluctuations or unsteadiness in a medium create sound wave. These fluctuations or unsteadiness may be of periodic nature or caused by random events. An example of periodic pressure fluctuation field is in rotating blades in turbomachinery and an example of random pressure fluctuation is the turbulence in the wake, within the freestream or in the boundary layer. Periodic unsteadiness exhibits characteristic frequencies (i.e. fundamental and its harmonics) whereas in random fluctuation, the emitted noise is of the broadband nature.

Sound propagation in a medium is an *infinitesimal pressure wave* propagation causing reversible and adiabatic (i.e. isentropic) changes to ambient pressure, temperature, and density of the fluid (i.e. p_0, T_0, and ρ_0, respectively). As an infinitesimal wave, all fluid property changes are very small compared to the unperturbed values. Namely, p', T', and ρ' are small perturbations in pressure, temperature and density, caused by sound

Table 3.12 Health and environmental effects of air pollutants.

Pollutant	Health effects	Environmental effects
Ozone	Lung function impairment, effects on exercise performance, increased airway responsiveness, increased susceptibility to respiratory infection, increased hospital admissions and emergency room visits, pulmonary inflammation, and lung structure damage (long term).	Crop damage, damage to trees, and decreased resistance to disease for both crops and ecosystems.
Carbon monoxide	Cardiovascular effects, especially in those persons with heart conditions.	Adverse health effects on animals similar to effects on humans.
Nitrogen oxides (NOx)	Lung irritation and lower resistance to respiratory infections.	Acid rain, visibility degradation, particle formation, contribute toward ozone formation, and greenhouse gas in the atmosphere, which may contribute to climate change.
Particulate matter (PM)	Premature mortality, aggravation of respiratory and cardiovascular disease, changes in lung function and increased respiratory symptoms, changes to lung tissues and structure, and altered respiratory defense mechanisms.	Visibility degradation, damage to monuments and buildings, safety concerns for aircraft from reduced visibility.
Volatile organic compounds (VOC)	Eye and respiratory tract irritation, headaches, dizziness, visual disorders, and memory impairment.	Contribute to ozone formation, odors, and have some damaging effect on buildings and plants.
Carbon dioxide, water vapor, and contrails	None	Act as greenhouse gases in the atmosphere and, therefore, may contribute to climate change.
Sulfur dioxide	Respiratory irritant. Aggravates lung problems, particularly for individuals with asthma.	Causes damage to crops and natural vegetation.in presence of moisture and oxygen, sulfur dioxide converts to sulfuric acid, which can damage marble, iron, and steel.

Table 3.13 National ambient air quality standards.

Standards pollutant	Primary/Secondary	Averaging time	Level	Form
Carbon monoxide (CO)1	Primary	8-hr	9 ppm	Not to be exceeded more than once per year
		1-hr	35 ppm	
Lead (Pb)2	Primary and secondary	Rolling 3 mo average	0.15 μg/m³,7	Not to be exceeded
Nitrogen dioxide (NO₂)3	Primary	1-hr	100 ppb	98th percentile, averaged over 3 years
	Primary and secondary	Annual	53 ppb8	Annual mean
Ozone (O₃)4	Primary and secondary	8-hr	0.075 ppm9	Annual fourth-highest daily maximum 8-hr concentration, averaged over 3 years
Particulate matter5 PM$_{2.5}$	Primary	Annual	12 μg/m³	Annual mean, averaged over 3 years
	Secondary	Annual	15 μg/m³	Annual mean, averaged over 3 years
	Primary and secondary	24-hr	35 μg/m³	98th percentile, averaged over three years
PM$_{10}$	Primary and secondary	24-hr	150 μg/m3	Not to be exceeded more than once per year on average over 3 years
Sulfur dioxide (SO₂)6	Primary	1-hr	75 ppb10	99th percentile of 1-hour daily maximum concentrations, averaged over 3 years
	Secondary	3-hr	0.5 ppm	Not to be exceeded more than once per year

Source: EPA, *National Ambient Air Quality Standards (NAAQS)*, 2013, http://www.epa.gov/air/criteria.html.

Notes: ppb = parts per billion, ppm = parts per million, μg/m³ = micrograms per cubic meter of air. Federal Registers: (1) 76 FR 54294, (2) 73 FR 66964, (3) 75 FR 6474 and 61 FR 52852, (4) 73 FR 16436, (5) 78 FR 3086, (6) 75 FR 35520, and 38 FR 25678. (7) Final rule signed October 15, 2008. The 1978 lead standard (1.5 μg/m³ as a quarterly average) remains in effect until 1 year after an area is designated for the 2008 standard, except that in areas designated nonattainment for the 1978, the 1978 standard remains in effect until implementation plans to attain or maintain the 2008 standard are approved. (8) The official level of the annual NO2 standard is 0.053 ppm, equal to 53 ppb, which is shown here for the purpose of clearer comparison to the 1-hour standard. (9) Final rule signed March 12, 2008. The 1997 ozone standard (0.08 ppm, annual fourth-highest daily maximum 8-hour concentration, averaged over 3 years) and related implementation rules remain in place. In 1997, EPA revoked the 1-hour ozone standard (0.12 ppm, not to be exceeded more than once per year) in all areas, although some areas have continued obligations under that standard ("anti-backsliding"). The 1-hour ozone standard is attained when the expected number of days per calendar year with maximum hourly average concentrations above 0.12 ppm is less than or equal to 1.(10) Final rule signed June 2, 2010. The 1971 annual and 24-hour SO₂ standards were revoked in that same rulemaking.

propagation. Mathematically, "very small" means at least two orders of magnitude i.e. 10^{-2}, but in acoustics, perturbations are yet smaller, typically 10^{-5} times the unperturbed quantities.

The homogeneous acoustic wave equation for p' is derived from the conservation laws of mass, momentum, and energy in fluids (e.g. see Kerrebrock 1992, for derivation):

$$\frac{1}{a_0^2}\frac{\partial^2 p'}{\partial t^2} = \nabla^2 p' \tag{3.18}$$

Where a_0 is the (acoustic) wave speed and p' is the acoustic pressure. For a plane wave propagating in the x-direction, the homogeneous wave equation simplifies to:

$$\frac{1}{a_0^2}\frac{\partial^2 p'}{\partial t^2} = \frac{\partial^2 p'}{\partial x^2} \tag{3.19}$$

From mathematical physics, we use separation of variables to solve the wave equation. We recall that any function of $(x - a_0 t)$, i.e. $f(x - a_0 t)$, and any function of $(x + a_0 t)$, i.e. $g(x + a_0 t)$, are solutions of the wave equation as they signify right and left-traveling waves, respectively. In terms of wave number, k, we may express a right-traveling wave of amplitude, P, as a sine wave:

$$p'(x,t) = P\sin(kx - \omega t) = P\sin[k(x - a_0 t)] \tag{3.20}$$

The wave number k is defined as $k \equiv \dfrac{2\pi}{\lambda}$ with λ representing the wavelength. The angular speed of the wave is related to its frequency, f, which is the inverse of its period, T, i.e. $\omega = \dfrac{2\pi}{f} = 2\pi T$. The wave propagation speed, a_0, is related to the wavelength and frequency, f (or its inverse, the period, T), according to:

$$a_0 = \lambda f = \lambda / T \tag{3.21}$$

For a general observer, the sine wave describing the acoustic pressure wave needs a phase angle, φ

$$p'(x,t) = P\sin(kx - \omega t + \phi) \tag{3.22a}$$

$$p'(x,t) = P\sin[k(x - a_0 t + \phi)] \tag{3.22b}$$

Equation (3.22a) or (3.22b) describes a wave of single frequency, f, propagating in the positive x-direction at a constant speed of a_0 and is measured by a general observer who introduces a "phase angle," ϕ. A single-frequency wave is called a *pure tone* in acoustics. When all the frequencies, or equivalently all wavelengths in the audible range are present in the measured sound wave, we refer to that as *broadband* noise. We also note that the sound wave described by Eq. (3.22a) or (3.22b) is a *plane wave*, in contrast to spherical or cylindrical wave. The amplitude of the plane wave, P, remains constant, whereas the amplitude of the acoustic wave of spherical shape drops inversely proportional to r, as we shall see in the "Sound Intensity" and "Pulsating Sphere" sections.

3.8.1.2 Sound Intensity

Sound intensity is defined as the time average power of the acoustic wave per unit area perpendicular to the sound propagation direction, therefore

$$I = \frac{1}{T}\int_0^T p'(r,t)u'(r,t)dt = \frac{1}{T}\int_0^T p'(r,t)\frac{p'(r,t)}{\rho_0 a_0}dt = \frac{1}{\rho_0 a_0}\frac{1}{T}\int_0^T p'^2(r,t)dt \qquad (3.23)$$

where T is the integration time ($>> T_{\text{period}}$). In Eq. (3.23), we replaced the induced particle speed, u', by $p'/\rho_0 a_0$, which is easily derived by a simple (steady) momentum balance in the wave frame of reference.

We define the root mean square (rms) of the fluctuating pressure as the *effective pressure*:

$$p_{\text{rms}} \equiv \sqrt{\frac{1}{T}\int_0^T p'^2(r,t)dt} \qquad (3.24)$$

Using the definition of p_{rms} in Eq. (3.24), we get:

$$I = \frac{p_{\text{rms}}^2}{\rho_0 a_0} \qquad (3.25)$$

Sound wave that is propagating spherically, has an expanding wave front (area, A_w) that grows as r^2. From conservation of energy, we conclude that $p'u'A_w$ remains constant. Therefore, both u' and p' must each vary as $1/r$ (recall that $u' = p'/\rho_0 a_0$). Since p' is inversely proportional to r and the intensity, I, is proportional to p'^2, we conclude that the intensity of a spherical wave drops off as $1/r^2$. This is known as *the inverse-square law*.

In contrast to spherical waves, the plane wave experiences undiminished intensity, as the pressure remains constant along the propagation path. Another way of explaining this is that there is no spreading of the plane wave, as compared to the spreading of the spherical wave. Hence, the acoustic intensity is undiminished in a plane wave. That is why we hear the heartbeat in a stethoscope or a whisper in a garden hose.

3.8.1.3 Acoustic Power

It is the total acoustic energy produced per unit time, which is related to acoustic intensity and the area of sphere (of radius, R) as:

$$W = \frac{p_{\text{rms}}^2}{\rho_0 a_0} 4\pi R^2 \quad (\text{Spherical wave}) \qquad (3.26)$$

A plane acoustic wave propagating in a constant-area tube of diameter D has an acoustic power

$$W = \frac{p_{\text{rms}}^2}{\rho_0 a_0}\frac{\pi D^2}{4} \quad \left(\text{Plane wave in a tube of diameter } D\right) \qquad (3.27)$$

3.8.1.4 Levels and Decibels

As the *perceived intensity of all sensations* ~ *log of external stimuli* in nature, we use the logarithmic scale to describe the intensity of a sensation, e.g. sound.

Level looks at the *logarithm* of the ratio of energy-related characteristics of the sound, e.g. W, I, p_{rms}^2, to a *reference value*. The *reference values* are established by an international agreement and are related to the *threshold of hearing at 1000-Hz pure tone* by a human ear. The American National Standard Institute (ANSI) recommends the following reference values:

$$W_{ref} = 10^{-12} \text{ W}$$

$$I_{ref} = 10^{-12} \text{ W / m}^2$$

$$p_{ref} = 2 \times 10^{-5} \text{ N / m}^2 (0.0002...\text{bars})$$

The logarithmic ratio, which we called the *level*, is called the *Bel* (in honor of Alexander Graham Bell). The more common unit is one tenth of the Bel, which is known as the *decibel*, dB.

3.8.1.5 Sound Power Level in *Decibels, dB*

$$L_W \equiv 10\log_{10}\left(\frac{W}{W_{ref}}\right) \quad \text{Re} 10^{-12} \text{ W} \tag{3.28}$$

3.8.1.6 Sound Intensity Level in *Decibels, dB*

$$L_I \equiv 10\log_{10}\left(\frac{I}{I_{ref}}\right) \quad \text{Re} 10^{-12} \text{ W / m}^2 \tag{3.29}$$

3.8.1.7 Sound Pressure Level in *Decibels, dB*

$$L_P \equiv 10\log_{10}\left(\frac{p}{p_{ref}}\right)^2 = 20\log_{10}\left(\frac{p}{p_{ref}}\right) \quad \text{Re} 2 \times 10^{-5} \text{ N / m}^2 = 20\,\mu Pa = 0.0002..\text{bar} \tag{3.30}$$

In Eq. (3.30), p is the measured rms pressure of sound.

3.8.1.8 Multiple Sources

It is logical to expect the overall intensity of several sound sources to be the sum of individual intensities, based on the conservation of energy principle, i.e.

$$I_{total} = I_1 + I_2 + I_3 + ... + I_n = \frac{p_{total}^2}{\rho_0 a_0} \tag{3.31}$$

Which, in turn, implies that the total effective (rms) pressure at any location is related to the individual effective pressures at that location, as follows:

$$p_{total}^2 = p_1^2 + p_2^2 + p_3^2 + \ldots + p_n^2 \tag{3.32}$$

3.8.1.9 Overall Sound Pressure Level in Decibels, dB

Environmental noise often contains all frequencies in the audible range (20–20000 Hz) and overall sound pressure level treats each frequency equally. Although it measures the total environmental noise level, it lacks the human factor with different perceptions to different frequencies.

3.8.1.10 Octave Band, One-Third Octave Band, and Tunable Filters

To measure noise and its frequency content, microphones with a variety of filters are used. For example, an *octave band filter* passes the highest frequency, which is twice the lowest frequency passed through the filter. The next octave band filter then has its lowest frequency set at the highest frequency of the previous filter, so as to cover the frequency range of interest. Now, the *one-third octave band filter* breaks down the octave band into three bands with $2^{1/3}$, or $\sqrt[3]{2}$, as the multiplier, thus provide a higher resolution than the octave band filter. If we start at 20 Hz, *which is the lowest audible frequency,* and then multiply by $\sqrt[3]{2}$, we get 25 Hz, and if we repeat this, we get, 31.5, 40, 50, 63, 80, and the rest of the frequencies in octave and one-third octave band filter, as shown in Figure 3.24. The measured data are plotted at the center frequency of the band. A tunable filter can provide a yet higher resolution than the one-third octave band filter.

A typical flyover noise spectrum, using a one-third octave band filter, is shown in Figure 3.25 (data from Pearson and Bennett 1974).

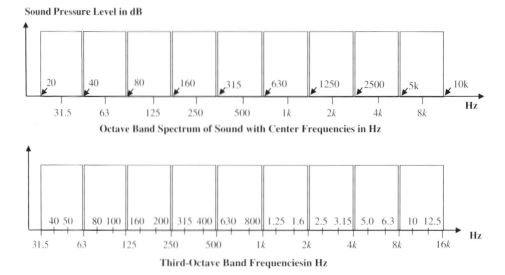

Figure 3.24 Center frequencies in octave band and one-third octave band filters.

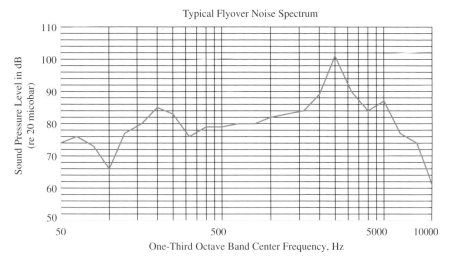

Figure 3.25 Typical flyover noise spectrum using one-third octave band filter. *Source:* Data from Pearsons and Bennett 1974.

3.8.1.11 Adding and Subtracting Noise Sources

In order to calculate the noise (intensity) level from several sources, i.e. to add or subtract noise sources, we first take the *antilog* of the noise level of each source to arrive at the power-related quantities (based on the conservation of energy), and then we add/subtract these energy terms. Finally, we take the logarithm of the sum (of the energy terms) to arrive at the new level. This is a rather cumbersome process; therefore, there are convenient charts, known as *adding levels* and *subtracting levels* charts, which can be used to simplify the task.

For example, we know that *doubling the source intensity* will add 3 dB to the noise intensity level, which is known as *the 3-dB Rule*:

$$10 \log 2 \sim 3 \, \mathrm{dB}$$

In addition, since sound meters typically have ±0.5 dB precision and the human ear cannot distinguish sound level differences less than 1 dB, we may construct convenient level charts based on sound meter measurements.

3.8.1.12 Weighting

There are several weighting networks used in a sound level meter that account for human sensitivity to loudness of a pure tone. The most common is the *A-weighted* network, which applies a prescribed correction to the sound pressure level as a function of frequency. The graph for the A-weighted correction for the octave and one-third octave band frequencies is shown in Figure 3.26 (from 50 to 12 500 Hz). As an example, a 77 dB pure tone at 100 Hz will be perceived as a 58 dBA sound by a human ear. With increased sensitivity of our ears to pure tones between 2000 and 5000 Hz, we experience an increased sensation, up to 1.3 dB at 2500 Hz.

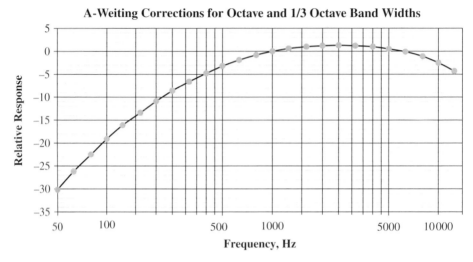

Figure 3.26 A-weighting corrections in dB for octave and one-third octave band frequencies.

3.8.1.13 Effective Perceived Noise Level (EPNL), dB, and Other Metrics

The human ear is most sensitive to pure tones, between 2000 and 5000 Hz. In addition to the tonal content of the noise, its duration impacts perceived noisiness as well, which is measured in NOYs. Effective Perceived Noise Level (EPNL) accounts for tonal corrections as well as their duration corrections. These corrections are rather involved and will not be reproduced here, for brevity. Pearsons and Bennett (1974) provide details in the calculations of all noise ratings, including EPNL. Noise regulations and certifications in the United States (FAR-36) and abroad (ICAO Annex 16) use EPNL in dB as their measure. However, we note that EPNL is a *single-event* measure of noise (e.g. for a single takeoff, or single landing) and not an accumulated over time effect of noise (as in total noise exposure in 24 hours).

Communities near major airports show different levels of tolerance to aviation noise during day and night. Sleep disorder may be caused by nighttime aviation noise, for example. In addition to the time of day, the exposure *duration*, the *noise level*, and the *tonal content* play an important role in sensitivities of people who live within airport boundaries. To address these and other community-level aviation noise concerns, new metrics and regulations are developed. Some of the key metrics are listed here for reference:

1) DNL is day-night average sound level (averaged over 24 hours)
2) SEL or LE is sound exposure level (flat-weighted)
3) CNEL is community noise equivalent level
4) LA_{eq} is equivalent sound pressure level (A-weighted)
5) LA_{max} is maximum sound pressure level (A-weighted)
6) A-weighted time above is metrics given as the time above a defined noise exposure
7) SELC or LCE is sound exposure level (C-weighted)
8) LC_{max} is maximum SPL (C-weighted)
9) NEF is noise exposure forecast
10) Maximum PNLT – Maximum tone corrected perceived noise level

11) WECPNL is weighted equivalent continuous perceived noise level
12) PNL time above metrics is Metrics, given as the time above a defined noise exposure

Airport Noise Compatibility Planning is the Federal Aviation Regulation document that is known as FAR Part 150. This document defines procedures and standards for developing airport noise exposure maps that are used for land use assessment. EPA in its "Levels Document," in 1974 identified DNL (day–night average sound level) as the best descriptor with an outdoor level of 55 dB (45 dB indoor level) as requisite to public health and welfare (see US Environmental Protection Agency 1974). For further reading, Pearsons and Bennett (1974) is recommended.

To study aerodynamics, we introduced source, sink, doublet, and vortex as elements that could be combined to create flows of interest around wings, bodies, and full aircraft. The superposition of the elementary solutions was a gift from linearized aerodynamics. The study of noise follows the same principle. i.e. we introduce elements of noise that may be combined to simulate the total engine noise emission. These elementary sources of sound are monopole, dipole, and quadrupole.

3.8.1.14 Pulsating Sphere: Model of a Monopole

Imagine there is a sphere of radius R_0 that is pulsating with certain amplitude and certain frequency. The pressure wave generated by the sphere will propagate in the medium in a spherical pattern according to the homogeneous wave equation written in spherical coordinate system:

$$\frac{\partial^2 p'}{\partial t^2} = a_0^2 \frac{1}{r^2} \frac{\partial}{\partial r}\left(r^2 \frac{\partial p'}{\partial r} \right) \tag{3.33}$$

The solution of the homogeneous wave Eq. (3.33) that satisfies the pressure boundary condition on the sphere, i.e. the pressure amplitude P, is:

$$p'(r,t) = \frac{R_0 P}{r} \sin\left[k(r - R_0) - \omega t + \varphi \right] \tag{3.34}$$

where P is the amplitude of the pressure at the surface of the sphere and r is the distance of the wave to the center of sphere.

For a point source, we shrink R_0 to zero while maintaining the product of (PR_0) constant. The acoustic pressure wave related to a pulsating point source is:

$$p'(r,t) = \frac{(PR_0)}{r} \sin\left[kr - \omega t + \varphi \right] \tag{3.35}$$

We note the amplitude of the pressure wave drops inversely proportional to the distance from the source, i.e. the sphere. However, the speed of propagation remains constant. The pulsating sphere describes a *monopole* with net mass flow of zero at any point in the fluid. The oscillating mass may be used to model a pulsating jet or the thickness or volume displacement effect of a vibrating blade in a compressor. Figure 3.27 shows a spherical (sound) wave propagation.

The acoustic power of the monopole is

$$\wp_m = \frac{2\pi (PR_0)^2}{\rho_0 a_0} \tag{3.36}$$

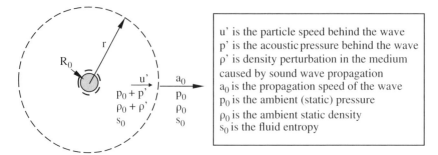

Figure 3.27 Spherical (sound) wave propagation in a medium at rest.

since the intensity of sound drops inversely proportional to the surface area of the sphere, i.e. $\sim r^2$, therefore, doubling the distance from a noise source causes intensity to drop to ¼. Hence, $10 \log (1/4) = -6\,dB$, which is known as *the 6-dB Rule*.

3.8.1.15 Two Monopoles: Model of a Dipole

Since acoustic equation is linear, solutions may be superimposed to create other solutions of interest. For example, if two monopoles of equal strength are placed at a distance d apart where one operates 180° out of phase with respect the other, they describe a dipole. This phase shift of π between the two monopoles creates an interesting flow pattern, which resembles the local flow pattern around a vibrating blade. Figure 3.28 shows the flow pattern for a dipole and a vibrating blade.

Dipole simulates the oscillating blade force since there is accelerating flowfield between the two poles. The fluctuating pressure field measured at (r,θ) position with respect to a dipole is approximated for an observer far away from the dipole, as

$$p' \approx \frac{PR_0}{r}(kd\sin\theta)\sin k(r - a_0 t) \quad kr \gg 1 \tag{3.37}$$

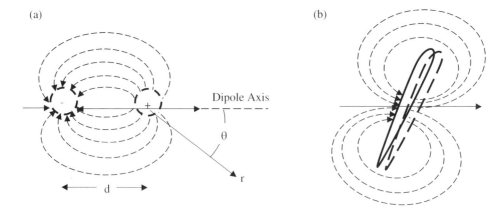

Figure 3.28 The flow pattern associated with a dipole and a vibrating blade: (a) flow pattern of a dipole; (b) flow pattern of a vibrating blade.

where r is the distance to the dipole and θ is the angle measured from the dipole axis. The dipole power is then expressed as:

$$\wp_d = \frac{2\pi \left(PR_0\right)^2}{\rho_0 a_0}\left(kd\cos\theta\right)^2 \tag{3.38}$$

where monopole lacks any directionality, and dipole exhibits directional character through $\cos^2\theta$ term in Eq. (3.38). Also, the acoustical power of the dipole is a function of wave number, k, thus the frequency. For derivation of Eqs. (3.37) and (3.38), see Kerrebrock 1992.

3.8.1.16   Two Dipoles: Model of Quadrupole

By placing two dipoles a distance d apart, as shown in Figure 3.29, the net mass flow acceleration and thus the net force cancels. This is the model of a quadrupole that is used to study radiated noise in turbulent jets. Lighthill (1952, 1954) laid the mathematical foundation of the general theory of aerodynamic sound (also known as aeroacoustics) by introducing a source term (in the form of turbulent stress tensor, T_{ij}) in the acoustic equation for density fluctuation. Lighthill's Acoustic Analogy (LAA) equation is:

$$\frac{\partial^2 \rho'}{\partial t^2} - a_0^2 \nabla^2 \rho' = \frac{\partial^2 T_{ij}}{\partial y_i \partial y_j} \tag{3.39}$$

Lighthill treated the nonlinear source term on the RHS of Eq. (3.39), as known, and related it to a distribution of quadrupoles of strength T_{ij} for calculating sound emission from a subsonic jet.

The acoustic power radiated from a quadrupole is derived by Kerrebrock (1992):

$$\wp_q = \frac{2\pi \left(PR_0\right)^2}{\rho_0 a_0}\left(kd\right)^4 \tag{3.40}$$

The power of the quadrupole is proportional to $(kd)^4$ and the power of the dipole is proportional to $(kd)^2$.

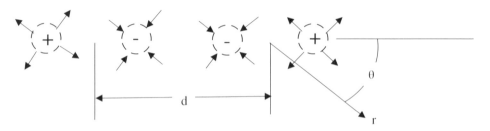

Figure 3.29 Two dipoles of equal strength are placed at a distance d apart to form a quadrupole.

3.8.2 Sources of Noise Near Airports

Communities in the vicinity of airports are mainly affected by aircraft noise in approach, landing, taxi, takeoff, climb, and flyover. However, there are other sources of noise, e.g. those from ground support vehicles, or even airport traffic, impact community noise. There are three aviation noise sources near airports:

1) Engine noise
   a) Fan, compressor, combustor and turbine noise
   b) Jet noise
2) Airframe noise, which originates on wings, landing gear, flaps, slats, and thrust reversers especially noticeable in approach and in landing phase are due to:
   a) Landing gear, or undercarriage noise due to massively separated turbulent wake originated at the landing gear/undercarriage
   b) Unsteady wake from wing, flaps/slats/high-lift devices, and radiated noise from flaps' edges
   c) Turbulent BL formation and noise radiation from eddies on fuselage and wings
   d) Panel vibration noise
   e) Noise from cavities, steps, wheel wells
   f) Interaction noise
3) Airport ground support equipment, e.g. towing, baggage handling, maintenance/repair, refueling, and food service.

Airframe noise due to slat is investigated by Khorrami (2003), who studied noise sources in slats. Figure 3.30 shows the fluctuating pressure field in a slat deployment due to vortex shedding at the slat trailing edge, TE, (see Khorrami 2003). Some passive treatments of slat TE are discussed by Khorrami and Choudhari 2003, that shows porous passive treatment at the slat TE has the potential of reducing the TE noise by ~20 dB, in the nearfield.

Another noise mitigation strategy targets flaps. A deflected flap creates turbulence and vortex shedding, which both contribute to airframe noise in LTO cycle. One

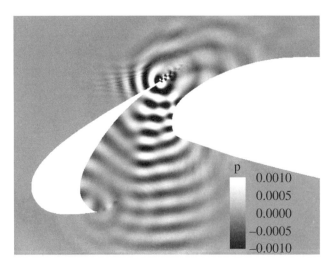

p	
	0.0010
	0.0005
	0.0000
	−0.0005
	−0.0010

Figure 3.30 Instantaneous fluctuating pressure field due to vortex shedding at the slat trailing edge. *Source:* from Khorrami 2003.

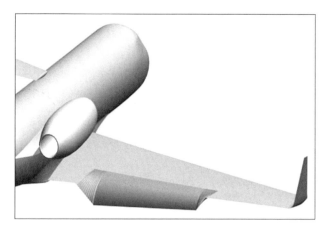

Figure 3.31 Adaptive compliant trailing edge technology. *Source:* Courtesy of FlexSys, Inc. https://www.flxsys.com/flexfoil/

solution is a hinge-free "continuous bendable flap" that eliminates the seam and the gap, which mitigates turbulence and vortex formation. Figure 3.31 shows the adaptive compliant trailing edge (ACTE) flap that is designed and patented by FlexSys, Inc. (see Adaptive Control Trailing Edge Flight Experiment, NASA website: https://www.nasa.gov/centers/armstrong/research/ACTE/index.html) and sponsored by the Air Force Research Lab (AFRL) (see Adaptive Compliant Trailing Edge, 2017). The flexible flap is made of composite material with variable geometry trailing edge structure. For the low-power adaptive system technology used in shape morphing wing, see the company's website (https://www.flxsys.com/flexfoil).

To elevate the technology readiness level (TRL) of Adaptive Compliant TE, flight test programs were conducted on board White Knight aircraft by AFRL. The next application was on board Gulfstream III aircraft, which was modified to perform flight tests at NASA-Armstrong with seamless, flexible flap technology from FlexSys, Inc. Figure 3.32 shows the Aerodynamic Research Test Bed aircraft that is used for bendable, seamless flap technology demonstration in flight at NASA-Armstrong Flight Research Center. The ACTE technology is also tested on board Boeing KC-135 aircraft. The future of quiet aviation inevitably has to target designs that mitigate turbulence production and vortex shedding. The broader potential of morphing wing clearly extends to stealth technology.

3.8.3 Engine Noise

High-performance gas turbine engines are very noisy machines, by design! Energy transfer in turbomachinery can only take place through *unsteady means* (see Kerrebrock 1992, Cumpsty 2003 or Farokhi 2014). Also to maximize energy transfer, modern engines use high-speed wheels of rotating and stationary or counter-rotating blade rows. Turbomachinery noise is thus dominating source of noise in the fore arc of the engine whereas jet noise dominates the aft arc. Figure 3.33 shows the noise sources in a separate-flow turbofan engine. Noise radiation produces directivity pattern in farfield.

Modern fans often involve relative tip speed that is *supersonic*, which create shocks and their associated *discrete tone* noise at blade passing frequency (BPF), which is the

Figure 3.32 Gulfstream III is modified to test the ACTE flexible-flap research project. *Source:* Courtesy of NASA/FlexSys, Inc. see https://www.nasa.gov/centers/armstrong/research/ACTE/index.html

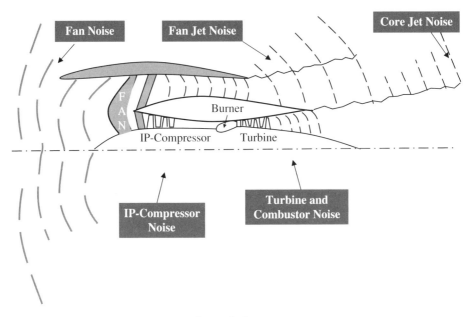

Figure 3.33 Noise sources in a separate-flow turbofan engine.

product of number of blades, B, times the rotor angular speed, ω. However, upstream propagating shocks that are not perfectly periodic due to manufacturing tolerances or blade vibration can coalesce and create combination-tone or *buzz-saw* noise. These characteristics are shown in Figure 3.34.

Since the energy transfer is unsteady, the blade wakes are unsteady with periodic vortex shedding structure. The unsteady wakes of a blade row are then chopped by the next row of blades, as they are in relative motion. *Rotor-stator interaction* creates a pressure pattern that rotates with the angular speed, ω_{eff}:

$$\omega_{\text{eff}} = \frac{B\omega_r}{V - B} \tag{3.41}$$

where B and V denote the number of rotor and stator blades, respectively, and ω_r is the rotor angular speed. It is apparent from Eq. (3.41) that V B has to be avoided, and indeed,

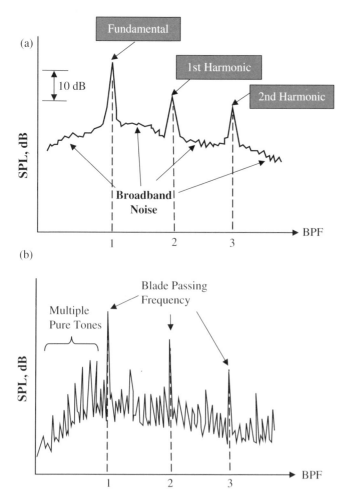

Figure 3.34 The effect of fan tip relative Mach number on spectral content of its radiated noise: (a) Noise spectra of subsonic tip fan showing harmonics of BPF and fan broadband noise; (b) noise spectra of a supersonic tip fan showing combination tone or buzz-saw noise.

modern design practices adopt $V = 2B \mp 1$ strategy to avoid high-intensity sound emission due to rotor-stator interaction.

An important lesson in the study of turbomachinery noise is the coupled nature of the turbomachinery, as a source of excitation and the duct, as a means of transmission. For example, the pressure disturbance modes that are created by high-speed fan blades are indeed coupled to duct modes. Under certain conditions, some acoustic disturbances propagate and some may decay. The condition for propagation of pressure disturbance modes is called *cutoff* condition. It may be shown that supersonic relative tip Mach number is required for propagating modes (see Kerrebrock 1992, Cumpsty 1977 and 1985 and Morfey 1972 and 1973). Therefore, for the propagating modes, i.e. the cutoff condition, we must have:

$$M_T^2 + M_z^2 > 1 \tag{3.42}$$

Due to blade elasticity and aerodynamic loads, the blades *vibrate* in various modes, e.g. first bending, second bending, first torsional mode, coupled bending and torsional mode, and higher–order modes. Turbulent boundary layers on blades, as well as end-wall regions that may contain flow separation, all contribute to radiated noise from turbomachinery. These areas of random unsteadiness produce the broadband noise between discrete tones of BPF.

Modern fan blades use thin profiles and incorporate blade sweep coupled with the acoustic liner in the inlet, fan duct, and nozzle, and core exit ducts offer successful solutions that alleviate the turbomachinery noise problem. For example, the acoustic duct liners, which are Helmholtz resonators embedded in the wall, attenuate the dominant frequency in the transonic fans' and low-pressure turbine (LPT) noise spectra, namely the discrete tone noise at the blade passing frequency. In addition, proper selection of blade numbers in successive blade rows and blade row spacing are the solutions that alleviate, by and large, the rotor-stator interaction problem. Modern fan stages use swept and leaned stator blades (and fan duct exit vanes) to reduce the rotor-stator inter-action noise (see Woodward et al. 1998, and Huff 2013).

Figure 3.35 shows partially assembled fan stage with swept-only stators (part (a)) and swept and leaned stators (part (b)), from Woodward et al. 1998. The acoustic benefit of the swept and leaned stator showed 4 dB reduction in broadband noise compared to the radial stators. The experiments at NASA were conducted on a 22-in. transonic fan designed by Allison with characteristics summarized in Table 3.14. Scaled results for a two-engine air-craft, on a prescribed flightpath, resulted in 3 EPNdB reduction, which is deemed significant. Figure 3.36 shows a baseline stage as compared to a swept and swept and leaned stator stage.

A promising passive technique used in inlet design redirects the downward propagat-ing noise upward through extended lower cowl lip, known as *scarf inlet* (see Clark et al. 1997). This noise-reduction technique does not attenuate the noise, such as acoustic liners;

(a)   (b)

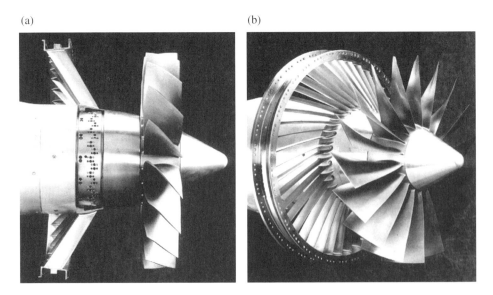

Figure 3.35 Concepts in fan noise reduction: (a) partially assembled fan stage with swept-only stator; (b) partially assembled fan stage with swept and leaned stator. *Source:* From Woodward et al. 1998.

Table 3.14 Fan design parameters (Allison).

Rotor diameter	55.9 cm (22 in.)
Rotor blade number	18
Rotor hub/tip ratio	0.30
Rotor aspect ratio	1.754
Stator vane number	(All modes) 42
Stator aspect ratio	3.073
Swept and leaned stator	30° lean/30° sweep
Swept-only stator	30° sweep
Design-stage pressure ratio	1.378 (1.45 tip–1.20 hub)
Design-specific weight flow	210.4 kg/s/m^2 (43.1 lbm/s/ft^2)
Design-corrected tangential tip speed	305 m s^{-1} (1000 ft s^{-1})
Design-tip relative Mach number	1.080

rather, it redirects the downward noise in the upward direction. The sideline noise measurements showed a reduction by as much as 8 dB for the scarf inlet, as compared to conventional inlet (see Clark et al. 1997). The noise shielding potential is utilized by over-the-wing installation of advanced turbofan engines, which uses the *wing barrier* concept, in approach, sideline and community noise reduction, as reported by Berton (2000). Over-the-wing installation of advanced turbofan engines resulted in the reduction of 96 EPNdB footprint from 0.96 to 0.57 mi^2, according to Berton. In terms of certification, 9.9 EPNdB cumulative noise reduction was attributed to the engine placement over the wing.

The *unsteady* mechanism of energy release in a combustor creates the bulk of combustion noise. Turbulent mixing of the flow with secondary and cooling jets contribute to the broadband noise. The combustor and turbine noise are lumped as *core noise* sources in a jet engine. Finally, we need to account for thrust reverser noise and directivity in landing. For additional discussion on thrust reversers, see von Glahn et al. 1972.

3.8.4 Subsonic Jet Noise

The pressure fluctuation in the exhaust jet due to *turbulent structure and mixing* is the source of *broadband* jet noise. In his classical paper, Lighthill (1952) showed that the acoustic power in a (cold) subsonic turbulent jet is proportional to the 8th power of the jet speed, V_j, i.e.

$$\wp_j \propto V_j^{\,8} \tag{3.43}$$

Therefore, to reduce the power of the radiated noise from an aircraft exhaust, the jet speed needs to be reduced. Interestingly, propulsive efficiency is also improved with reduced exhaust speed. The development of high and ultra-high bypass ratio turbofan engines accomplish both goals, i.e. higher propulsive efficiency (with reduced thrust specific fuel consumption [TSFC]) and reduced jet noise, simultaneously. Also, the concept of variable cycle engine (VCE) promises the capability of higher bypass ratio at

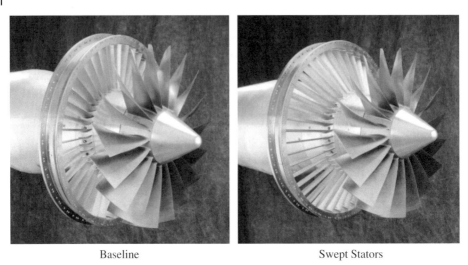

Baseline   Swept Stators

Swept/Leaned Stators

Figure 3.36 Fan noise reduction research.

takeoff and landing and reduced bypass ratio at cruise condition, which is advantageous in the reduction of the airport, i.e. community noise. Since the kinetic power in a jet is proportional to V_j^3, an acoustic efficiency parameter may be defined as the ratio of the radiated acoustic to the jet kinetic power (see Kerrebrock 1992,), which is then proportional to the fifth power of the jet Mach number, M_{j0}, which in this derivation is defined as the ratio of V_j and the ambient speed of sound, a_0, i.e.

$$\eta_{\text{jet noise}} \propto \left(\frac{\rho_j}{\rho_0} \right) M_{j0}^{5} \tag{3.44}$$

This efficiency parameter is used to compare different noise suppression concepts. The noise suppressors of the 1950s that were developed originally for turbojets added significant weight to the engine and caused significant loss of gross thrust. The turbofan

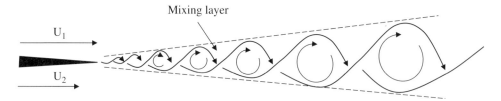

Figure 3.37 The formation of large-scale vortical structures in a subsonic free shear layer in the nearfield.

engines of the 1960s and 1970s alleviated the jet noise problem based on their reduced jet speed, also their blade and duct treatments reduced the turbine noise; but still more stringent noise regulations were on the horizon. These limitations are set by the Federal Aviation Regulation–Part 36 Stage 3 and the new FAA Stage 4 Aircraft Noise Standards (that is used for the certification of turbofan-powered transport aircraft).

The dynamics and structure of a subsonic turbulent jet as it emerges from a nozzle is initially governed by the Kelvin-Helmholtz (K-H) instability wave at the interface between the jet and the ambient surrounding. The interface is indeed a vortex sheet that, through spatial growth of the instability wave, creates large-scale vortical structures (also known as coherent structures) in the initial shear layer through entrainment (see Brown and Roshko 1974). Figure 3.37 shows the roll up of the shear layer and the formation of large-scale vortical structures. The flow visualization images of large-scale vortical structures in a free shear layer at random times are captured by Brown and Roshko 1974, among others. The size of these eddies spans across the mixing layer, and they are convected downstream at the average speed of the jet and the ambient air. The mixing layer growth reaches the centerline in four to six jet diameters, thus ending the potential core region of the jet. Eventually, these large-scale vortical structures break up into smaller eddies (through a transition process) and the jet reaches a fully developed state (in 10–12 diameters). The appearance of multiscale structures in turbulent flowfield of a free jet, namely, large scales in the initial mixing layer followed by fine scale in the fully developed jet, changes the scale of the quadrupole along the jet, causing high-frequency sound emission from the mixing layer to low-frequency sound from the fully developed region. Kerrebrock (1992) elucidates this point by slicing the subsonic jet in thin segments, from the initial mixing layer to fully developed state, and gives a rationale for the subsonic jet acoustic signature.

3.8.5 Supersonic Jet Noise

Supersonic jets are invariably turbulent (due to high Reynolds number) and often contain shocks, e.g. in case of underexpanded nozzles. Since shocks are formed by coalescence of compression Mach waves that are reflected from a fluctuating jet boundary, they are inherently unsteady. The radiated noise due to the interaction of the periodic shock cell structure and turbulence in underexpanded supersonic jets is known as the broadband shock-associated noise (BBSAN). The large-scale structures in the mixing layer that are convicted at the mean speed of the jet and the freestream may attain supersonic speeds and thus radiate Mach waves. In addition to turbulent mixing broadband noise, supersonic jets exhibit prominent discrete frequencies in their spectra, known as screech tone. The screech tone is related to a feedback mechanism between growing instability waves and the shock cell structure that propagates in the upstream direction and excites new instability wave at the nozzle lip, thus closing the loop. A comprehensive review article by Tam (1995) is recommended for discussion of screech and the physics of supersonic jet noise. Figure 3.38 shows the narrow-band spectrum of a supersonic jet (from Seiner and Yu 1984) where the three basic sources of supersonic jet noise, turbulent mixing noise, screech, and broadband shock- associated noise, are identified. The frequency scale is the nondimensional Strouhal number, St. These are the essential characteristics that exist in a supersonic jet, which are mainly absent in their subsonic counterpart. Finally, LAA that we cited earlier is only applicable to subsonic jets and its result that the radiated acoustic power from an unheated jet is proportional to M_j^8, in a supersonic jet turns into M_j^3. The relationship between acoustic power and jet speed is shown, on a log-log scale, in Figure 3.39 (from Powell 1959). The initial slope, on the log plot, for the low-speed (unheated) jets shows proportionality with M_j^8 power, whereas the slope drops, corresponding to M_j^3, for the supersonic jets. For heated jets, density ratio, or equivalently total temperature ratio (TTR), impacts sound radiation, as demonstrated by Ahuja and Bushell (1973). The velocity exponent in the jet drops from 8 for the unheated jet to 5–6 for the heated jet.

A comprehensive review article on supersonic jet noise by Tam (1995) is recommended for further reading. The hot jets and sources of jet noise are discussed by Khavaran et al. (2010). The exponent on the jet velocity for the heated jets is 5–6 instead of 8 for cold jet. The acoustic radiation from turbofan engines, which involves dualstream nozzles, is discussed by Khavaran and Dahl (2012).

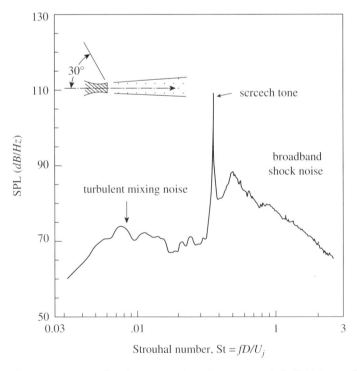

Figure 3.38 Narrow-band supersonic jet noise spectrum in farfield. *Source:* from Seiner and Yu (1984).

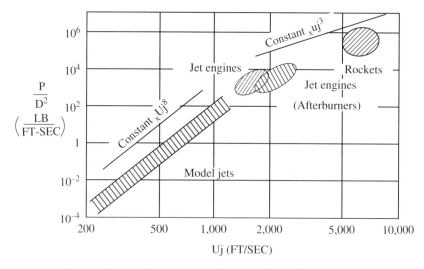

Figure 3.39 Acoustic power dependency on jet speed. *Source:* from Powell 1959.

3.9 Engine Noise Directivity Pattern

The radiated sound from an aircraft engine is characterized both by its intensity as well as its directivity pattern. The directivity of radiated sound from an inlet in the farfield is shown in Figure 3.40 (adapted from Tam et al. 2013). The measurements were taken from static engine tests of JT-15D. The maximum SPL is measured at 60° from the inlet axis.

In the case of a hard-wall inlet, the radiated sound directivity is shown in Figure 3.40a whereas an inlet with an acoustic liner has a directivity pattern as shown in Figure 3.40b. Directivity of the radiated noise from an inlet is unaffected by the acoustic liner, but the radiated noise level is reduced by 20 dB (which is very significant based on the logarithmic nature of decibel).

Directivity of radiated noise from a subsonic jet in farfield is shown in Figure 3.41 (from Powell 1959). Note that the radial scale of relative SPL is graphed linearly in Figure 3.41. It shows that 35° is the angle from the jet axis where noise reaches its peak level.

To examine the absolute scale of the radiated jet noise and its directivity pattern, Figure 3.42 (from Treager 1979), is plotted. The graph shows equal sound level contours surrounding a jet engine. Higher-level contours are in the aft region of the engine where the jet noise dominates (note the 140 dB contour in the nearfield). Also, we note that the peak noise at 100 ft radius downstream of the engine is reached at an angle of ~40° from the jet axis.

Two dominant parameters that affect noise emission in supersonic jets are the jet Mach number and the jet total temperature. Figure 3.43 shows the measured directivity at selected Strouhal numbers in a Mach-2 jet at 500 K total temperature (from Seiner et al. 1992). The peak levels are measured in the 25°–45° range from the jet axis (or in terms of χ: $135° < \chi < 155°$).

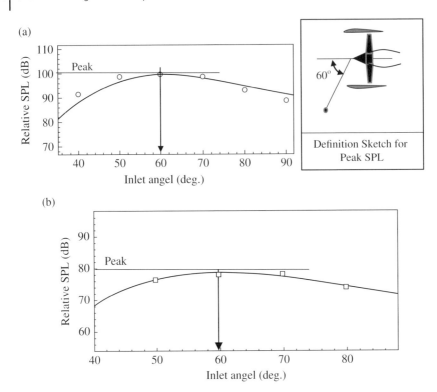

Figure 3.40 Directivity of the radiated sound from an aircraft engine (JT-15D) inlet: (a) computed and measured directivity of sound in the farfield (inlet with hard wall); (b) computed and measured directivity of sound in the farfield (inlet with liner). *Source:* Adapted from Tam et al. 2013.

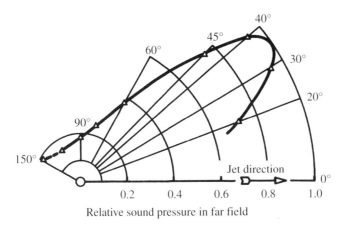

Figure 3.41 Directional distribution of radiated sound (relative to peak) from a subsonic jet. *Source:* From Powell 1959.

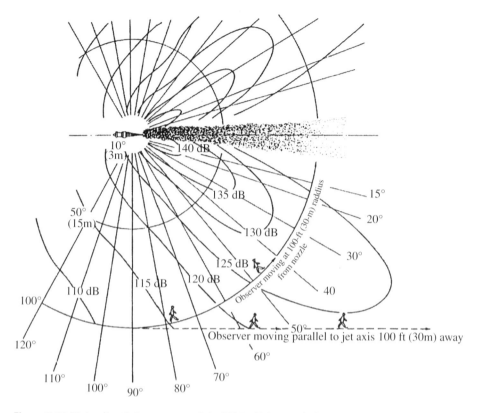

Figure 3.42 Noise directivity pattern and the SPL in dB for a turbulent (round) jet in near- and farfield. *Source:* From Treager 1979.

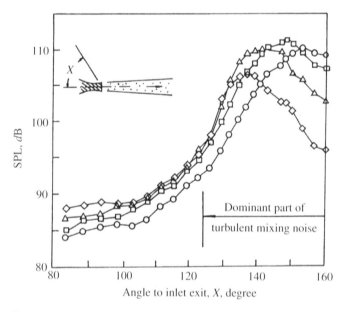

Figure 3.43 Measured noise directivities at selected Strouhal numbers in a Mach-2 jet. At 500 K total temperature. *Source:* From Seiner et al. 1992.

3.10 Noise Reduction at the Source

In this section, we briefly present the technologies that have successfully mitigated engine noise at its source. The first technology uses wing shielding for engine noise mitigation in landing and takeoff (LTO) cycle.

3.10.1 Wing Shielding

To quantify the effect of wing shielding on noise reduction on a large commercial transport, Berton (2000) studied a notional long-haul commercial quadjet transport with advanced high bypass ratio turbofan engines mounted over the wing. This configuration is shown in Figure 3.44a (from Berton 2000). The advanced engine in this study is chosen for entry into service of 2020 at 55000 lbs. thrust class. The fan has a design pressure ratio of 1.28, bypass ratio of 13.5, and tip speed of 850 fps ($260\,\mathrm{m\,s^{-1}}$). The full-scale fan has a diameter of 130 in. (3.3 m). The tone-weighted perceived noise levels (in EPNdB) and their sources are shown in Figure 3.44b–d for the sideline (certification) observer, for the community noise and the approach noise, respectively. The total noise mitigation in sideline (certification point) is 4.3 EPNdB, followed by the same level of reduction for the community noise, and the approach noise shows the least impact of wing shielding with 1.3 EPNdB reduction. The cumulative effect of wing shielding in this study is 9.9 EPNdB.

Before we leave this subject, we point out that any noise mitigation study that involves airframe propulsion system integration is strongly configuration-dependent, as expected. However, we recognize that noise mitigation through aircraft wing, empennage shielding is certainly a viable and strong option in the development of future quiet aviation.

3.10.2 Fan Noise Reduction

The fan bypass ratio has played a decisive role in engine noise reduction in commercial aviation. First, the core jet speed is reduced, since its power is drained and supplied to a larger fan. Second, the fan nozzle velocity is lower than the core, which mitigates noise. The use of acoustic liners in the fan inlet and exit ducts has proven effective in fan noise emission. The modern fan blades that use sweep and reduce thickness contribute to the fan noise reduction as well. The combination of sweep and lean that are used on fan exit vanes in the bypass duct have reduced rotor-stator interaction noise. Introduction of geared turbofan (GTF), in service since 2014 (GTF-1000 by P&W), is a technology demonstrator for ultra-high bypass (UHB) turbofan engines that are designed with bypass ratios in excess of 12. This feature alone is responsible for engine noise reduction by −20 dB margin with respect to FAA Stage 4 Aircraft Noise Standards.

Acoustic liners are Helmholtz (cavity) resonators, which may be integrated in an inlet duct or fan exit duct to attenuate fan noise at its dominant frequency, e.g. BPF. The perforated plate on a honeycomb is the acoustic liner configuration for a single layer. Other liner configurations may use porous layer over a honeycomb, or use bulk absorbers instead of honeycomb or use double or triple-layer honeycomb. The use of triple-layer, called three degrees of freedom (3 DoF), offer the capability of a broader frequency attenuation (see Leylekian et al. 2014).

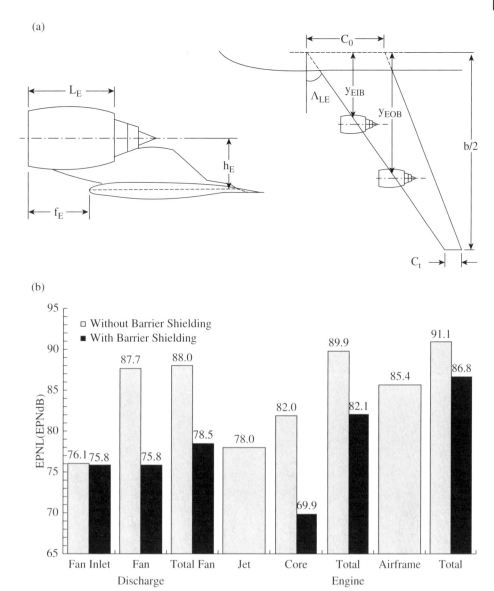

Figure 3.44 Impact of wing shielding on aircraft noise mitigation: (a) configuration of the four-engine (Quad) installation over the wing; (b) certification observer at the sideline; (c) community noise; (d) approach noise. *Source:* Courtesy of NASA.

The acoustic liner design concepts and their performance in sound attenuation, in addition to sonic inlets, are presented by Feiler et al. (1972) and Klujber (1972). These features are quantified in Figure 3.45 (adapted from Smith 1986).

The new advances in design and manufacturing of low-drag acoustic liner (LDAL) is shown in Figure 3.46 (from Norris 2018). The promising acoustic liner design incorporates an embedded mesh within the honeycomb and a slotted face sheet to reduce liner drag. The embedded mesh helps increase the frequency bandwidth of acoustic absorption

(c)

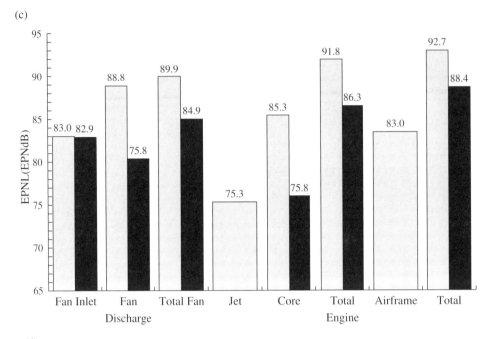

(d)

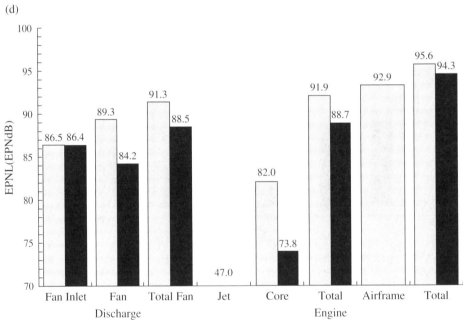

Figure 3.44 (Continued)

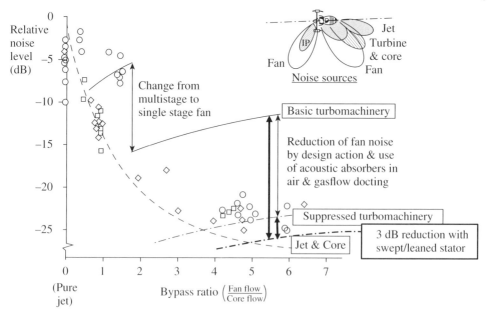

Figure 3.45 Noise-reduction technologies. *Source:* Adapted from Smith 1986.

Figure 3.46 New low-drag acoustic liner with slotted face sheet undergoing flight test at NASA.
Source: Courtesy of NASA/Boeing.

by the liner. The early results are promising, which prompts the incorporation of LDAL in the aft-bypass duct of separate flow turbofan engines with flight test demonstration.

3.10.3   Subsonic Jet Noise Mitigation

Jet noise is a perfect candidate for mitigation since it scales as V_j^8, in acoustic power emission. Beyond a reduction in jet speed, as in high bypass ratio turbofan engines, are there any other means of reducing jet noise without paying a severe penalty on gross thrust? The jet noise reduction depends on many parameters, but predominantly on the nozzle pressure ratio (NPR) and the jet Mach number. In this section, the most promising development in transonic jet noise reduction is briefly presented. For the past two decades, research and development by NASA-Glenn and aircraft engine and airframe industry on jet noise reduction has led to the development of a new winning technology: chevron nozzle. An excellent review article by Zaman et al. (2011) that chronicles the development of mixing devices, mainly tabs and chevrons, is suggested for further reading (also see Bridges et al. 2003).

3.10.3.1   Chevron Nozzle

Chevrons are triangular serrations at the nozzle exit plane that have the potential of "calming" turbulent jets (see Zaman et al. 2011). The *calming* effect stems from a redistribution of azimuthal vorticity at the nozzle exit that attains certain streamwise component. For optimal redistribution of vorticity, the chevron tips have to slightly penetrate the jet. Excessive penetration gives rise to thrust loss and an insignificant dipping into the jet does not produce the desired sound pressure level (SPL) reduction. Figure 3.47 shows a baseline nozzle that is used to compare the aeroacoustic and propulsive performance of various chevron configurations on the core and/or the fan nozzles in a separate flow turbofan engine.

The most promising of the chevron configurations in static test were then flown to demonstrate takeoff, landing, and the flyover noise-reduction effectiveness. Naturally,

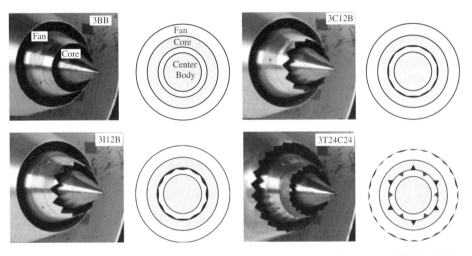

Figure 3.47 Baseline nozzle (3BB) and examples of chevron and tab configurations. (3C12B, 3I12B and 3T24C24, respectively) in static test. *Source:* From Saiyed et al. 2000.

on the propulsion side, thrust loss coefficient at cruise conditions is the measure of viability of any noise reduction device.

The benefits in noise reduction of chevron-equipped aircraft in flyovers, i.e. community noise, is first measured (using logarithmic scale, known as Bel or more commonly the decibel, dB) in the frequency range where human ear is sensitive (i.e. between 50 and 10000 Hz) and subsequently weighted toward the frequencies where the human ear is most sensitive, i.e. 20000–50000 Hz.

Other corrections contribute to the calculation of what is known as the *effective perceived noise level* (EPNL) in dB, which is the basis of FAR-36 certification limits. Kerrebrock (1992) offers a clear methodology to EPNL calculations and is suggested for further reading on engine noise. There is ample data that supports a reduction in turbulent kinetic energy in a jet where streamwise vorticity is injected at the nozzle exit, as produced by chevrons. The significance of chevron technology is in its light weight and negligible impact (i.e. loss) on nozzle gross thrust, which is measured at ¼%.

At the outset, the goal of jet noise reduction was to achieve 2.5 EPNL (dB) in flyover, i.e. community, noise reduction while limiting the thrust penalty in cruise to less than 0.5%. As evidenced in Table 3.15 where the noise benefit and cruise thrust loss data from Zaman et al. (2011) is shown, several chevron configurations can achieve the desired performance (see Brausch et al. 2002).

The success of the chevron nozzle has made it into a standard component on several new aircraft such as Boeing 787. It is noteworthy to mention that the earlier jet noise suppressors in the form of multiple and corrugated nozzle schemes exhibited 1% thrust loss per dB reduction. In contrast, we note from Table 3.15 that the best chevron configuration produced a 2.71 dB reduction while only incurring a thrust loss of 0.06% at cruise.

3.10.3.2 Acoustic Liner in Exhaust Core

In addition to chevron, special acoustic liners for high-temperature applications are used in the engine exhaust core. The new LDAL technology discussed in Section 3.10.1 finds applications in the aft bypass duct (see Norris 2018).

Table 3.15 Chevron noise benefit and cruise thrust loss data.

Configuration	Noise benefit EPNdB	% loss in thrust coefficient at cruise
3C12B	1.36	0.55
3I12B	2.18	0.32
3I12C24	2.71	0.06
3T24B	2.37	0.99
3T48B	2.09	0.77
3T24C24	–	0.43
3T48C24	–	0.51
3A12B	–	0.34
3A12C24	–	0.49

3.10.4 Supersonic Jet Noise Reduction

We start this section with a brief overview of the concepts used in supersonic jet noise reduction.

The concepts are divided into five categories: (i) plasma actuators, (ii) chevrons, (iii) fluidic injection, (iv) 2-D nozzles, and (v) twin jet research"

1) *Plasma actuators research.* Plasma actuators are used to excite instability waves in supersonic shear layers. The research was pursued on a scale model in a laboratory as well as computational fluid dynamics (CFD) and large-eddy simulation (LES). Active control technique was used in an open and closed loop (see Samimy et al. 2009).

2) *Chevron research for supersonic jets.* Following the promise of chevrons in subsonic jet noise reduction (2–3 dB), researchers proposed to use these in under- and over-expanded supersonic jets. A variant of the mechanical chevron approach used fluid chevrons (via microjet injection) for active noise control. The application of these concepts to military aircraft are detailed by Morris (2011), and in a separate report by Kailasanath et al. (2011). Frate (2011) gives a detail aero-acoustic assessment of convergent-divergent (CD) nozzles with chevrons.

3) *Fluid injection research.* Recent work at NASA focused on single and dual-stream jets in subsonic and supersonic flow. Water, water–air, and air injection were used as noise mitigation techniques. The dual-stream jets with fan and/or core nozzles operating with supersonic jets were investigated. The results were mixed, i.e. when low-frequency noise was suppressed, high-frequency noise was excited. When BBSAN was unaffected, slight noise reduction at peak jet noise angle ($60°$) was recorded. Mass flow ratio for the injection compared to the jet was in 1–2% range. Henderson (2009) details the history and current status of fluidic research in supersonic jet noise mitigation.

4) *2-D nozzle research.* Innovating nozzle concepts for low-noise/high-thrust applications to commercial supersonic transport for the 2020 time frame are studied. 2D nozzles with and without chevrons, including geometrical variants, such as bevel, extended edge, and aspect ratio were included in the CFD studies. Frate (2011) gives an overview of 2-D nozzle research for supersonic applications.

5) *Twin-jet interaction research*: This area of research is the closest to the statement of need (SON) for the noise mitigation in twin-engine military aircraft (e.g. F-18). Corrugated seals in the divergent flap and the twin-jet coupling effects were investigated. Nozzle configurations that incorporated bevel and canting angle were studied. Seiner et al. (2003) may be consulted for direction study of twin-engine fighter aircraft.

Unfortunately, there is no simple solution to supersonic jet noise reduction, beyond a reduction in jet speed. There are, however, three advanced propulsion system designs that hold a promise for supersonic civil transports:

1) Variable-cycle engine (VCE)
2) Mid-tandem fan (MTF)
3) Mixed-nozzle ejector (MNE)

These advanced propulsion concepts are presented, in the context of suitability for supersonic transport, by Whurr (2004). Other supersonic cruise propulsion concepts are discussed by Fournier (2004). The noise at takeoff due to high-speed jet for supersonic transports such as Concorde, severely hampered the use and economic viability of the aircraft around the globe. Indeed Concorde average noise (over its landing, sideline, and flyover phases) was 20 dB above Stage 3! However, LTO noise is not the only problem supersonic vehicles pose to communities. Indeed, there is the problem of sonic boom in flyover.

3.11 Sonic Boom

Aircraft in supersonic flight create waves to make room for the vehicle volume. The first wave is an oblique or conical shock wave that forms ahead of the aircraft and is followed by a series of weak expansion and compression waves around the vehicle, and finally, shock waves appear near the aft portions of the aircraft to recover parallel streams downstream of the vehicle. These waves that are originated at the aircraft propagate through a nonhomogeneous atmosphere (which is weather-related as well as the environment with temperature, density pressure, and humidity gradients) and the wave strength is diminished. The initial flow compression through the shock wave followed by the continuous expansion and finally sudden compression at the aft produces a characteristic N-shaped signature on the ground. The pressure field on the ground due to the shock wave is known as the sonic boom. Figure 3.48 shows the schematic of N-shaped signature (or N-wave) of sonic boom, and the corresponding overpressure, on the ground. N-wave is the signature of supersonic aircraft at constant altitude and cruise Mach number. In maneuvering aircraft, the shape of the boom signature on the ground changes to U-wave. The observer on the ground is exposed to the aircraft shock overpressure twice (since there is a leading and an aft shock wave, as shown), which sweep over the observer within a fraction of a second.

Sonic boom overpressure is the measure of ground shock strength (in lbf ft^{-2}), as shown in Figure 3.48. The overpressure on the ground corresponding to Concorde cruising at Mach 2 at 52000-ft altitude, is about 1.94 lb ft^{-2}. For a real sonic boom N-wave, Figure 3.49

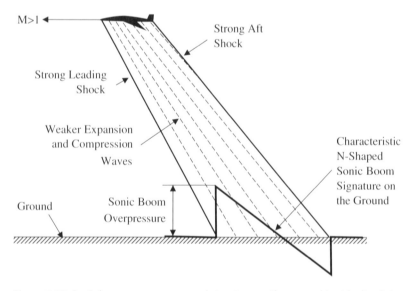

Figure 3.48 Sonic boom overpressure and signature on the ground (an idealized view of boom signature; N wave).

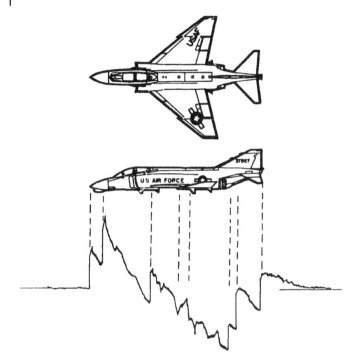

Figure 3.49 Supersonic fighter, F-4C and its sonic boom pressure time history in low-altitude flight shows multiple shock waves and expansion fans in its boom signature. *Source:* From Nixon et al. 1968.

(from Nixon et al. 1968) shows the F-4C in extreme low-altitude supersonic flight, where a series of shocks and expansion waves spread across the vehicle in its sonic boom signature.

Typical ground overpressures by various high-speed aircraft flying at different altitude and Mach number are listed here for perspective and comparison:

- SR-71 : 0.9 psf ($lbf\,ft^{-2}$) at Mach 3, 80 000 ft
- Concorde: 1.94 psf, at Mach 2, 52 000 ft
- F-104 : 0.8 psf, at Mach 1.93, 48 000 ft
- Space Shuttle: 1.25 psf, at Mach 1.5, 60 000 ft, landing approach

From the theory of sonic boom, the ground signature depends on aircraft *size and weight* (at the same flight altitude). Alternatively, the sonic boom overpressure depends on aircraft *volume* and *lift*. This feature is captured well in Figure 3.50 from Fisher et al. 2004) where HSCT, Concorde, and Supersonic Business Jet (SBJ) are compared in their ground sonic boom signature. For a review of sonic boom theory, see Plotkin (1989).

Recent advances in sonic boom reduction has made the case for the quiet SBJ that is allowed to fly over land at Mach 1.4–1.8. The concept of extendable nose spike transforms the characteristic N-shape sonic boom signature on the ground to a smooth sinusoidal "thump," known as *shaped sonic boom*. Figure 3.51 shows a concept NASA-X plane, X-59 QueSST (Quiet Supersonic Transport), designed by Lockheed-Martin as a supersonic low-boom configuration. Figure 3.52 shows another low-boom concept plane from NASA (courtesy of NASA).

Figure 3.53 shows a supersonic airliner designed by Boom Supersonic to cruise at Mach 2.2 with 55 passengers (see Norris and Warwick 2018, and Boom Technology website, https://boomsupersonic.com/).

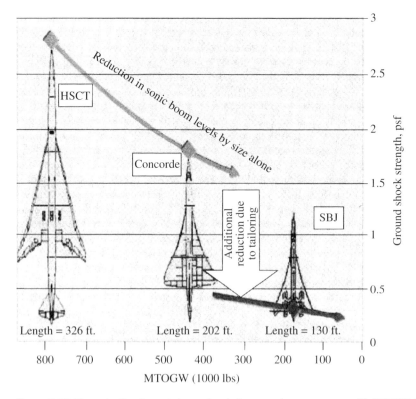

Figure 3.50 The reduction in sonic boom levels, in ground overpressure, with MTOGW (in 1000 lbs) and size at the same altitude. *Source:* From Fisher et al. 2004.

Figure 3.51 NASA X-59 *QueSST* aircraft designed for low-boom commercial supersonic flight over land. *Source:* From NASA and Lockheed Martin.

Figure 3.52 NASA's low-boom supersonic test case. *Source:* Courtesy of NASA.

Figure 3.53 55-seat supersonic airliner under development at Boom Supersonic (for Mach 2.2). (https://boomsupersonic.com.)

Figure 3.54 shows the three-view of a supersonic business jet designed by Aerion Supersonic for Mach 1.4 cruise (see Aerion Supersonic website, https://www.aerionsupersonic.com) and Table 3.16 shows the preliminary performance data for the aircraft.

A quiet supersonic jet, *Spike S-512*, is designed by Spike Aerospace (http://www.spikeaerospace.com) with supersonic overland flight capability. The cruise Mach number is 1.6 and the company predicts the perceived loudness level, on the ground, to be less than 75 dB (see Spike Aerospace website). The aircraft is sized for 12–18 passengers.

FAA regulations currently prohibit supersonic commercial flights over land in the United States. Similarly, supersonic flight over land is currently prohibited in Europe.

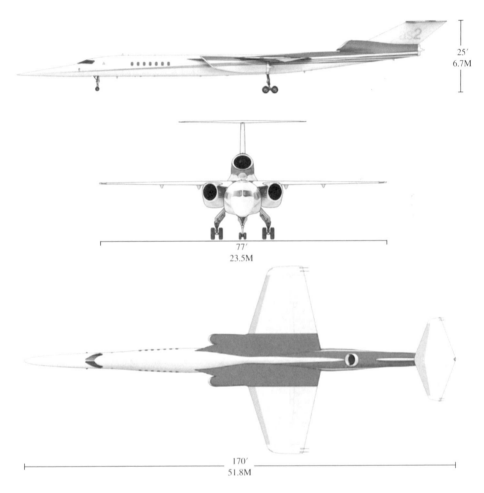

25′
6.7M

77′
23.5M

170′
51.8M

Figure 3.54 AS2 supersonic business jet designed by Aerion Corporation for Mach 1.4 cruise. *Source:* Courtesy of Aerion Supersonic. (https://www.aerionsupersonic.com.)

Table 3.16 Preliminary performance data.

SuperCruise	1.4 Mach
BOOMLESS CRUISE[sm]	1.1–1.2 Mach
LRC, subsonic	0.95 Mach
Max range IFR, Mach 1.4	4200 nm/7780 km
Max range IFR, Mach 0.95	5400 nm/10 000 km
Balanced field length at ISA S.L.	7500 ft./2286 m
MTOW	133 000 lbs./60 328 kg
Fuel	59 084 lbs./26 800 kg
Wing area	1511 ft^2/140 m^2

3.12   Aircraft Noise Certification

The guidelines on noise regulations for turbofan-powered transport (FAA FAR-36 Stage 4) are shown in Figure 3.55. The metric of choice in noise certification is *effective perceived noise* measured in dB, EPNdB. The aircraft takeoff gross weight is used in approach and sideline regulation. The number of engines (two, three, or four) regulate the takeoff noise (see Figure 3.55).

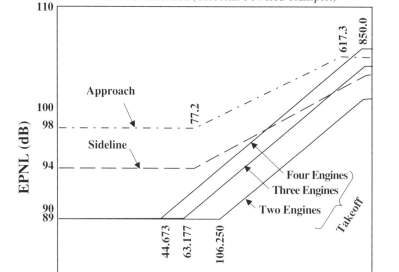

Figure 3.55 FAA Stage-4 aircraft certification guidelines for turbofan powered transport.

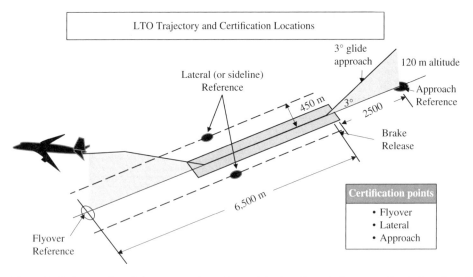

Figure 3.56 Approach and departure trajectories, and the location of reference points (microphone).

Figure 3.56 shows the trajectory and certification locations that are used in the United States and by ICAO. We note three certification points on approach, lateral, and flyover (see ICAO Annex 16 Standards, Appendix 2). Noise is sampled at 44.1 kHz and all microphones are positioned 1.2 m above ground level (in open field). Equations for the calculation of noise certification limits at takeoff, sideline, and approach certification are summarized in ICAO Annex 16 Appendix B.

Advancements in noise reduction show an 80% improvement since 1960 (see Figure 3.57a). Average aircraft engine noise reduction relative to Stage 3 is shown in Figure 3.57b), from Huff 2013. The *average* noise reported in Figure 3.57b is taken over three reference points, namely approach, sideline, and flyover. The military aircraft and the supersonic civil transport, Concorde, are shown for reference on top of the graph with 20 dB over Stage 3. The main reason for excessive noise of these aircraft lies in their high jet speed, as compared to modern turbofan engines.

The relative contribution of engine and airframe components to overall noise in approach and takeoff is shown in Figure 3.58 (from NASA 1999). The takeoff is dominated by the fan exhaust jet noise and the core exhaust jet noise. The contributions from the fan inlet are small to takeoff noise as compared to the exhaust jets. As expected, the opposite of these trends exist in approach, where the fan inlet makes dominant contribution to the overall noise, followed by airframe noise.

3.13 NASA's Vision: Quiet Green Transport Technology

Table 3.17 is the summary of NASA's vision on subsonic transport system level metrics. At the time of this writing, we still have three more years to reach N + 2 (2020) goals and eight more years to N + 3. The striking lack of metrics on CO_2 emissions (or other GHG emissions) is evident in Table 3.17.

Cumulative EPNdB is a NASA metric that represents the sum of sideline, flyover, and approach.

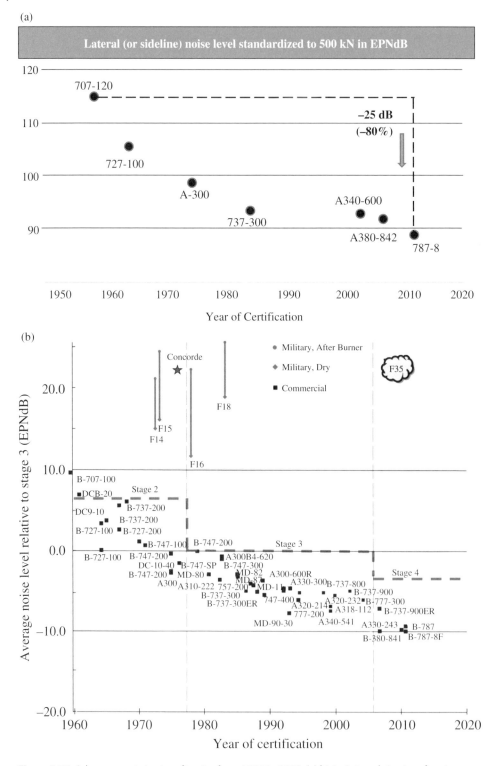

Figure 3.57 Advancements in aircraft noise from 1960 to 2020: (a) historic trends in aircraft noise reduction; (b) average aircraft engine noise relative to stage 3. *Source:* From Huff 2013.

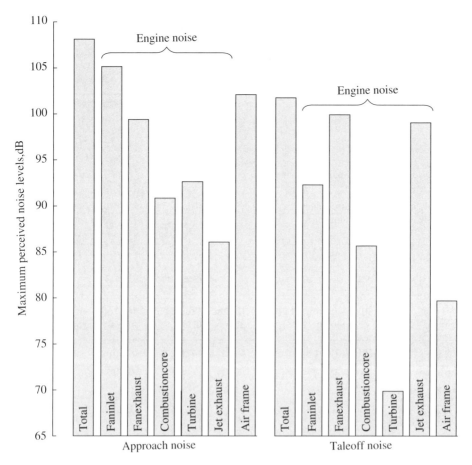

Figure 3.58 Noise components of an aircraft with turbofan engine during takeoff, approach, and landing. *Source:* From NASA, Glenn Research Center 1999.

Table 3.17 NASA subsonic transport system level metrics.

Technology benefits	Technology generations (Technology readiness level = 4–6)		
	N + 1 (2015)	N + 2 (2020)	N + 3 (2025)
Noise (Cum. margin rel. to Stage 4)	−32 dB	−42 dB	−71 dB
LTO NOx emissions (Rel. to CAEP 6)	−60%	−75%	−80%
Cruise NOx emissions (Rel. to 2005 best in class)	−55%	−70%	−80%
Aircraft fuel/energy consumption (Rel. to 2005 best in class)	−33%	−50%	−60%

Table 3.18 NASA's environmental goals for supersonic aircraft (see Huff 2013).

Environmental Goals	N+1 supersonic business class aircraft (2015)	N+2 small supersonic airliner (2020)	N+3 efficient multi-mach aircraft (beyond 2030)
Sonic Boom	65–70 PLdB	65–70 PLdB	65–70 PLdB Low-boom flight 75–80 PLdB Over water flight
Airport noise (cum below Stage 4)	Meet with margin	10 EPNdB	10–20 EPNdB
Cruise emissions (cruise NOx in $g\,kg^{-1}$ of fuel)	Equivalent to current subsonic	<10	<5 and particulate and water vapor mitigation

Table 3.19 NASA's performance goals for supersonic aircraft (see Huff 2013).

Performance Goals	N+1 supersonic business class aircraft (2015)	N+2 small supersonic airliner (2020)	N+3 efficient multi-Mach aircraft (beyond 2030)
Cruise speed	Mach 1.6–1.8	Mach 1.6–1.8	Mach 1.3–2.0
Range (n mi)	4000	4000	4000–5500
Payload (passengers)	6–20	35–70	100–200
Fuel efficiency (passenger-miles per lb fuel)	1.0	3.0	3.5–4.5

NASA's research goals for supersonic aircraft is focused on airport noise and overland sonic boom reduction. These goals and cruise NOx emission are summarized in Table 3.18 (Huff 2013). Sonic boom pressure level reduction is measured in perceived noise level in dB (PLdB). Table 3.19 shows NASA's research performance goals for future supersonic aircraft (Huff 2013).

3.14 FAA's Vision: *NextGen* Technology

In response to environmental concerns, energy efficiency, sustainability, and security, FAA announced the *NextGen* Program, in 2004. Five pillars are identified in the environmental approach in *NextGen*:

1) Better scientific understanding and improved tools for integrated environmental analysis, with the key aspects:
   a) Characterize problem.
   b) Assess risk.
   c) Inform development of mitigation solutions.

2) Mature new aircraft technologies, with the key aspects:
   a) Accelerate aircraft and engine technology development.
   b) Assess the technology.
3) Develop aviation alternative fuels, with the following emphasis:
   a) Fuel certification.
   b) Promote environmental and economic sustainability.
   c) Facilitate feedstock and fuel production.
   d) Support collaboration.
4) Develop and implement clean, quiet, and energy-efficient operational procedure through:
   a) New air traffic management capabilities.
   b) Gate-to-gate and surface operational procedures.
5) Policies, environmental standards, market-based measures, and environmental management system with the following components:
   a) Domestic policy (e.g. *NextGen* deployment).
   b) International policy (e.g. ICAO, CAEP).

The operation of the National Airspace System using piecewise continuous management of the flight paths was an artifact of the legacy human-centric communication, navigation, and surveillance systems. The next-generation air transportation system is the seamless integration of automation and optimization objectives for flight path management from gate-to-gate (i.e. integrated en route, arrival, departure, and surface management). *NextGen* proposes initiatives to advance today's national airspace system – in particular, ATM. The program started in 2004 with a short-term plan (2004–2012), mid-term plan (2012–2020), and long-term plan (2020–2030). The scope of the *NextGen* program and its benefits are summarized as follows:

- A better travel experience for passengers and operators
- Fuel savings for aircraft operators
- A reduction in emissions due to more direct and efficient routes and approaches
- Reduced separation minimums
- Reduced congestion
- Better communications across the airspace system and its users
- Standardized access to weather information
- Improved on-board technology

It is estimated that ATM improvements and other operational procedures could reduce fuel burn by 8–18%, according to an IPCC Special Report on Aviation and the Global Atmosphere (1999). Although FAA's vision does not address directly CO_2 and other GHG emissions, the indirect impact on aviation emissions comes from broader goals of energy efficiency through "fuel savings" and "efficient routes and approaches," as well as development of "aviation alternative fuels."

3.15 The European Vision for Sustainable Aviation

The European counterpart of *NextGen* is called *Single European Sky ATM Research* (SESAR). Their vision for air transportation in 2050 is formulated as *Flightpath 2050*

Table 3.20 Emission goals in EU's *Flightpath 2050*.

Emissions	Goals for 2050
CO_2 emissions per passenger-km	−75% (compared to 2000)
NO_x emissions reduction	−90% (compared to 2000)
Noise reduction	−65% (compared to 2000)
Emission at taxiing	0

(see European Commission Report 2011). The goals stated in Flightpath 2050 ensure Europe with sustainable aviation that is achievable by 2050. The emission goals for 2050 are summarized in Table 3.20.

The broader goals of Flightpath 2050 are passenger-centric, in timeliness of the flights, efficiently integrated ground transportation that enables door-to-door travel time of less than four hours (within Europe) and integration of advanced technology in ATM. For sustainable aviation, the Flightpath 2050 outlines the critical areas of research and development, e.g. alternative jet fuel, advanced propulsion, and power concepts and electrification that help achieve zero-carbon aviation.

3.16 Summary

This chapter viewed aviation's impact on climate through the lens of combustion-generated pollutants of fossil fuels and noise. On the combustion side, we learned that fossil-fuels in aviation contain heavy dosage of carbon (on mass basis), which create carbon monoxide, which is poisonous, carbon dioxide, which is a greenhouse gas, or soot that impacts surface air quality and the associated health risks. Since air contains 79% N_2 (by volume), its combustion with jet fuel creates oxides of nitrogen, known as NO_x, that impact surface air quality through ozone formation with health implications as well as impact on ozone depletion in the upper atmosphere. Ozone depletion in the upper atmosphere leads to harmful levels of UV radiation on Earth. Another oxide of nitrogen, N_2O, nitrous oxide, acts as a GHG with the same harmful impact on climate change as carbon dioxide. The other byproduct of combustion is water vapor, which is a greenhouse gas and may lead to formation of persistent contrail. These, in turn, impact radiative forcing in the atmosphere and cause an imbalance of day–night temperature levels on a regional basis.

The advances in combustor design have drastically reduced NO_x, but carbon dioxide and water vapor are still left intact in the emissions. Regulated combustion emissions are carbon monoxide and unburned hydrocarbons, NO_x, SO_x, particulate matter, and soot/smoke. No GHG are included in the regulations. We briefly introduced lifecycle concepts of CO_2 emissions from *well-to-tank* (WTT) and *tank-to-wake (TTW)*. *WTT* introduced a two to three times multiplier effect on the *TTW* contributions to CO_2 emissions, which is easily quantifiable. GHG absorb IR radiation from Earth, transmit a portion back toward Earth and reflect the balance (of IR radiation) into space. With exponentially increasing concentration levels of GHG in the upper atmosphere, the IR radiation based on anthropogenic interference in GHG causes an increase in temperature on the Earth, which in climate research circles is referred to as *global warming*. Melting of the polar ice caps and rising water levels causes extreme weather conditions, such as catastrophic floods and droughts, which is collectively and benignly called *climate change*. The electric power industry and transportation sectors are the top polluters in terms of GHG emissions and NO_x. Within the transportation sector, aviation contributes about 10% to carbon dioxide national inventory. Aviation generation of NO_x and its impact on air quality is estimated at less than 3% by 2020 (measured as percentage of national inventory). Road transportation sector, i.e. cars and trucks, are, however adopting advanced emission standards in hybrid and electric vehicles. Thus, aviation impact on environment, as a percent of transportation, is expected to grow. In addition to aircraft engines emission of GHG and NO_x, the gate and ground support equipment, ground access vehicles, construction activities, electric power generating plants, and maintenance operations at the airports also contribute to environmental pollution, mainly in carbon monoxide, VOC, and smoke/soot emissions at the airports.

After a brief introduction to noise, we learned that EPNL has emerged as the dominant scale to measure human objection to noise. It corrects for the tonal content of the noise and the duration of the exposure to noise. Prolonged exposure measure of noise emerged as a concern in airport communities and thus prompted the definition of new levels, such as DNL. Objection to noise has resulted in the prohibition of night flights in major airports around the world. International regulations from ICAO and FAA created certification standards for commercial aviation for approach-landing, takeoff, and flyover noise.

Noise sources in an aircraft engine are related to the engine itself, e.g. fan, intermediate-pressure compressor (IPC), combustor, turbine, and jet as well as airframe noise in landing and takeoff. The landing gear and high-lift devices are two examples of the airframe noise. On the engine side, the noise that emanates from the inlet is dominated by the fan noise. On the aft-side (or downstream) of the engine, jet noise dominates. Rotor-stator interaction caused buzz-saw noise in supersonic tip Mach number fans. A passive solution was in acoustic liners, or Helmholtz (cavity) resonators, that were effective in attenuating the power of the radiated noise from the fan blades through the inlet duct. The acoustic liners were also integrated in the fan exit duct as well as in the duct downstream of low-pressure turbine. Another remedy in rotor-stator noise mitigation was in swept-leaned stator blades as well as wider spacing between the rotors and stators. Reducing the relative tip Mach number in the fan stages also help reduce the radiated noise. The acoustic power in a subsonic jet was shown by Lighthill to be proportional to the eight-power of the jet speed, whereas in a supersonic jet, the dependency drops to the third-power of the jet speed. These suggest that jet noise reduction in best achieved through a reduction in jet speed. At the subsonic level, the jet noise attenuation was achieved through chevrons, which introduced streamwise vortices in the jet shear layer. These, in turn, interacted with and altered the dynamics of the large-scale structures and attenuated radiated sound, by nearly 3 dB. On the supersonic jet noise, the phenomenon of screech was added to the noise sources in an underexpanded supersonic turbulent jet. Supersonic flight creates sonic boom on the ground, which led to the prohibition of such flights over land. *Boom-shaping* technology offered a promising approach to soften the sonic boom, N-shaped wave, and signature on the ground, which turned it into a sinusoidal "thump" signature. The smaller-volume supersonic business jet, with boom-shaping technology, is the obvious first candidate to fly supersonically over land. There is exciting new research at NASA and industry in low-boom technology.

References

Adaptive Compliant Trailing Edge (ACTE) 2017. Flight Experiment. March 10, 2014, Updated August 7, at NASA website: https://www.nasa.gov/centers/armstrong/research/ACTE/index.html.

Aerion Supersonic website, n.d., https://www.aerionsupersonic.com.

Ahuja, K.K. and Bushell, K.W. (1973). An experimental study of subsonic jet noise and comparison with theory. *Journal of Sound and Vibration* 30 (3): 317–341.

Anderson, D. N.1974. Effect of Equivalence Ratio and Dwell Time on Exhaust Emissions from an Experimental Premixing Pre-vaporizing Burner. *NASA TM-X-71592.*

Berton, J.J. 2000. Noise Reduction Potential of Large, Over-the-Wing Mounted, Advanced Turbofan Engine. *NASA TM-2000-210025.*

Biofuel Atlas from Department of Energy, National Renewable Energy Laboratory, n.d., https://maps.nrel.gov/biofuel-atlas.

Blazowski, W.S. (1985). Fundamentals of combustion. In: Aerothermodynamics of Aircraft Engine Components, AIAA Education Series (ed. G.C. Oates). Washington, DC: AIAA Inc.

Blazowski, W.S. and Henderson, R.E. 1974. Aircraft Exhaust Pollution and Its Effects on U.S. Air Force. Air Force Aero Propulsion Lab, Wright-Patterson AFB, Ohio. *Report AFAL-TR-74-64.*

Boom Technology website, n.d.: https://boomsupersonic.com.

Bradley, M.K. and Droney, C.K.2011. Subsonic Ultra Green Aircraft Research: Phase I Final Report. *NASA/CR-2011-216847.*

Brausch, J. F., Bangalore, A.J., Barter, J.W. and Hoff, G.E. 2002. Chevron exhaust nozzle for gas turbine engine. US patent 6360528 B1.

Bridges, J.E., Wernet, M., and Brown, C.A.2003. Control of jet noise through mixing enhancement. Presented at Noise-Con, Cleveland, Ohio, *NASA TM*2003- 212335, Washington, D.C. (June 23–25, 2003).

Brown, J.L. and Roshko, A. (1974). On density effects and large structures in turbulent mixing layers. *Journal of Fluid Mechanics* 64, Part 4,: 775–816.

Chang, C.T. (2012). Low-Emissions Combustors Development and Testing. NASA-Glenn Research Center.

Clark, L.R., Thomas, R.H., Dougherty, R.P., Farrassat, F., and Gerhold, C.H.1997. Inlet Shape Effects on the Far-Field Sound of a Model Fan. AIAA Paper Number 97–1589.

Collier, F. 2012. NASA Aeronautics: Environmentally Responsible Aviation Project – Real Solutions for Environmental Challenges Facing Aviation. AIAA Aerospace Sciences Meeting.

Cumpsty, N.A. (1977). Critical review of turbomachinery noise. *Journal of Fluids Engineering, Transactions of ASME* 99, Ser. I, No. 2,: 278–293.

Cumpsty, N.A. (1985). Engine Noise. In: Aerothermodynamics of Aircraft Engine Components, AIAA Education Series (ed. G.C. Oates). New York.

Cumpsty, N.A. (2003). Jet Propulsion: A Simple Guide to the Aerodynamic and Thermodynamic Design and Performance of Jet Engines, 2nde. Cambridge, UK: Cambridge University Press.

Environmental Protection Agency. 1973. *Control of air pollution from aircraft and aircraft engines.* Federal Register, Vol. 38, No. 136.

Environmental Protection Agency. 1978. *Control of air pollution from aircraft and aircraft engines.* Federal Register, Vol. 43, No. 58.

Environmental Protection Agency. 1982. *Control of air pollution from aircraft and aircraft engines.* Federal Register, Vol. 47, No. 251.

Environmental Protection Agency. 2013. National Ambient Air Quality Standards (NAAQS). http://www.epa.gov/air/criteria.html.

Environmental Protection Agency. EPA 2016 Report, 1990-2014. Inventory of US Greenhouse Gas Emissions and Sinks.

European Commission. 2011. Flightpath 2050, Europe's Vision for Aviation, Report of the Hugh Level Group on Aviation Research, Luxemburg; see https://ec.europa.eu/transport/sites/transport/files/modes/air/doc/flightpath2050.pdf.

European Environment Agency (2016). Air Pollutant Emission Inventory Guidebook. EEA.

European Environment Agency report. (n.d.). http://reports.eea.eu.int/EMEPCORINAIR3/en/B851vs2.4.pdf.

Expert Group on Future Transport Fuels. 2015. State of the Art in Alternative Fuels Transport Systems. Final Report, the European Commission, EU, July.

Farokhi, S. (2014). Aircraft Propulsion, 2e. Chichester (UK): Wiley Ltd.

Federal Aviation Administration. 1979. *Airport Noise Compatibility Planning.* 14 CFR Part 150. Washington, D.C.,.

Federal Aviation Administration. 2005. Aviation & Emissions: A Primer, Washington, D.C.

Feiler, C.E., Groeneweg, J.F., Rice, E.J. et al. (1972). Fan Noise Suppression. In: Aircraft Engine Noise Reduction, vol. III. Washington, D.C: NASA SP-311.

Fetterer, F., Knowles, K., Meier, W., and Savoie, M. (2016). Sea Ice Index, Version 2. Boulder, Colorado: National Snow and Ice Data Center.

Fisher, L., Liu, S., Maurice, L.Q., and Shepherd, K.P. (2004). Supersonic aircraft: balancing fast, affordable, and green. *International Journal of Aeroacoustics* 3 (3): 181–197.

FlexSys, Inc.n.d. https://www.flxsys.com/flexfoil.

Flightpath 2050 n.d. Goals. http://www.acare4europe.org/sria/flightpath-2050-goals.

Fournier, G.F. (2004). Supersonic-transport takeoff silencing. *International Journal of Aeroacoustics* 3 (3): 249–258.

Frate, F.C. 2011. An aerodynamic and acoustic assessment of convergent-divergent nozzles with chevrons. AIAA Paper Number 2011-976, presented at the Aerospace Sciences Meeting in Orlando, Florida (January 2011).

Frate, F.C. (2011). Supersonic Nozzles for Low-Noise/High-Thrust at Takeoff. NASA Publication, https://ntrs.nasa.gov/archive/nasa/casi.ntrs.nasa.gov/20150010334.pdf.

Goldstein, M.E. (1974). Aeroacoustics. Washington, D.C.: NASA SP-346.

GREET Tool. n.d. Argonne National Laboratory. http://www.transportation.anl.gov/modeling_simulation/GREET.

Henderson, B. (2009). Fifty years of fluidic injection for jet noise reduction. *International Journal of Aeroacoustics* 9: 91–122.

Henderson, R.E. and Blazowski, W.S. (1989). Turbopropulsion combustion technology. In: Aircraft Propulsion Systems Technology and Design, AIAA Education Series (ed. G.C. Oates). Washington, DC: AIAA Inc.

Huff, D.L.2013. NASA Glenn's Contributions to Aircraft Engine Noise Reduction. *NASA/TP-2013-217818*.

ICAO Worldwide Communications Conference, Montreal, Canada, April 1981. https://www.icao.int/publications/pages/publication.aspx?docnum=9082.

Intergovernmental Panel on Climate Change. 1999. IPCC Special Report on Aviation and the Global Atmosphere.

International Civil Aviation Organization (ICAO) Committee on Aviation Environmental Protection (CAEP). 2003. *Operational Opportunities to Minimize Fuel Use and Reduce Emissions*, Circular 303-AN/176.

IPCC (2007). *Climate change 2007*, The physical science basis. In: Contribution of Working Group I to the Fourth Assessment Report of the Intergovernmental Panel on Climate Change (eds. S. Solomon, D. Qin, M. Manning, et al.). Cambridge, UK: Cambridge University Press.

Jardine, C.N. (2008). Calculating the Environmental Impact of Aviation Emission, 2e. Oxford, UK: Environmental Change Institute, Oxford University.

Kailasanath, K., Gutmark, E., and Martens, S. 2011. Mechanical Chevrons and Fluidics for Advanced Military Aircraft Noise Reduction. *Final Report on SERDP Project WP-1584.*

Kerrebrock, J.L. (1992). Aircraft Engines and Gas Turbines, 2e. Cambridge, Mass: MIT Press.

Khavaran, A. and Dahl, M.D. 2012. Acoustic Investigation of Jet Mixing Noise in Dual Stream Nozzles. *NASA/TM-2012-217226.*

Khavaran, A., Kenzakowski, D.C., and Mielke-Fagan, A.F. (2010). Hot Jets and Sources of Jet Noise. *International Journal of Aeroacoustics* 9 (4, 5): 491–532.

Khorrami, M.R.2003. Understanding slat noise sources. Paper presented at EUROMECH 449, Chamonix, France (December 9-12, 2003).

Khorrami, M.H and Choudhari, M. 2003. Application of Passive Porous Treatment to Slat Trailing Edge Noise. *NASA/TM-2003-212416*, Washington, DC,.

Klujber, F. (1972). Sonic Inlet Development for Turbofan Engines. In: Aircraft Engine Noise Reduction. Proceedings of the conference, 305. Cleveland, OH (May 16-17, 1972): NASA SP-311.

Lee, D.S., Fahey, D.W., Forster, P.M. et al. (2009). Aviation and global climate change in the 21st century. *Atmospheric Environment* 43: 3520–3537.

Lee, D.G., Pitari, G., Grewe, V. et al. (2010). Transport impacts on atmosphere and climate: aviation. *Atmospheric Environment* 44 (37): 4678–4734.

Lefebvre, A.H. (1983). Gas Turbine Combustion. New York: Hemisphere Publishing.

Leylekian, L., Lebrun, M., and Lempereur, P. (2014). An overview of aircraft noise reduction technologies. *Aerospace Lab Journal* (7).

Lighthill, M.J. (1952). On sound generated aerodynamically, I: general theory. *Proceedings of the Royal Society, Series A* 211: 564–587.

Lighthill, M.J. (1954). On sound generated aerodynamically. II, turbulence as a source of sound. *Proceedings of the Royal Society of London, A* 222: 1–32.

Lipfert, F.W.1972. Correlations of Gas Turbine Emissions Data. *ASME Paper 72-GT-60.*

MacKay, D.J.C. (2009). Sustainable Energy – Without the Hot Air. Cambridge, (UK): UIT Cambridge Ltd.

Merkur, R.A.1996. Propulsion System Considerations for Future Supersonic Transports – A Global Perspective. *ASME Paper 96-GT-245.*

Minnis, P., Ayers, J.K., Palikonda, R., and Phan, D. (2004). Contrails, cirrus trends, and climate. *Journal of Climate* 17: 1671–1685.

MIT. 2004. Report to the US Congress: Aviation and the Environment. Cambridge, Massachusetts, December.

Morfey, C.L. (1972). The acoustics of axial flow machines. *Journal of Sound and Vibration* 22 (4): 445–466.

Morfey, C.L. (1973). Rotating blades and aerodynamic sound. *Journal of Sound and Vibration* 28 (3): 587–617.

Morris, P.J. 2011. The Reduction of the Advanced Military Aircraft Noise. *Final Report on SERDP Project WP-1583.*

NASA. 1987. *Aeropropulsion '87. NASA CP-3049.*

NASA. 1999. Making Future Commercial Aircraft Quieter. Glenn Research Center. *FS-1999-07-003-GRC.* Cleveland, Ohio.

National Research Council (2002). For Greener Skies: Reducing Environmental Impacts of Aviation. Washington, D.C.: National Academy Press.

Nixon, C.W., Hille, H.K., Sommer, H.C., and Guild, E. 1968. Sonic Booms Resulting From Extremely Low-Altitude Supersonic Flight: Measurements and Observations on Houses, Livestock and People. *Aerospace Medical Research Laboratory Report AMRL-TR-68-52,* WPAFB, Ohio.

NOAA, *Trends in Atmospheric Carbon Dioxide – Monthly Average Mauna Loa CO_2*; see https://www.esrl.noaa.gov/gmd/ccgg/trends/

Norris, G. (2018). Boeing-NASA Low Drag, Low-Noise Engine Feature Beats Expectations. *Aviation Week & Space Technology* (August 28).

Norris, G. and Warwick, G. (2018). Noise and Emissions Central Issues for Resurgent Supersonics. *Aviation Week and Space Technology* (July): 31–32.

Pearsons, K.S. and Bennett, R.L. 1974. Handbook of Noise Ratings, *NASA CR-2376.*

Penner, J.E. (1999). Aviation and Global Atmosphere. Cambridge, UK: Cambridge University Press.

Plotkin, K.J. 1989. Review of Sonic Boom Theory. AIAA Paper Number 89–1105. Paper presented at AIAA Aeroacoustics Conference, San Antonio, Texas (April 1989).

Powell, A. 1959. On the Generation of Noise by Turbulent Jets. ASME Paper Number 59-AV-53. Paper presented at the Aviation Conference, Los Angeles, CA (March 1959).

Ragozin, N.A. (1962). Jet Propulsion Fuels. Oxford: Pergamon Press.

Royal Commission on Environmental Pollution 2002. The Environmental Effects of Civil Aircraft in Flight, London, UK. www.rcep.org.uk/avreport.htm.

Rogelj, J., Schaeffer, M., Meinshausen, M. et al. (2015). Zero emission targets as long-term global goals for climate protection. *Environmental Research Letter* 10.

Saiyed, N., Mikkelsen, K.L., and Bridges, J. 2000. Acoustics and Thrust of Separate-Flow Exhaust Nozzles with Mixing Devices for High-Bypass-Ratio Engines, *NASA TM 2000-209948*.

Samimy, M., Kim, J.-H., and Kearney-Fischer, M. 2009. Active Control of Noise in Supersonic Jets Using Plasma Actuators. Paper Number GT2009-59456 Presented at the ASME Turbo Expo, Orlando, Florida (June 2009).

Sausen, R., Isaksen, I., Hauglustaine, D. et al. (2005). Aviation radiative forcing in 2000: An update on IPCC (1999). *Meteorologische Zeitschrift* 14 (4): 555–561.

Seiner, J.M. and Yu, J.C. (1984). Acoustic near-field properties associated with broadband shock noise. *AIAA Journal* 22: 1207–1215.

Seiner, J.M., Ponton, M.K., Jansen, B.J., and Lagen, N.T. 1992. The Effect of Temperature on Supersonic Jet Noise Emission. *AIAA Paper Number 92-02-046*.

Seiner, J.M., Jansen, B.J., and Ukeiley, L.S. 2003. Acoustic Fly-Over Studies of F/A-18E/F Aircraft during FCLP Mission. *AIAA Paper Number, 2003-3330*.

Smith, M.J.T. (1986). Aircraft Noise. Cambridge, UK: Cambridge University Press.

Society of Automotive Engineers. 1970. (SAE) document - ARP 1179.

Spike Aerospace website. n.d. http://www.spikeaerospace.com.

Strack, W.C. 1987. Propulsion Challenges and Opportunities for High-Speed Transport Aircraft.

Strahan, S.E. and Douglass, A.R. (2018). Decline in Antarctic ozone depletion and lower stratospheric chlorine determined from aura microwave limb sounder observations. *Geophysical Research Letters* 45 (1): 382–390.

Tam, C.K.W. (1995). Supersonic jet noise. *Annual Review of Fluid Mechanics* 27: 17–43.

Tam, C.K.W., Parrish, S.A., Envia, E., and Chien, E.W. (2013). Physical processes influencing acoustic radiation from jet engine inlets. *Journal of Fluid Mechanics* 725: 152–194.

Treager, I.E. (1979). Aircraft Gas Turbine Engine Technology, 2e. New York: McGraw-Hill.

Tschudi, M., Fowler, C., Maslanik, J. et al. (2016). EASE-Grid Sea Ice Age, Version 3. NASA (ed.). Boulder, Colorado: National Snow and Ice Data Center Distributed Active Archive Center.

US Environmental Protection Agency (1974). Information on Levels of Environmental Noise Requisite to Protect Public Health and Welfare with an Adequate Margin of Safety. Washington, D.C.: US EPA.

US General Accounting Office (GAO) Report. 2003. Aviation and the Environment, Strategic Framework Needed to Address Challenges Posed by Aircraft Emissions. *GAO-03-252*, Washington, D.C.

US Federal Aviation Administration. 2006. Voluntary Airport Low Emissions Program (VALE): Eligible Airports, Washington, DC.

Von Glahn, U.H., Gray, V.H., Krejsa, E.A. et al. (1972). Jet Noise. In: Aircraft Engine Noise Reduction, vol. IV. Washington, D.C: NASA SP-311.

Wang, M. 2011. *Greenhouse Gasses, Regulated Emissions, and Energy Use in Transportation (GREET) Model*, Version GREET1_2011.Center for Transportation Research, Argonne National Laboratory, Argonne, Illinois.

Whitefield, P.D., Lobo, P., Hagen, D.E., et al. 2008. Summarizing and Interpreting Aircraft Gaseous and Particulate Emissions Data. *Airport Cooperative Research Program Report Number 9*, Transportation Research Board.

Whurr, J. (2004). Propulsion system concepts and technology requirements for quiet supersonic transport. *International Journal of Aeroacoustics* 3 (3): 259–270.

Woodward, R.P., Elliot, D.M., and Berton, J.J.1998. Benefits of Swept and Leaned Stators for Fan Noise Reduction. *NASA TM-1998-208661*.

Zaman, K.B., Bridges, J.E., and Huff, D.L. (2011). Evolution from 'tabs' to 'chevron technology'-a review. *International Journal of Aeroacoustics* 10 (5 & 6): 685–710.

Zumdahl, S.S. and Zumdahl, S.A. (2000). Chemistry, 5e. Boston: Houghton Mifflin Co.

4

Future Fuels and Energy Sources in Sustainable Aviation

4.1 Introduction

Sustainable aviation demands renewable jet fuels (RJFs) to make future air travel environmentally friendly. Sustainable aviation also demands reliability in fuel supply and stability in fuel price as fossil fuels are replaced. The history of crude oil price fluctuation alone makes it an unreliable and unsustainable fuel. The price of crude oil per barrel on the world market has seen a nearly tenfold variation in the last 10 years (2008–2017), i.e. crude oil was sold between ~\$15 and ~\$145 in the last decade! This wild swing in crude oil price, in a relatively short time span, is the textbook definition of fuel *price volatility*. In short, fossil fuels are not renewable, lack supply reliability, lack stability in price, and lack the requisite longevity that transportation and power industry needs. On the pollution side, combustion of fossil fuels in power plants, aircraft engines, ships, and road vehicles produce greenhouse gas (GHG) emissions, destroy ozone in the upper atmosphere, and create harmful effects for humans through surface air quality. Moreover, GHG emissions alter radiative forcing in the upper atmosphere and contribute to climate change through rising temperatures on the Earth. In contrast, alternative jet fuels (AJFs) are renewable and can help reduce GHG emissions, as well as soot and particulate matter (PM) emissions from jet engines. In addition, the absence of aromatics hydrocarbons and sulfur in AJF, as compared to conventional fossil fuels, is considered a significant environmental advantage of AJF, which positively impacts surface air quality. Ultimately, it is in the reduced lifecycle emissions assessment of AJFs, as compared to the conventional fossil fuels, that AJF finds strong support in the aviation industry today. For a detailed exposition and assessment of the twenty-first century needs for the aeropropulsion industry; see Epstein (2014).

To define specific goals in the development of sustainable aviation, we may use the *Technology Readiness Level* (TRL) scale that NASA used to define the next three generations of commercial aircraft, $N+1$, $N+2$, and $N+3$. We propose the following ambitious timelines toward achieving sustainable aviation:

1) Develop and implement promising options in the *near- to medium-term*, i.e. up to 2040.
2) Advance long-term promising options in the *transition period*, 2040–2060.
3) Develop/demonstrate technologies for the *long-term solutions*, 2060–2100.

Future Propulsion Systems and Energy Sources in Sustainable Aviation, First Edition. Saeed Farokhi.
© 2020 John Wiley & Sons Ltd. Published 2020 by John Wiley & Sons Ltd.
Companion website: www.wiley.com/go/farokhi/power

The concept of TRL is borrowed from the Department of Defense and NASA, which measures the maturity of technology (from 1–9) needed to take a promising concept from inception/basic research to mission-proven technology in successful mission operations. Briefly, TRLs, in general terms, are defined as:

- TRL 1 is the transition from basic to applied research.
- TRL 2 is for technology concept or application formulated.
- TRL 3 is for proof-of-concept validation.
- TRL 4 is for laboratory validation of component/subsystem.
- TRL 5 is for prototyping in representative environment.
- TRL 6 is for prototype demonstration in a relevant end-to-end environment.
- TRL 7 is the full system prototype demonstration in an operational environment.
- TRL 8 is the actual (full-scale) system completed and "mission-qualified."
- TRL 9 is for the actual system "mission-proven" through successful mission operations.

The TRL definitions are detailed in www.nasa.gov/pdf/458490main_TRL_Definitions. pdf. In the near term, where TRL is 6 for key technologies, we have the option (and opportunity) of replacing carbon-intensive fossil fuels in jet engines, e.g. Jet-A, by renewable fuels. We also have strong incentives to continue developing renewable AJFs that are *drop-in* biofuels that are suitable for current fleet of aircraft, with only minor modifications. These options are feasible in the near-to-medium term and could serve as a bridge to long-term solutions in propulsion and power systems that could eliminate harmful GHG emissions altogether. In addition to sustainable biofuels, we have the option of using liquefied natural gas (LNG), which is approximately 95–99% methane, CH_4. It is clearly a low-carbon replacement of Jet-A, which has an approximate chemical formula of $CH_{1.92}$, i.e. predominantly $C_{12}H_{23}$. Since LNG is a cryogenic fuel that requires special handling and storage, it cannot be used as a *drop-in fuel* in the current fleet without modification. The use of LNG thus introduces new design features and requirements in commercial aircraft fuel systems, at extra cost. However, if there is a collective (i.e. global) will and financial investments/resources, the costs and the technical challenge of new aircraft design that incorporates cryogenic fuel could be met in the near-to-medium term (i.e. 2030–2040).

In the transition period, i.e. 2030–2050, the key technologies have lower TRL by the year 2020, e.g. 4/5. In this category, we find *hybrid-electric propulsion systems* that take on parallel or serial architectures:

- Parallel hybrid-electric propulsion where the fan is driven through the gas turbine core by the low-pressure (or power) turbine and an electric motor shaft power. Figure 4.1a shows the schematic drawing of a parallel hybrid-electric propulsion architecture.
- Serial hybrid-electric propulsion system uses a turboshaft engine that powers an electric generator. The ducted fan is driven by an electric motor that is powered by the generator. Figure 4.1b shows the building blocks of a serial hybrid-electric propulsion system.

Figure 4.2 shows one of NASA's aircraft concepts that utilizes a serial hybrid-electric propulsion architecture. In this design, the aft fan is driven by an electric motor that is powered by the generators that are integrated with the twin turbofan wing-mounted engines. The inlet of the aft, i.e. fuselage, electric fan is located in the fuselage boundary layer. This turboelectric commercial aircraft is slated for 2035 technology maturity time frame.

(a)

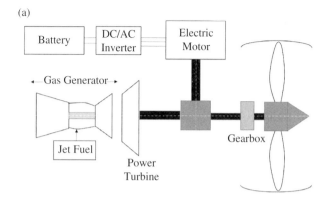

(b)

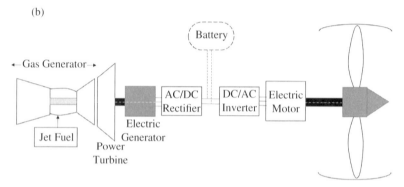

Figure 4.1 Schematic drawing of hybrid electric propulsion systems: (a) parallel hybrid-electric propulsion concept; (b) serial hybrid-electric propulsion concept.

Figure 4.2 Hybrid electric propulsion on a conventional (tube and wing) configuration with aft electric fan ingesting fuselage boundary layer. *Source:* Courtesy of NASA.

The hybrid electric propulsion system may be extended to multiple electric ducted fans in a distributed propulsion (DP) concept. The electric fans in DP are powered by gas-turbine-based turbo-generator and are best suited in an unconventional aircraft configuration, e.g. hybrid wing–body (HWB), or blended wing–body (BWB). Figure 4.3 shows an aircraft configuration (known as N3-X) with distributed turboelectric propulsion system in a HWB configuration (courtesy of NASA). This concept utilizes turboshaft engines with superconducting generators at the wingtips that provide electric power to 14 superconducting motor-driven distributed fans through high-power electric transmission lines. DP (14 electric fans and 2 wingtip turbo-generators) allow for the reduction or elimination of the vertical tail, since it is often sized for *One Engine Inoperative* (OEI) case (see Kim et al. 2013).

The placement of the turbine-driven generators at the wingtips offers several advantages; namely, they relieve the wing (root) bending stress in cruise, they receive undisrupted airflow to the engines, and their jet exhausts disrupt wingtip vortices, thus reduce drag due to lift, or induced drag. Therefore, in the transition period, we develop advanced technologies in electric power generation, high-energy density batteries and superconducting electric motors/generators, energy storage, thermal management system, and high-power superconducting transmission lines that serve as the cornerstones of electrification in commercial aviation propulsion and power systems. Advanced

Figure 4.3 N3-X Turboelectric distributed propulsion (TeDP) system with two gas turbine-driven electric generators at the wingtips and 14 boundary-layer ingestion, BLI, central distributed electric propulsor design. *Source:* Courtesy of NASA.

hybrid propulsion technology offers an opportunity for efficient electric taxiing on runways, known as *eTaxi*.

Use of pure alcohol, e.g. ethanol, as jet fuel is impractical because of its low energy density as compared to conventional jet fuels (i.e. lower heating value [LHV] of ethanol is only 62% of Jet-A). Therefore, pure alcohol is not a *drop-in* jet fuel. However, ethanol may be used as a blending ingredient in synthetic jet fuel (see Spear 2018). In the long-term, i.e. beyond 2050, the only carbon-free, zero-GHG emission solution in terms of environmental impact seems to be hydrogen fuel cell–based electric propulsion or an advanced form of low-energy nuclear propulsion. Net-zero emission through human activities is widely accepted as the foundation of sustainable future on the Earth.

Forecasting the technology maturity and commercial adoption in civil aviation is complicated by the myriad of factors that affect its realization. The best-educated guess, by industry experts, stakeholders and the decision makers, however, is valuable since it shows the potential for breakthroughs. One such team is called AHEAD for *Advanced Hybrid Engines for Aircraft Development* with headquarter in Europe (see Rao 2016). The projected time frames, as seen by the AHEAD team, is shown in Figure 4.4 (adapted from Rao 2016). It is noted that the energy sources in controlled fusion reaction is excluded from the forecast in Figure 4.4. This forecast also excludes general aviation (GA) and the regional jets that are expected to enter the market using AJFs. With current intense activity in both aviation sectors, i.e. electric propulsion in the GA sector and the hybrid-electric propulsion in the regional jet/commuter market, we anticipate accelerated development, certification and confidence in these technologies and in both sectors. They are expected to reach TRL 9 and wide use by 2030–2040.

In this chapter, we address challenges and opportunities in sustainable aviation fuels (SAFs) from renewable AJFs (biofuels), to LNG, liquid hydrogen, LH_2, and fuel for compact fusion reactor. The ultimate goal is to identify a sustainable energy mix, mainly in the power and transportation sector, that supports growth in aviation without pollution.

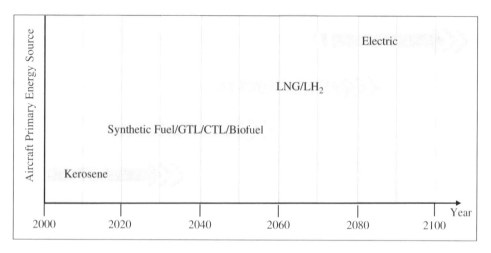

Figure 4.4 Forecast on alternative jet fuels adoption in civil aviation. *Source:* Adapted from Rao 2016.

4.2 Alternative Jet Fuels (AJFs)

The second pillar of *sustainable aviation* is the renewability of energy sources. Renewable energy sources will allow us *to grow* our energy needs and distance ourselves from fossil-fuel dependency. To accomplish this, we need biofuels of diverse origins, i.e. feedstocks, to achieve independence from fossil fuels. The fundamental challenges of biofuel feedstocks are their adequate availability and supply, the change in land-use and increased strain on water resources, the economies of scale to enable competitive costs, financing for commercial production, and the societal acceptance of large-scale processing plants. Indeed the pathways from feedstock to *drop-in* jet fuels go through many complex processes that require additional research and technological breakthroughs to reach commercial viability. Under the best circumstances, AJFs from renewable sources can only provide a small fraction of commercial aviation fuel/energy needs. To quantify the percent market share of AJF in aviation, we use the forecasts in the broader transportation sector, as reported in the European Union for the use of biofuels in transportation. It states that biofuels can reach 5% market share in the transport sector by 2020 and 10% market share by 2030. Aviation energy consumption is about 12% of transport sector; therefore, the market share of AJF in aviation could be around 1%. For a European perspective on future transport fuels, see the Final Report of Expert Group on Future Transport Fuels, "State of the Art in Alternative Fuels Transport Systems," from the European Commission 2015. To assess the full environmental impact of AJF, however we need to conduct lifecycle analysis (LCA) of GHG emissions (see Stratton, Wong, and Hileman 2010).

International Air Transport Association (IATA) in 2015 reports that 1700 commercial flights with passengers have flown using SAF since 2011 (see IATA 2015). Many international airlines, including United, Lufthansa, Air France, KLM, JAL, Air China, Aeroméxico, Finnair, Qatar, Etihad, Virgin Atlantic, and Iberia, among others, have flown commercial passenger flights using 50% blend Jet and hydro-processed esters and fatty acids (HEFA) fuel. The biofuel feedstock and the airlines that have flown on sustainable jet fuel are shown in Figure 4.5 (data from IATA 2015). Figure 4.6 shows the Airbus A320 Advanced Technology Research Aircraft (ATRA) before its flight on blended biofuel (courtesy of German DLR).

AJFs, e.g. 50/50 mixture of JP-8/HEFA produced from Camelina-based plant oil, produced 40–60% less soot than conventional jet fuels at cruising altitude (see Anderson, NASA ACCESS-II Program, 2014). The lower soot level and PM generated by biofuels is expected to impact ice crystal formation in contrails that are formed in upper troposphere at cruise. The net positive radiative force, i.e. greater warming influence, due to contrail cirrus clouds poses a challenge to aviation-induced climate change (see Burkhardt and Kärcher 2011). AJFs contain no aromatics hydrocarbons and sulfur, which reduces soot and PM emissions at cruise. Soot and PM are deemed as the pathway to ice crystal formation in ice-saturated contrails. The same harmful emissions impact surface air quality near airports.

Switching to biofuels does not have a major impact on CO_2 and NO_x production when burned; their benefit is in the absorption of CO_2 during plant growth. They can even create negative GHG emissions in some biomethane pathways. For reference, Table 4.1 shows WTT (well-to-tank), TTWs (tank-to-wheels), and WTW (well-to-wheel) carbon dioxide emissions from four fuel sources used in road vehicles (Joint European Commission 2014). The four fuels are: CNG (compressed natural gas),

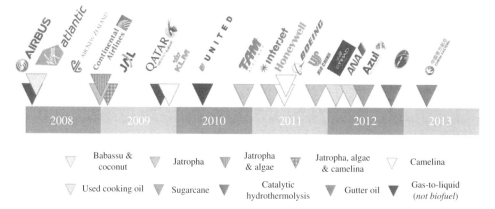

Figure 4.5 Test flights (2008–2013) using sustainable jet fuels. *Source:* From IATA Report 2015.

Figure 4.6 Airbus A320 Advanced Technology Research Aircraft (ATRA) before flight tests on 48% biofuel and 52% Jet A-1 blend, 2015. *Source:* Courtesy of German DLR (see Deutsches Zentrum für Luft-und Raumfahrt, 2015).

Table 4.1 The range of GHG emissions for CNG and biomethane (road vehicles).

Alternative fuel	WTT (g CO_2) km^{-1}	TTW (g CO_2) km^{-1}	WTW (g CO_2 km^{-1})
CNG, EU-mix	30	132	163
Biomethane	−290 to −33	132	−158 to 99
Conventional gasoline	29	156	185
Conventional diesel	25	120	145

biomethane, conventional gasoline, and conventional diesel. Note that WTW is the sum of WTT and TTW. The combustion column, i.e. TTW, for the four fuels shows nearly equal CO_2 emissions (i.e. minimum of $120\,g\,CO_2\,km^{-1}$ for conventional diesel to $156\,g\,CO_2\,km^{-1}$ for conventional gasoline). The biomethane creates $132\,g\,CO_2\,km^{-1}$, which is within 10% of conventional gasoline.

Based on the feedstock, we may define three generations of liquid renewable biofuels (from Del Rosario 2011, Expert Group on Future Transport Fuels, European Commission 2015):

1) *First generation.* From food crops, e.g. ethanol from sugar; biodiesel (FAME) from vegetable oils, or soybeans. The production of the first-generation biofuels is well-established and mature.
2) *Second generation.* From nonfood sources, algae, Jatropha, halophytes. Production technologies of the second-generation biofuels is more complex and costly. Since these are from nonfood sources, they are more sustainable and with a potential to reduce GHG emissions more than the first-generation biofuels.
3) *Third generation.* From bioengineered bacteria, e.g. biofuels from bioengineered algae, hydrogen from biomass, or synthetic methane. Production technology is under intense research and development (R&D) and not near commercialization.

The nonrenewable liquid jet fuels are from natural gas, GTL (gas to liquid) and CTL (coal to liquid) synthetic jet fuel, or oil-derived jet fuel from oil sands and shale oil. Alcohols, e.g. ethanol and butanol, do not offer benefits to aviation, as drop-in jet fuel. The high vapor pressure of alcohols pose problems to high-altitude flight, and the low energy density (Q_R) of alcohols directly affects aircraft range through the Breguet range equation:

$$R = \eta_0 \cdot \frac{Q_R}{g_0} \cdot \frac{L}{D} \cdot \ell n \frac{W_i}{W_f} \tag{4.1}$$

With full tank and maximum takeoff weight, Hileman, Stratton, and Donohoo (2010) show that depending on aircraft type, alcohol causes a reduction in range, which for ethanol is 52–66% and is 33–54% for butanol. Table 4.2 shows a comparison of important properties of two alcohols with kerosene with a nominal formula $CH_{1.953}$ (property data from Levy et al. 2015).

Table 4.2 Properties of methanol and ethanol and comparison to kerosene.

Property	Kerosene $CH_{1.953}$	Methanol CH_3OH	Ethanol C_2H_5OH
Lower heating value, LHV (MJ kg^{-1})	43.1	21	26.952
Heat of evaporation (kJ kg^{-1})	251	1100	841
Auto-ignition temperature (°C)	220	385	363
Density (kg m^{-3})	780–810	790	789
Dynamic viscosity (kg m^{-1}s^{-1})	0.00164	0.00056	0.00104
Surface tension (N m^{-1})	0.023–0.032	0.022	0.02239

Figure 4.7 Supersonic fighter aircraft, *Gripen*, undergoing flight test on 100% biofuel on April 2017. *Source:* SAAB 2017.

Figure 4.7 (from Saab AB) shows the attractiveness of alternative fuels for military jet aircraft. In this figure, Saab Gripen fighter aircraft is shown undergoing flight test program on 100% biofuel in Linköping, Sweden. The tested biofuel is known as CHCJ-5, which is made of rapeseed oil. The flight test results were completely satisfactory, based on Saab's report. No engine change or modifications were necessary for these flight tests. Similar efforts by the US Air Force on biofuels started in 2006. The flight test program used a 50–50 blend of hydroprocessed renewable jet (HRJ) fuel and JP-8. The HRJ fuel was made up of animal fat and cooking oil. The F-15 Eagle was used as the flying testbed and technology demonstrator. According to the pilot's report, no change in thrust or performance degradation was observed. For more details, see Cuttita 2010. These are small, but valuable steps toward reducing the environmental impact of military aviation.

4.2.1 Choice of Feedstock

The first generation of feedstocks, having their origin in food crops, may not be the best choice for AJF, as it competes with food, land use, and water resources. The renewable feedstocks that are deemed acceptable generally utilize waste sources (e.g. algae grown using wastewater, lignocellulosic crop residues, municipal solid waste, wastewater, and renewable fats, oils, and grease), or existing infrastructure (e.g. woody biomass), or promising new energy crops (e.g. oil-based plants such as Jatropha) and perennial grasses. These belong to the second and third generation of feedstocks. Their impact on GHG emissions is highly diverse and, as shown in Table 4.3 (Joint European Commission 2014), they are quantified for surface vehicles.

Table 4.3 The range of WTT (well-to-tank), TTWs (tank-to-wheels), and WTW (well-to-wheel) GHG emissions for different synthetic fuels.

Alternative fuel	WTT (g CO_2) km^{-1}	TTW (g CO_2) km^{-1}	WTW (g CO_2 km^{-1})
HVO (hydro-treated vegetable oil)	−111 to −22	116	5 to 94
GTL (gas-to-liquid)	22 to 38	116	138 to 154
CTL (coal-to-liquid)	65 to 211	116	181 to 328
BTL (biomass-to-liquid), woody biomass (syndiesel)	−104 to −111	116	5 to 12
DME (dimethyl ether) natural gas/coal/wood	38/218/−104	117	154/334/12
Conventional gasoline	29	156	185
Conventional diesel	25	120	145

From the host of alternative (synthetic) fuels listed in Table 4.3, only hydro-treated vegetable oil (HVO) and biomass-to-liquid, BTL, woody biomass (syndiesel), reduce GHG emissions, as compared to the conventional fossil fuels. HVO (that is 100% pure, i.e. unblended) offers 40–90% and BTL (that is 100% pure, i.e. unblended) offers 60–90% reduction in GHG emissions. Note that the assumption of 100% pure HVO and BTL fuels may not meet all the conventional fossil fuel specifications and properties and are listed here for reference. The GTL pathway creates comparable GHG emissions, whereas CTL shows higher carbon dioxide emission than conventional diesel. However, for the CTL fuels, the balance of CO_2 emissions can be improved by CO_2 capture in the plant.

Although BTL is considered to be a promising pathway to alternative biofuels, biomass feedstock is of low-energy density and widely dispersed, which pose an economical/technological challenge for efficient, cost-effective mass production facilities.

4.2.2   Conversion Pathways to Jet Fuel

There are several conversion pathways for AJFs. Among these, three dominant pathways are:

1) Thermochemical, i.e. HRJ fuel
2) Fischer-Tropsch liquids (FTLs) from biomass, natural gas, and coal
3) Sugar-to-jet (STJ) fuel, with sugar extraction from feedstock for direct conversion to hydrocarbon or catalytic conversion to drop-in jet fuel

HRJ fuel is also known as HEFA fuel, i.e. hydroprocessed esters and fatty acids, or bio-SPK (synthetic paraffinic kerosene). Triglycerides and free fatty acids (from plant oil and animal fat) are used as feedstock in a process to produce jet fuel. The near-term application is in 50/50 blend with conventional jet fuel (see Edwards, Shafer, and Klein 2012).

Fischer-Tropsch synthesis is a catalytic chemical process that converts a mixture of carbon monoxide and hydrogen (called synthesis gas or syngas) into liquid hydrocarbon fuels. This process is a key component of GTL biofuel technology, or CTL technology when coal gasification is used as the syngas in F-T liquid synthesis. Depending on the catalyst, temperature, and the process, a range of hydrocarbon fuels are produced, with the desirable carbon chain ranging from 5 to 25.

STJ fuel is based on corn and corn stover as feedstock that use two conversion pathways, namely biological and catalytic conversion (see Han, Tao, and Wang 2017).

4.2.3 AJF Evaluation and Certification/Qualification

TRL is also used to describe the stages in the new fuel evaluation and certification. The description of TRL, as applied to new biofuels, is shown in Table 4.4 (Edwards, Shafer, and Klein 2012).

ASTM International sets the commercial aviation standards for qualification and approval of AJFs. The ASTM D4054 is the test program for the new aviation turbine fuels and fuel additives. On the military side, more stringent fuel evaluation and certification is needed, e.g. for aerial refueling, afterburner use, auxiliary power unit (APU) cold start, high-altitude operation, which is described in MIL-HDBK-510.

In Table 4.4, *Fit for Purpose* properties, TRL 3, include fuel properties such as fuel density as a function of temperature (from −40 to 90 °C), fuel viscosity, and vapor pressure, dielectric constant and fuel electrical conductivity versus temperature, among other fuel properties. *Extended Laboratory Fuel Property Testing* (TRL 4) include material compatibility (soak) test – 28 days, dynamic seal testing, turbine hot section materials compatibility, emissions and endurance testing, among others. In *Component Rig Testing* (TRL 5), fuel pump durability for 500 hours, combustor section performance, combustor nozzle coking evaluation, full annular combustor evaluation, dielectric constant/fuel tank gauging, among others, are conducted. In *Small Engine Testing* (TRL 6) demonstration, 150-hour engine testing on a small engine, e.g. T63, among other engine tests, is conducted. Upon completion of TRL 6, *On-Aircraft Evaluation* (TRL 7) begins. For *Validation/Certification* (TRL 8), aircraft performance with respect to range is evaluated. The final stage is the *Field Service Evaluation* (TRL 9). For details of these tests, see Edwards, Shafer, and Klein 2012, or MIL-HDBK-510.

The following biofuels, pathways and the date of their certification are noted here for reference:

1) F-T Kerosene (up to 50% blend) was certified in 2009 under ASTM D7566 for semi-synthetic jet fuel (fossil-biofuel); see ASTM D7566 2016.
2) F-T SPK (synthetic paraffinic kerosene) is certified in both commercial and military specifications in 2010 (see Kinder and Rahmes 2010).

Table 4.4 Technology readiness level for alternative jet fuel (AJF).

TRL 1	Basic fuel properties observed and reported
TRL 2	Fuel specification properties
TRL 3	Fit for purpose
TRL 4	Extended laboratory fuel property testing
TRL 5	Component rig testing
TRL 6	Small engine demonstration
TRL 7	Pathfinder: APU & on-aircraft evaluation, afterburning engine test
TRL 8	Validation/certification
TRL 9	Field service evaluations

Table 4.5 Biofuels and pathways to certification (as of 2015).

Certified	Fischer-Tropsch (any feedstock)	2009
Certified	HEFA (vegetable oils, animal fat)	2011
Certified	Synthetic-Iso-paraffin (direct sugar) (SIP)	2014
Under review	Alcohol-to-jet	2015
Under review	FT synthetic paraffinic kerosene with aromatics (SKA)	Exp. 2016
Under review	Hydro-processed depolymerized cellulosic jet	Exp. 2016
Testing	Alcohol-to-jet SKA	—
Testing	Catalytic hydro-thermolysis	—
Testing	Synthetic aromatic kerosene by catalytic conversion of sugars	—
Testing	Synthetic (paraffinic) kerosene by catalytic conversion of sugars	—

3) HVO-Kerosene (called HEFA) for up to 50% blend, under alternative fuel specification D7566, was certified in 2011.
4) Sugar-to-Jet (STJ) fuel was approved under ASTM D7566 in 2014.

Table 4.5 shows the various stages of the biofuels certification process (from IATA Report 2015). For a complete list of the biofuels that are either under review or in data compilation phase for their certification by ASTM International, see IATA report 2015.

4.2.4 Impact of Biofuel on Emissions

During flight tests in 2013 and 2014 near NASA's Armstrong Flight Research Center in Edwards, California, data was collected on the effects of alternative fuels on engine performance, emissions, and aircraft-generated contrails at altitudes flown by commercial airliners (35–37 kft). The test series was part of the *Alternative Fuel Effects on Contrails and Cruise Emissions Study*, or ACCESS.

The DC-8's four engines burned either JP-8 jet fuel or a 50/50 blend of JP-8 and renewable AJF of HEFA produced from Camelina plant oil.

Figure 4.8 (from NASA) shows a DC-8 aircraft in flight to assess the impact of drop-in AJF on contrail clouds and cruise emissions using HRJ biofuel 50/50 blend with JP-8 (known as HRJ8-Camelina). The impact of reduced soot in AJF on ice crystal formation in contrail clouds formation is showing statistical deviation between the laboratory and the flight tests. Flying a research aircraft in the wake of a lead aircraft to measure the emissions of its engines' exhaust is both dangerous and highly complex. Statistical spread between different flights, as a function of altitude, speed, atmospheric turbulence, engine setting, and data obtained in the laboratory testing can be significant.

With no land-use change (LUC), HEFA fuel derived from soybean oil feedstock produces 31–67% in lifecycle CO_2 emissions (with a median estimate of 42%) as compared to conventional aviation fossil fuel (see Stratton, Wong, and Hileman 2010). STJ synthesized via biological conversion has the potential to reduce GHG emissions by 59% as compared to conventional Jet-A.

Figure 4.9 (adapted from Penner 1999) shows the relative CO_2 emissions (relative to crude oil conversion to jet fuel) for the manufacture and use of AJFs. We note in Figure 4.9

Figure 4.8 NASA DC-8 contrail and emissions flight test using JP-8 and renewable jet fuel blend (50/50) composed of HRJ fuel from Camelina plant oil and JP-8. *Source:* Courtesy NASA.

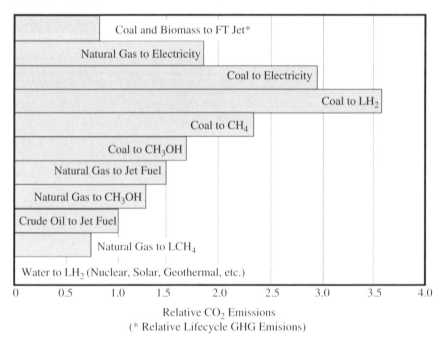

Figure 4.9 Comparison of carbon-dioxide production in the manufacture and use of alternative aviation fuels. *Source:* Adapted from Penner 1999.

that coal produces the highest net CO_2 in conversions to electricity, or conversion to liquid hydrogen or to liquid methane. Two processes produce a net reduction of CO_2 and those are in natural gas conversion to LNG and water conversion to LH_2, using nonfossil source of energy, e.g. nuclear, solar, geothermal. Lifecycle GHG emissions for biofuels as compared to petroleum-based jet fuel is discussed in Section 4.2.6. The lifecycle example of coal and biomass conversion to FT fuel is shown in Figure 4.9 (*).

4.2.5 Advanced Biofuel Production

FAA stated (in 2012) the production goal of 10^9 (or billion) gallons of advanced biofuels by 2018. To put this number in perspective, the United States consumed 13×10^9 gallon of Jet-A in 1985. Therefore, 10^9 gallons of AJF produced in 2018 is equivalent to 7% of the jet fuel consumed in 1985. Moving forward, US demand for aviation fuel is about 20 billion gallons in 2012, which with one billion biofuel goal in 2018, the share of AJF is about 5% of the US consumption. In 2015, US consumption of jet fuel increased to nearly 23 billion gallons, which gives 4.4% share to advanced biofuels in 2018 production goal. The percent share of one billion gallons of RJF represents *only 1.7%* of predicted total fuel consumption in the United States in 2018 (see Winchester et al. 2013). Global aviation fuel consumption in 2006 was 190 Mt (megaton is 10^6 ton, and ton is 1000 kg), which is equivalent to 6.28×10^{10} US gallons of Jet-A. In the global context, 10^9 gallons of AJF amounts to about 1.6% of the global aviation fuel consumption in 2006. At the global level, aviation fuel consumption in 2012 is nearly 377 billion liters, equivalent to 13.1 trillion MJ (i.e. 13.1×10^{12} MJ), in energy terms.

Multiplier	Prefix	Abbreviation
10^{15}	Peta-	P
10^{12}	Tera-	T
10^9	Giga-	G
10^6	Mega-	M
10^3	Kilo-	k
10^2	Hecto-	h
10^1	Deca-	da
10^{-1}	Deci	d
10^{-2}	Centi-	c
10^{-3}	Milli-	m
10^{-6}	Micro-	μ
10^{-9}	Nano-	n
10^{-12}	Pico-	p
10^{-15}	Femto-	f
10^{-18}	Atto-	a

To estimate land-use requirement to produce biofuel(s) from different feedstock, in liters/hectare/year, we refer to Table 4.6 (where *hectare* is $10000\,m^2$), with data from Stratton, Wong, and Hileman (2010). For example, soybeans can produce 400 l of biofuel in one hectare per year. Salicornia produces 1200 l of jet fuel and 1700 l of biodiesel (note that hydro-processed renewable diesel, HRD, with additional hydro-processing can be converted into jet fuel, HRJ, as shown in Figure 4.10). Palm produces 3300 l of jet fuel per hectare per year. Finally, conversion of algae to HEFA fuel produces 17000 l of fuel, per hectare per year.

Table 4.6 Biofuel yield from feedstock.

Synthetic fuel yield	l/ha/yr
Soy HEFA	400
Salicornia HEFA/Diesel	1 200/1 700
Palm HEFA	3 300
Algae HEFA	17 000

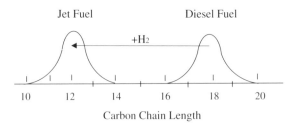

Figure 4.10 Schematic of converting biodiesel (HRD) to jet fuel (HRJ) through additional hydro-processing.

Taking the global aviation fuel need in 2030 as 5.8×10^{12} l yr^{-1}, based on *The Energy Information Administration* (EIA) projections, we may convert the data in Table 4.5 into areas (in hectare) and diameter of a circle of equal area. Table 4.7 shows the surface area needed to produce jet biofuel and the corresponding circular diameter to supply unblended biofuel for aviation in 2030.

The smallest diameter in Table 4.7 is ~2000 km, corresponding to an algae pond! This diameter engulfs several US Midwestern states, since the horizontal width of the United States is about 2680 mi, or 4300 km. Now, what about 50/50 blend of the biofuel and Jet-A?

Table 4.8 is the area requirements based on 50/50 blend of biofuel and fossil fuel (Jet-A). The area reduction for the 50/50 blend biofuel production is one half (½), and the diameter scales as the square root of the area, which is $1/\sqrt{2}$, or about 70%. In these estimates, the biomass crop yield in 2006 was applied to the forecast of aviation fuel demand in 2030. The purpose here is to learn from an *order of magnitude* estimate of the land-use requirements and also recognize that the numbers are not exact.

In addition to the land-use requirement for advanced biofuel production, there are still additional critical questions that need to be asked and answered:

1) Does biofuel feedstock production cause LUC either directly or indirectly? Is biofuel feedstock produced in an active farmland, a new farmland, in a converted peatland rainforest, in a marginal land, or in a degraded land?
2) What are the water (and climate) requirements and *water resources* for biomass crop development? How much is the water requirement, what is the effect of climate on biofuel feedstock yield, and what are the sources of water, e.g. irrigation, rainfall, surface water, underground aquifer, other?
3) What is the impact of feedstock and biofuel composition on *lifecycle GHG emissions*? Included in the GHG emissions, besides carbon dioxide, CO_2, are nitrous

Table 4.7 Land use needed for (global 2030) biofuel production from feedstock.

Biofuel (unblended)	Land use area (ha)	Diameter of a circle (km)
Soy HEFA	1.45E + 10	1.36E + 04
Salicornia HEFA	4.84E + 09	7.85E + 03
Palm HEFA	1.76E + 09	4.73E + 03
Algae HEFA	3.41E + 08	2.08E + 03

Table 4.8 Land use requirement for (50/50 blend) biofuel production (global 2030).

Biofuel (50/50 blend)	Land use area (ha)	Diameter of a circle (km)
Soy HEFA	7.25E + 09	9.61E + 03
Salicornia HEFA	2.42E + 09	5.55E + 03
Palm HEFA	8.79E + 08	3.35E + 03
Algae HEFA	1.71E + 08	1.47E + 03

oxide, N_2O, and methane, CH_4. Which biomass crop (feedstock) and fuel composition lead to negative or positive lifecycle GHG? Transportation of sustainable feedstock to production facilities is an element of lifecycle GHG emissions. In addition, transportation cost impacts economic viability, which poses the ultimate question: *What is the biofuel cost* at the pump?

4) *Economic viability*, i.e. can biofuel feedstock cultivation be scaled and efficiently commercialized, and will the price of AJF be competitive with Jet A?

5) Are policy makers supportive of AJF development? Is there regulatory requirement and/or financial (incentive) support from the governments, e.g. *carbon tax, or subsidies*?

On the question of LUC, we note that trees absorb carbon dioxide; therefore, clearing the trees (say in a forest) for biofuel production purposes almost invariably results in net positive lifecycle GHG emissions (Stratton, Wong, and Hileman 2010). Therefore, the impact of land use change on GHG emissions depends on the type of *converted land* and the type of bio-crop among other parameters such as agricultural practices. The estimation of GHG emissions due to LUC is based on a range of converted land scenarios that are used in the lifecycle model (e.g. GREET). Some examples of the LUC scenarios are (from Stratton, Wong, and Hileman 2010):

Scenario 1. Carbon-depleted soils converted to switchgrass cultivation
Scenario 2. Tropical rainforest conversion to soybean field
Scenario 3. Peatland rainforest conversion to palm plantation field

Marginal lands have little to no competition in land use, which makes them ideal for renewable energy development. Figure 4.11 (from Milbrandt et al. 2014) shows the *availability* of marginal lands in the United States, by county. Light color-coded counties show

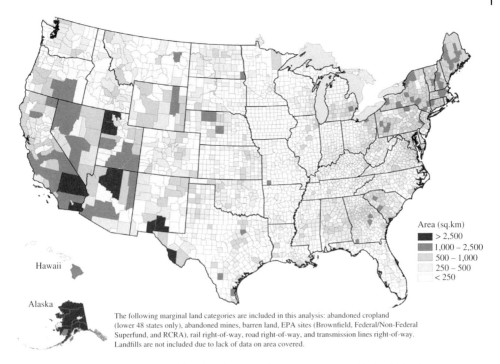

The following marginal land categories are included in this analysis: abandoned cropland (lower 48 states only), abandoned mines, barren land, EPA sites (Brownfield, Federal/Non-Federal Superfund, and RCRA), rail right-of-way, road right-of-way, and transmission lines right-of-way. Landfills are not included due to lack of data on area covered.

Figure 4.11 Availability of marginal lands with renewable energy potential in the United States. *Source:* From Milbrandt et al. 2014.

less than 250 km^2 availability, whereas dark color-coded counties may offer higher than 2500 km^2 of availability. The marginal lands comprise nearly 11% of the US mainland in area (i.e. 865 000 km^2) and have the potential to produce electricity from renewable resources such as wind, solar, biomass, landfill gas (LFG), geothermal, and hydrothermal. In the category of biomass, only herbaceous crops such as miscanthus, switchgrass, and sorghum were considered, and no woody crops were included. Using these energy crops, 1.9 PWh of electricity may be produced (P in PWh stands for Peta-, which is 10^{15}). To estimate the crop yield, Biofuel Ecophysiological Traits and Yields Database (see https:// www.betydb.org) was used.

The seven categories of renewable energy, i.e. wind, geothermal photovoltaic, concentrating solar power (CSP), hydropower, biomass, and LFG sources are mapped in Figure 4.12 that shows the renewable energy harvesting potential in the United States (from Milbrandt et al. 2014).

Naturally, *climate* impacts the optimal choice of feedstock as well. For example, rapeseed (or canola) is produced mainly in China, Canada, India, the United Kingdom, and the European Union. Palm fresh fruit bunches (palm FFBs) come from Southeast Asia (e.g. Malaysia). Jatropha comes from southwestern United States. The map of the continental United States is shown in Figure 4.13 where strong miscanthus yield (tons/acre) is produced from NREL Biofuel Atlas (see: https://maps.nrel.gov/bioenergyatlas). Figure 4.14 shows the location and capacity of biofuel plants in the continental United States (data courtesy of NREL, Map export on 2017-08-23).

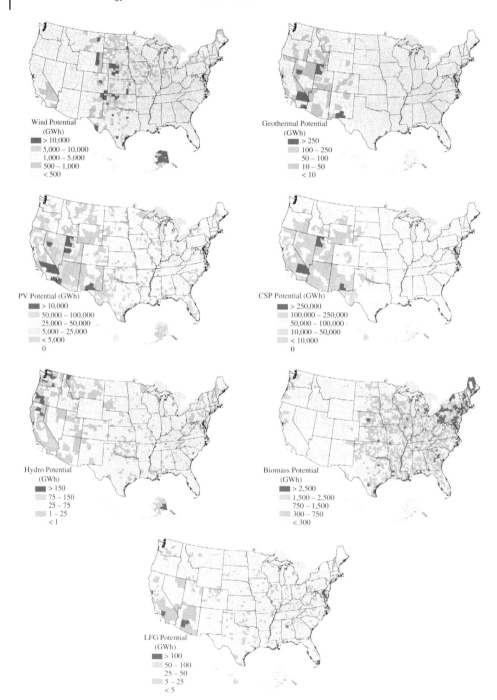

Figure 4.12 Renewable energy potential on marginal lands in the United States. *Source:* From Milbrandt et al. 2014.

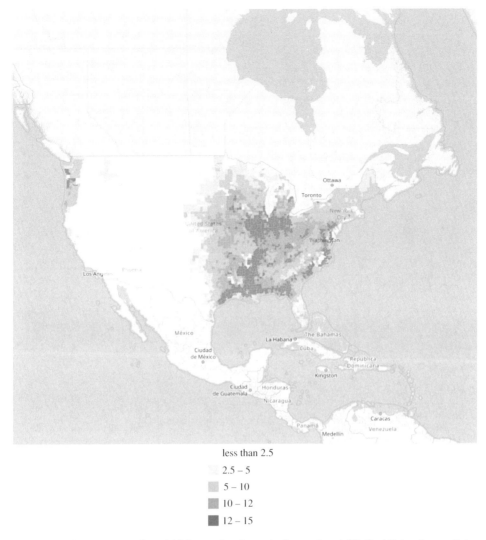

less than 2.5

2.5 – 5

5 – 10

10 – 12

12 – 15

Figure 4.13 Strong miscanthus yield (in tons/acre) map in the continental United States. *Source:* Data courtesy of NREL.

The question of water requirements and resources depends on the bio-crop and the region. For example, Jatropha is well adapted to semi-arid conditions, whereas some feedstocks require arable land for cultivation (e.g. soy or palm). Some feedstock are more resilient and may use marginal land (e.g. herbaceous crops). Freshwater consumption in the United States is dominated by irrigation. Runoff from agricultural activity may cause *eutrophication* and reduce water quality. Nitrogen loading caused by fertilizer in agricultural runoff causes an increase in algae growth near the surface water. Beneath the surface water, algal decomposition takes place that causes a reduction in oxygen concentration in bottom waters. The bacteria release in decaying algae uses oxygen in the water. Low dissolved oxygen in the water adversely impacts marine organisms. This process is called *eutrophication.*

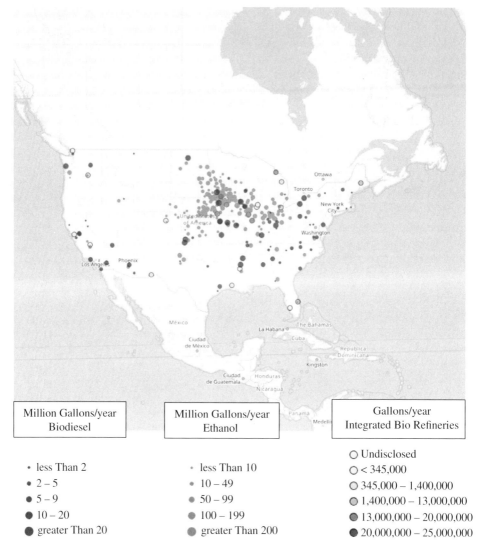

Million Gallons/year Biodiesel	Million Gallons/year Ethanol	Gallons/year Integrated Bio Refineries
• less Than 2	• less Than 10	○ Undisclosed
• 2 – 5	• 10 – 49	○ < 345,000
• 5 – 9	• 50 – 99	○ 345,000 – 1,400,000
• 10 – 20	• 100 – 199	◐ 1,400,000 – 13,000,000
• greater Than 20	• greater Than 200	● 13,000,000 – 20,000,000
		● 20,000,000 – 25,000,000

Figure 4.14 Biofuel plants and their capacity in the continental United States. *Source:* Data from NREL.

Lifecycle CO_2 emissions, by stages, are described in Stratton, Wong, and Hileman (2010):

- Biomass credit (gCO_2/MJ)
- Recovery of feedstock (gCO_2/MJ)
- Transportation of feedstock (gCO_2/MJ)
- Processing of feedstock to fuel (gCO_2/MJ)
- Transportation of jet fuel (gCO_2/MJ)
- Combustion CO_2 (gCO_2/MJ)

Feedstock byproducts and processing are important in AJF environmental impact studies. As a cautionary note, the processing of some co-products may be toxic. For example, Jatropha has co-products that, if used as fertilizer in food crops, raises biosafety concerns and warnings (see Achten et al. 2008).

On commercial-scale economic viability, developing supply chains for AJF production is necessary, but still insufficient, for competitively priced AJF and petroleum-based fuel (e.g. Jet-A). Commercial Aviation Alternative Fuels Initiative (CAAFI), as an aviation and AJF industry coalition, defines useful tools, such as *fuel readiness level* (FRL) and *feedstock readiness level* (FSRL) that tie feedstock and FRL to economic viability (www.caafi.org). FRL 1–5 signifies leading R&D phase, FRL 6–7 is for certification phase and FRL 8–9 for the business and economics phase. FSRL addresses the development and availability of raw material (feedstock) for AJF production. The commercial end-user of AJF is the airlines that require AJF to meet:

- Fuel certification, i.e. compliance with ASTM certifications
- Drop-in criteria, i.e. total compatibility with the storage, transportation, handling infrastructure, engine, aircraft, and other equipment
- Reliability of supply and on-time delivery
- Environmental benefit
- Economic viability, i.e. price competitiveness

For an example of the feedstock and the size of the land to support a biofuel plant, we refer to Lane (2015). Planting nearly 500 000 ac (or nearly 200 000 ha) of canola in Oklahoma–Kansas region produces almost double the feedstock required to support a GTL-HEFA hybrid plant. The circular diameter of nearly 51 km (~32 mi) produces the same plantation area as 500 000 ac.

Finally, regulatory requirements, at the national and international level, such as carbon tax and financial incentives, such as government subsidies, impact commercial production of AJF. In a recent market study of RJF adoption in the United States, Winchester et al. (2013) concluded that a RJF subsidy in the amount of $0.35–$2.69 per gallon is required to meet FAA biofuel goal of one billion gallons of AJF consumption by the US aviation market from 2018. The wide range of subsidy is based on the feedstock, e.g. oilseed rotation crops grown on fallow land require $0.35 per gallon subsidy, whereas using soybean oil as feedstock requires $2.69 per gallon. The biofuel used in the study was produced from HEFA from renewable oils. Consequently, HEFA fuel produced from soybean oil feedstock is not deemed to be cost-competitive with conventional jet fuel for the next decade, or longer (see Winchester et al. 2013). Among promising oilseed rotation crops in the US midwestern states, camelina, rapeseed, and pennycress are cited in Winchester et al. (2013), which utilize the land between rotations that otherwise would be left fallow.

Regulation on carbon tax plays a critical role in market share of AJF in commercial aviation. For example, under a high carbon price model and optimistic AJF development technology, Bauen et al. (2009) conclude that 100% of global aviation fuel can be supplied by AJF by 2040. Under no carbon price and slow development, AJF will have 3% share in 2030, and grows to 37% by 2050.

4.2.6 Lifecycle Assessment of Bio-Based Aviation Fuel

The figure of merit used in environmental studies for selecting an alternative bio-based jet fuel to replace petroleum-based jet fuel is based on the *lifecycle analysis*, or *well-to-wake* assessment, of equivalent CO_2 emissions. The unit of measurement is gram of (equivalent) CO_2 produced per megajoule of fuel energy. Figure 4.15 shows lifecycle GHG emissions of petroleum jet as well as six biofuels. Five of the six biofuels are HRJ fuels from various feedstocks. The sixth biofuel is BTL jet fuel from Stover feedstock. The data used in producing Figure 4.15 comes from Han et al. 2013. To show the percent reduction in lifecycle GHG emissions of biofuels as compared to petroleum jet fuel, Figure 4.16 is produced from the data of Figure 4.15. The reductions range from 41% for HRJ based on Jatropha or rapeseed to 89% for BTL based on Stover feedstock.

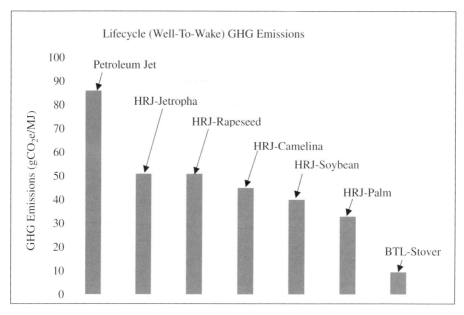

Figure 4.15 Lifecycle GHG emissions for drop-in (bio) jet fuels compared to petroleum-based jet fuel. *Source:* Data from Han et al. 2013.

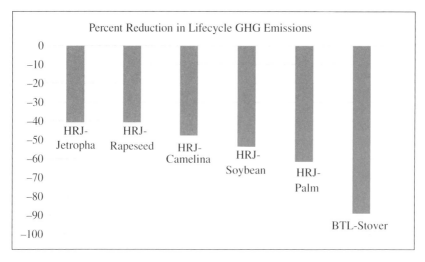

Figure 4.16 Percent reduction of lifecycle GHG emissions of bio-jet fuels compared to the conventional petroleum jet fuel ($C_{12}H_{23}$).

4.2.7 Conversion of Bio-Crops to Electricity

As a source of energy, bio-crops can be used to generate electricity. The next generation of support vehicles at the airports, batteries onboard aircraft, among others, all need electricity. Non-fossil-based electric power generation, therefore, is an integral part of carbon footprint reduction of aviation. Table 4.9 lists sustainable crop conversion to electricity. The data in Table 4.9 is extracted from MacKay (2009), where the power densities of plants in W/m^2 are listed as a range that depends on irrigation and fertilization levels. The average column in Table 4.9 is the mean value. As an example, to support a 30 MW plant, we need 6000 ha (i.e. about 15 000 ac) of energy crop with an average power density (yield) of $0.5\,W\,m^{-2}$.

4.3 Liquefied Natural Gas, LNG

The simplest hydrocarbon compound is methane, CH_4, which is the main ingredient in natural gas. It has the highest hydrogen fraction (by mass) in a hydrocarbon fuel at 25%. (e.g. by comparison, hydrogen content in Jet-A, by mass, is 13.8%). Unfortunately, it is still a fossil fuel, which produces carbon dioxide in combustion with air. The nitrogen in air still produces NOx in combustion with methane. Due to its low carbon content, it is less polluting than the conventional jet fuel, e.g. Jet-A. The mass of carbon dioxide produced per kilogram of methane is 2.75 kg, whereas Jet-A produces 3.16 kg of CO_2 per kilogram of Jet-A (see Table 3.8). The natural state of methane is gaseous and thus it is volumetrically inefficient for use as aircraft fuel. To liquefy the natural gas, we need to condense it, which is achieved at $-160\,°C$ ($-256\,°F$) at atmospheric pressure. The liquefaction of natural gas causes its volume to shrink 600-fold as compared to the gaseous form at atmospheric temperature and pressure. Figure 4.17 shows the relative volume of natural gas in gaseous and liquid state. This demonstrates the need to transport natural gas from the production site to the user as well as storage of natural gas (e.g. in the fuel tank of an aircraft) demands liquefaction of natural gas, i.e. LNG.

Table 4.9 Land use requirement for biocrop conversion to electricity.

Sustainable crop yield (to generate electricity)	$W\,m^{-2}$	Avg. ($W\,m^{-2}$)
Willow, miscanthus, poplar	0.15–0.3	0.22
Oilseed rape	0.22–0.42	0.32
Sugar beet		0.40
Switchgrass		0.20
Maize		0.10
Jatropha	0.07–0.19	0.13
Tropical plantations (eucalyptus)	0.1–0.58	0.34
Energy crops	0.2–0.8	0.50
Wood (forestry)	0.1–0.25	0.18

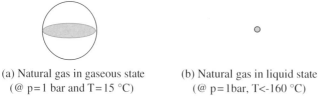

(a) Natural gas in gaseous state
   (@ p=1 bar and T=15 °C)

(b) Natural gas in liquid state
   (@ p=1bar, T<-160 °C)

Figure 4.17 Relative volume of natural gas at atmospheric temperate and pressure and in liquefied (cryogenic) state: (a) natural gas in gaseous state (@ p = 1 bar and T = 15 °C); (b) natural gas in liquid state (@ p = 1 bar, T < −160 °C).

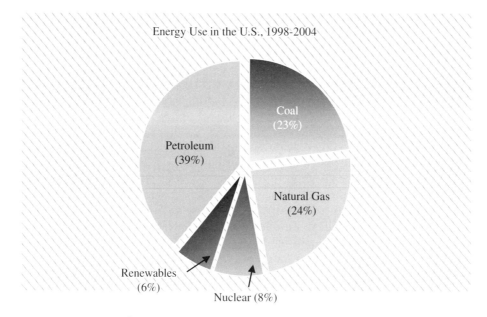

Figure 4.18 Statistics of the US energy use, 1998–2004. *Source:* Data from Energy Information Administration 2005.

The heating value of LNG is higher than typical jet fuel, i.e. nearly 50.2 MJ kg^{-1} LHV for LNG as compared to Jet-A with LHV of 43.4 MJ kg^{-1}. With boiling point temperature of −160 °C (at atmospheric pressure), LNG is a cryogenic fuel that requires special handling in transportation, storage (i.e. double-walled fuel tank), regasification, and use. Adding the weight of the cryotanks to LNG, its energy density (per unit mass basis) is lower than kerosene (Jet A). The volumetric energy density of LNG is about one half of Jet A, i.e. 19 MJl^{-1} for LNG vs 36 MJl^{-1} for Jet A. This, in essence, doubles the fuel-storage volume requirement for the LNG as compared to Jet A. Based on the volumetric needs as well as the special fuel handling requirements for cryogenics, the use of LNG will certainly impact aircraft design, and thus the fleet mix in the future.

Natural gas, in various forms, are produced and widely used in power generation industry (utilities), home heating/cooling, cooking, and transportation industry. It accounts for nearly a quarter of all US energy use (see Figure 4.18 from US Department of Energy, Energy Information Administration (EIA), *Energy Outlook 2005*).

Large reserves of petroleum oil and gas fields around the globe are the main suppliers of natural gas, and LNG. Global LNG business is known as the supply, or value chain. The LNG value chain is divided into four categories:

1) Exploration and production
2) Liquefaction capacity
3) Shipping
4) Storage and regasification

These are used for cost calculations and price projections in economic forecasts. For the US energy market and outlook to 2030, see US Energy Information Administration *Annual Energy Outlook 2007 with Projections to 2030* (2007). The full impact of LNG as aviation fuel on climate, i.e. LCA, has to account for all four categories in the value chain. A comprehensive approach is detailed in the *well-to-wake* analysis (WTWa) module for aviation fuels in GREET1 (Wang 2011).

4.3.1 Composition of Natural Gas and LNG

Natural gas composition extracted from an oil and gas field reservoir is as diverse as the geographic location of the oil and gas fields around the globe. Hence, we may only attribute *typical* compositions, based on their statistical significance, to the natural gas and LNG. Figure 4.19 shows the *typical* composition of natural gas and Figure 4.20 is the typical LNG composition.

The liquefaction plants are required to remove contaminants from the natural gas that could form solid particles when cooled to LNG temperatures. These are mainly water and carbon dioxide. The liquefaction processing plants can be designed to produce nearly 100% methane, as aviation fuel.

Based on the global abundance, i.e. proven reserves of natural gas, competitive price, and the capacity to build additional LNG liquefaction plants to meet aviation needs, LNG can safely enter the mix of energy sources under consideration for sustainable aircraft fuel in the near- to medium term. It certainly is an available and cleaner alternative to the conventional fossil fuels used in commercial aviation today, but with the drawbacks of being cryogenic and carrying less energy per unit volume, as compared to conventional jet fuel.

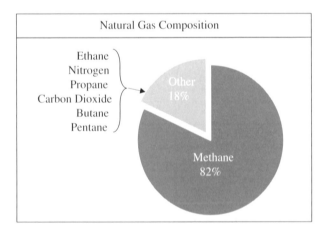

Figure 4.19 Typical composition of natural gas.

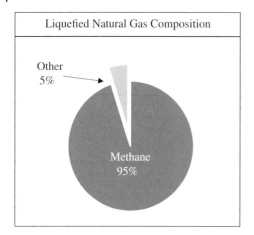

Figure 4.20 Typical composition of LNG.

Table 4.10 Properties of LNG and methane.

LNG boiling point	$-160\,°C$ ($-256\,°F$)
LNG density	Between 430–$470\,kg\,m^{-3}$ (typical value $456\,kg\,m^{-3}$)
LNG specific gravity	Typical value 0.456
LNG lower heating value	$50.2\,MJ\,kg^{-1}$
LNG flammability range (as liquid)	Nonflammable
LNG/methane flammability range in air	5–15% concentration in air
Methane auto-ignition temp	$540\,°C$ ($1004\,°F$)
LNG flashpoint[a]	$-188\,°C$ ($-306\,°F$)
LNG physical properties	Odorless, colorless, noncorrosive, nontoxic, noncarcinogenic

[a] Definition of flashpoint is the minimum temperature at which a liquid produces vapor in a test vessel in concentrations that is flammable (i.e. ignitable mixture with air), based on OSHA 2008.

The volumetric efficiency (i.e. energy/volume) of LNG is about 62% of Jet-A fuel, therefore it requires about 1.5 times the fuel tankage volume to deliver the same fuel energy as the conventional jet fuel. Table 4.10 is a summary of the physical and chemical properties of LNG and methane (data from The International Group of Liquefied Natural Gas Importers, LNG Information Paper No. 1, www.giignl.org).

LNG/methane exhibit a narrow flammability range in air, as shown in Figure 4.21 (Foss 2003).

4.4 Hydrogen

Hydrogen is the most abundant element in the universe, followed by helium. The most abundant element on Earth is oxygen. Hydrogen is an energy carrier that is known as *the clean fuel*; thus, it is deemed to be the fuel of choice for energy and transportation

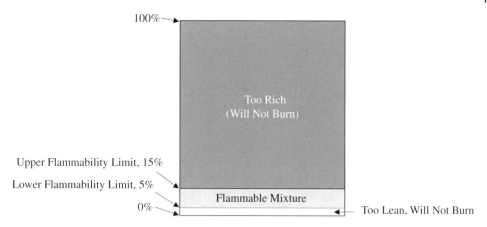

Figure 4.21 Flammability range for methane/LNG (mixture with air). *Source:* Data from Foss 2003.

industry. The main advantage of hydrogen is that it is not a fossil fuel; thus, it contains no carbon! Also, its reaction with oxygen produces water vapor, heat, and/or electricity, which are desirable outputs. Furthermore, its long use in the space program as rocket fuel has produced valuable experience base in aerospace industry. In addition, as a fuel, it contains the highest-energy density, *per unit mass*, of all fossil fuels. In fact, its LHV is nearly three times that of Jet-A, since hydrogen has a LHV of $\sim$120 MJ kg^{-1} whereas Jet-A carries $\sim$43 MJ kg^{-1} specific energy. The challenges with hydrogen are significant, too. For example:

1) Hydrogen is not found in abundance in pure (or elemental) form, rather, it is in compounds such as water and methane.
2) Hydrogen production requires energy expenditure, like electricity, and the process may involve net positive GHG emissions in lifecycle analysis; this is in addition to being economically uncompetitive as compared to Jet-A at the current technology level.
3) Hydrogen has low density in gaseous form, thus for volumetric efficiency, it needs to be in liquid state (LH$_2$), which then pushes it into cryogenic state, at temperature of 20 K and 1 atm. pressure that requires special insulation and thermal management systems. Due to extreme low cryogenic temperature state ($\sim$−253 °C), liquid hydrogen poses a more severe challenge than LNG (at $\sim$−160 °C) in insulation and thermal management.
4) Liquid hydrogen is still a low-density liquid (i.e. $\sim$71 kg m^{-3} at boiling point) as compared to conventional jet fuel ($\sim$800 kg m^{-3}), which carries about a quarter of energy *per unit volume* as compared to aviation fuel (i.e. 34 560 MJ m^{-3}, for Jet-A versus 8520 MJ m^{-3} for LH$_2$, with a ratio of 4.06). This means that hydrogen-fueled aircraft need about four times the fuel tankage and thus requires a totally new design (Guynn and Olson 2002).
5) Hydrogen is not a *drop-in* fuel to be used in the current fleet of aircraft. LH$_2$ requires special storage, handling, and delivery system/regasification; thus, a new fleet of hydrogen-fueled aircraft need to be designed.
6) Combustion of hydrogen in air will create NOx that impacts ozone and surface air quality and the associated health hazard.

7) Hydrogen is highly flammable, and thus strict safety issues, codes, and standards are required.
8) Lack of supporting infrastructure at the airports for liquid hydrogen limits its broad utility in broad markets.

The elements of hydrogen energy system are (see Department of Energy *National Vision* report 2002):

1) *Production*: Hydrogen production from fossil fuel, biomass, or water
2) *Delivery*: Distribution of hydrogen from production and storage sites
3) *Storage*: The confinement of hydrogen for delivery, conversion, and use
4) *Conversion*: The making of electricity and/or thermal energy
5) *End-use energy applications*: Use of hydrogen for portable power, mobile phones/ laptop computers, or fuel-cell vehicles, among others

4.4.1 Hydrogen Production

Hydrogen is currently produced and used in large quantities in oil-refining processes (i.e. hydro-treating crude oil) to improve hydrogen-to-carbon ratio in the fuel, in food production industry (e.g. hydrogenation), in fertilizer industry (for ammonia production), and in other industrial uses.

According to Lipman (2011), 10–11 million metric tons of hydrogen are produced in the United States, each year.

Hydrogen production may use fossil fuel, biomass, or through dissociation of water by electrolysis. On the fossil-fuel side, the prominent hydrogen production method uses *steam methane reforming* (SMR) reaction, which is considered a mature technology. In this process, the natural gas, i.e. up to 95% methane, is used to produce hydrogen through steam reformation and partial oxidation.

The SMR reaction is an *endothermic* reaction modeled by:

$$H_2O + CH_4 + Heat \rightarrow CO + 3H_2 \tag{4.2}$$

Steam (700–1000 °C) at 3–25 bar pressure reacts with natural gas through a catalyst to produce carbon monoxide and hydrogen. Steam reforming can be used with other fuels, such as ethanol, to produce hydrogen.

Water-gas shift reaction is:

$$CO + H_2O \rightarrow CO_2 + H_2 + Small\ amount\ of\ heat \tag{4.3}$$

Carbon monoxide and steam react in the presence of a catalyst to produce carbon dioxide and hydrogen.

Other (fossil) fuels may be used in lieu of methane.

Renewable hydrogen from nonfood biomass may be produced from:

- Municipal solid waste
- Energy crops
- Short rotation woody crops
- Forestry
- Mill and wood waste
- Agricultural residues, e.g. corn stover

Heating biomass feedstock in a gasification process yields low-to medium calorific value syngas, which upon cleaning and processing converts into hydrogen.

On rare occasions, nature produces hydrogen in elemental form. "In one of nature's quirks, certain algae and cyanobacteria photoproduce hydrogen for short times as a way to get rid of excess energy," according to National Renewable Energy laboratory (NREL) researchers (see NREL fact sheet 2007). Algal and bacterial hydrogen production fall under the category of photobiological hydrogen production systems. Liquid or solid nutrients, in the form of nitrogen, potassium, and phosphorous fertilizer are used in aquacultures, e.g. fish/algae ponds to promote organism growth and hydrogen production. With promising results that are obtained in large ponds, commercial-scale hydrogen production through photobiological pathways are deemed to be feasible in the medium to long-term. In a recent NREL-sponsored project, James et al. address different biological pathways to hydrogen production (2009), with promising results. The cost of hydrogen production poses a major hurdle.

At the present, the (truck) delivered price of hydrogen from biomass is $5–7 per kilogram. Large-scale production and a pipeline is estimated to reduce the delivered price to $1.50–3.50 per kilogram. For high-purity hydrogen (99.95%+), there are costs associated with the additional purification and by adding transport cost we get another $1/kg of additional cost.

Purified hydrogen also may be produced as byproduct of high-temperature fuel cells running on methane, wastewater treatment, or LFG. These fuel cells use molten carbonate (MCFC) or solid oxide (SOFC) technology.

Electrolysis of water is another process to produce hydrogen. Electrolysis uses electricity and an electrolyzer to split water molecule into O_2 and H_2:

$$e^- + H_2O \rightarrow \frac{1}{2}O_2 + H_2 \tag{4.4}$$

There are two common types of electrolyzers:

1) Alkaline uses potassium hydroxide electrolyte.
2) PEM (proton exchange membrane) uses a solid polymer membrane as electrolyte.

Electrical sources used in the electrolysis of water in hydrogen production may come from conventional electric power grid or from renewable sources, i.e.:

- Utility grid power
- Solar photovoltaic (PV)
- Wind power
- Hydropower
- Nuclear power

The scale of production varies between a few kW to 2000 kW per electrolyzer (see Lipman 2011). The grid cost of hydrogen production and delivery is $6–$7 per kilogram, with a potential reduction to $4/kg. The cost increases for wind-power-driven electrolysis, which is currently $7–$11/kg with a potential of $3–4 per kilogram in the future. Solar hydrogen power stands at $10–30 per kilogram at present with a future delivered cost of $3–4. The source of these estimated hydrogen production costs is National Academy of Science and National Academy of Engineering 2004 and US Energy Information Administration (EIA) 2008. Milbrandt and Mann address the potential for hydrogen production from renewable (2007) and other sources (2009).

4.4.2 Hydrogen Delivery and Storage

Hydrogen is an industrial gas that uses the codes and standards developed by ASME and DOT. These are ASME Boiler and Pressure Vessel Codes (for stationary use) and 49 Code of Federal Regulations for transportation/delivery uses. There are additional codes governing the piping, vent, and storage systems. The delivery system is either in the form of compressed gas or cryogenic liquid. For details on various storage systems and their associated safety, cost, weight, and rate of energy transfer, see: www1.eere. energy.gov/hydrogenandfuelcells/storage/current_technology.html.

4.4.3 Gravimetric and Volumetric Energy Density and Liquid Fuel Cost

Liquid hydrogen and LNG are both cryogenic fuels that require special handling, tankage, and fuel-delivery systems. The energy-density costs of liquid fuels are summarized in Table 4.11 (data from Tennekes 2009).

The cost per unit energy of the cryogenic liquid fuel, LH_2, is an order of magnitude higher than Jet-A. The LNG's specific cost is lower than Jet A, but its volumetric efficiency (expressed in MJ per liter) is nearly half of the conventional fossil fuel, thus requiring double the fuel tankage, as noted earlier. Including the cryotank weight for LNG, we note that its GED (expressed in MJ per kilogram) is also lower than the conventional Jet A fuel. The cryogenic fuels are still not competitive in GED, VED, and cost per unit energy. The safety of cryogenic fuel delivery to commercial aircraft, especially the liquid hydrogen case, remains a continuing concern in need of technology development.

4.5 Battery Systems

The use of batteries in aviation is not new. We use batteries to start the APU and power up the aircraft systems before the APU or the engines are started. Batteries also support ground operations such as refueling (see Figure 4.22, courtesy of Boeing). Batteries provide backup power in the unlikely event of power failure on the aircraft. Lithium-ion batteries are selected for the modern More-Electric commercial airplane, B787 for their

Table 4.11 Gravimetric (GED) and volumetric energy density (VED) and cost of liquid fuels.

Fuel type	GED (MJ kg^{-1})	VED (MJ l^{-1})	Cost (\$ MJ^{-1})
Li battery (rechargeable)	0.3	0.3	0.03
Li battery (primary)	0.6	0.6	170
Honey	14	20	0.29
Goose fat	38	35	0.26
Kerosene (Jet A)	44	36	0.018
Natural gas	45	19[a]	0.005
Hydrogen	117	8.3[a]	0.44

[a] Volume of liquid only, not accounting for cryotank.
Source: Data from Tennekes 2009.

(a)

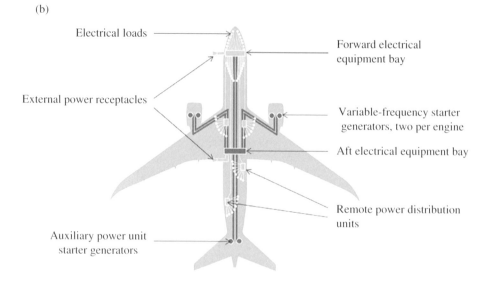

787 Batteries

APU Battery

- Located in the aft electrical equipment bay
- Powers the APU, which can power generators to start engines, if needed; also powers navigation light

Main Battery

- Located in the forward electrical equipment bay
- Powers up airplane systems, before APU or engines are started

In flight, the airplane is powered by the engine generators. The batteries are part of the multiple layers of redundancy that would ensure power in the extremely unlikely event of a power failure.

(b)

Electrical loads

Forward electrical equipment bay

External power receptacles

Variable-frequency starter generators, two per engine

Aft electrical equipment bay

Remote power distribution units

Auxiliary power unit starter generators

Figure 4.22 Batteries and the electric power production and distribution system in B787 aircraft: (a) batteries on board B787; (b) generator locations and electrical power production and distribution system on board B787. *Source:* Courtesy of Boeing.

high-voltage, high-current production, improved power quality, quick recharge-ability, weighing 30% less than the equivalent NiCd batteries and their compact size.

Once the APU and the engines start, electric generators on the main engines and the generator on APU takeover the task of electric power production. Traditionally, aircraft have one generator on each of the two engines and one on the APU. In more electric aircraft, i.e. B787, there are six generators, two on each engine and two generators on the APU. In flight, the four engine generators are the primary and the two APU generators are the secondary sources of electric power.

4.5.1 Battery Energy Density

The current energy density in various batteries is listed in Figure 4.23. Those who doubt the utility of electric propulsion, note that the Achilles' heel of batteries is exactly this parameter, which is hardly competing with the fossil fuel (Jet-A) when it packs more than $12\,000\,\mathrm{Wh\,kg^{-1}}$ and the nonrechargeable lithium battery offers $400\,\mathrm{Wh\,kg^{-1}}$. However, there is more to electrification of aircraft than just this single parameter, i.e. the battery energy density. New aircraft designs and capabilities are made possible courtesy of electric propulsion (see Moore and Fredericks 2014). For example, Stoll et al. (2014) describes the drag reduction made possible through distributed electric propulsion, among other design advantages offered through electrification.

The battery technology is rapidly accelerating based on the demands of electric mobility. For example, Table 4.12 shows a 10–20 years forecast of battery technology according to industry and prominent researchers.

In addition to the battery, we have wiring and extra cooling weight. These weights are estimated to be around 65% of the battery weight. The chemical reaction in batteries create heat, which require cooling. Table 4.13 shows the operating temperature and cooling air requirement for current batteries and forecast of 20 years.

The energy densities of batteries, even the advanced lithium-air open-cycle batteries, are a fraction of aviation jet fuel. Kerosene has an energy density higher than of $12\,000\,\mathrm{Wh\,kg^{-1}}$, which assuming 33% conversion efficiency, still one-third of the energy is available to the propulsor. The MIT forecast puts the battery energy density at

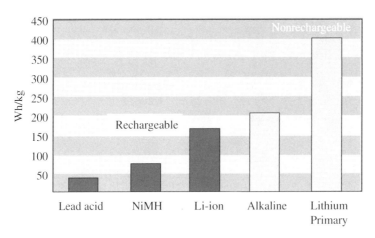

Figure 4.23 Energy density, in Wh/kg, for different batteries.

Table 4.12 Forecast of battery energy density in 10–20 yr.

Battery energy density	Wh kg^{-1}
Today[a]	120–200
Boeing (SUGAR Volt)	750
Rolls-Royce	750
In 20 yr	1000
MIT forecast in 10–15 yr	1000–1500

[a] 2018

Table 4.13 Cooling requirements for lithium-ion batteries.

Temperature	°F	°C
Battery max temp allowed (today)	140	60
Battery max temp allowed (in 20 yr)	212	100
Cooling air temperature	100	38

1–1.5 kWh kg^{-1} in 10–15 years, which will still be about a quarter to a third of the kerosene. However, as noted by Moore and Fredericks (2014), it is misleading to compare the energy density of kerosene and batteries and draw false conclusions that electric propulsion will not be feasible until energy densities are comparable. Electric propulsion enables a totally new design of aircraft that is impossible by the conventional kerosene-fueled system. This promising technology is addressed in Chapter 5.

4.5.2 Open-Cycle Battery Systems

There are two promising open-cycle battery systems:

1) Zinc-air
2) Lithium-air

The lithium-air open-cycle battery offers the highest storage density (see Stückl et al. 2012). The lithium-air reaction consumes oxygen from air or from a supply tank according to:

$$2Li + O_2 \rightarrow Li_2O_2; G_0 = -145 \, kCal \tag{4.5}$$

According to Friedrich et al. (2012), the theoretical energy density of lithium-air reaction is 5200 Wh kg^{-1}, including the mass of oxygen. On the cell level basis, however, where the weight of electrolytes and housing is included, the energy density of lithium-air battery is significantly reduced. For example, the predicted energy density at the cell level is 750–2000 Wh kg^{-1} (see Visco et al. (2006), Girishkumar et al. (2010), and Johnson (2010)). The lithium-air open-cycle battery technology is anticipated to be market-ready in 2030, according to Thielmann et al. (2010).

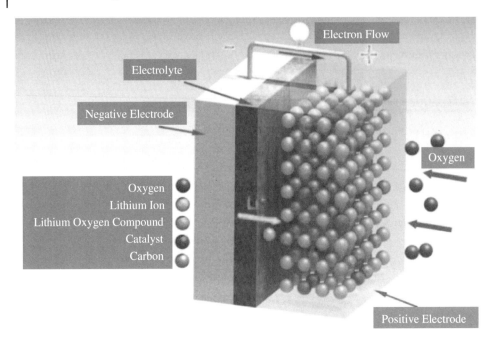

Figure 4.24 Lithium-air open-cycle battery anticipated to reach 2000 Wh kg^{-1} by 2030. *Source:* From Stückl et al. 2012.

Figure 4.24 shows the schematic drawing of lithium-air open-cycle battery system (from Stückl et al. 2012).

4.5.3 Charging Batteries in Flight: Two Examples

Battery-assisted takeoff and climb in hybrid-electric aircraft saves fuel and, in turn, requires charging of the battery system by the gas turbine engine in cruise. Two examples serve the purpose of demonstrating the battery-assist in takeoff and climb as well as charging of battery system in cruise. Table 4.14 is the example where 1500 kWh of battery needs to be recharged by a gas turbine engine. Table 4.15 is the example of electric assist in takeoff and climb with the assumption of 10% fuel burn saving in takeoff and climb due to electric assist. The numerical examples are used here to help the reader with similar calculations.

4.5.4 All-Electric Aircraft: Voltair Concept Platform

Voltair, an all-electric, regional category aircraft, is shown in Figure 4.25 (from Stückl et al. 2012) that was designed based on superconducting electric motors and a battery system with energy density of 1000 Wh kg^{-1}. Transmission voltage was assumed at 1 kV DC. The choice of DC transmission from the battery system to the motors is based on minimizing the electromagnetic interference as well as cable weight. The propulsion integration in Voltair takes advantage of the aft fuselage BLI configuration. The electric motor weight in the aft fuselage is balanced by the battery weight in front, to avoid backward shift of the center of gravity (see Stückl et al. 2012). The Voltair

Table 4.14 Example of recharging a 1500 kWh battery in flight.

Jet-A fuel needed to charge 1500 kWh of battery	
Energy in Jet-A (in kJ kg^{-1})	43 500
Conversion efficiency (in %)	30
Battery energy (in kJ)	=1 500 * 3 600 = 5 400 000
Fuel mass	Battery energy/(LHV of the fuel × Efficiency)
Jet A mass (in kg)	414

Table 4.15 Example of electric-assist in takeoff and climb

Concept of electric assist in TO and climb		10% saving in fuel
Fuel burned in TO and climb		6 000 lb
10% saving in fuel		600 lb
600 lb of fuel carries energy in kJ (kWh)		1.18E + 07
		3 284
Taking 30% efficiency in Jet-A conversion (kWh)		985
Weight (in kg) of batteries for 985 kW-h	Used today's tech	8 211
Battery weight (in lb)		18 125
Extra cooling and wiring weight (lb)		11 781
Total battery system weight (lb)		29 906

Figure 4.25 Voltair configuration. *Source:* From Stückl et al. 2012.

drag-and-weight reduction compared to conventional aircraft are summarized in four categories:

1) Zero lift drag:
   a) 60% natural laminar flow wing: → −15%
   b) No wheel well fairing: → −5%
2) Induced drag:
   a) Addition of winglets: → −10%
3) Propulsive efficiency:
   a) Boundary-layer ingestion: → +5%
4) Empty weight:
   a) Low slenderness ratio fuselage: → −12%

The net effect of these gains is 25% improvement in energy efficiency and 30% reduction in maximum gross weight, according to Stückl et al. (2012).

4.6 Fuel Cell

Fuel cells have been the primary source of electric power in spacecraft since the Gemini program. Fuel cell converts chemical energy from hydrogen-rich fuels into electrical power. A fuel cell consists of individual cells. Each cell contains an anode, a cathode, and an electrolyte layer. A group of individual cells form a fuel cell stack. Hydrogen-rich fuel enters the fuel cell stack and reacts with the oxygen in the ambient air. This reaction produces an electric current as well as heat and water. Figure 4.26 shows the schematic of fuel cell.

Types of fuel cells are classified based on their electrolyte solution:

1) Alkaline fuel cells
   Solution of potassium hydroxide

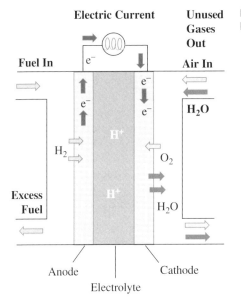

Figure 4.26 Schematic drawing of a fuel cell using hydrogen fuel.

2) Molten carbonate fuel cells
   High-temperature compounds of salt carbonates
3) Phosphoric acid fuel cells
   Phosphoric acid
4) PEM fuel cell
   Thin, permeable sheet of a polymer electrolyte
5) Solid-oxide fuel cell (SOFC)
   Hard, ceramic compound of metal oxides

Table 4.16 shows the operating temperature, power output, and efficiency of different fuel cells.

Fuel cell power density is expected to rise to $10 \, kW \, kg^{-1}$ by 2020 according to Sehra and Shin (2003). The authors note that $10 \, kW \, kg^{-1}$ is $20 \times SOA$ in 2003 (SOA stands for State-of-the-Art). Also, fuel-cell charging is equivalent to replacing the fuel consumed, unlike batteries that require rapid charging technology.

The most attractive and the best candidate for the hydrogen fuel cell is replacing the kerosene-fueled APU in an aircraft that supplies electricity and compressed air for the onboard air conditioning and starting the engines. Kerosene APU has a very low efficiency of 20% and even lower efficiency at idle (10%). Kerosene APU is noisy and highly polluting. For these reasons, airports limit the use of APUs only for short periods to start the engines. At other times, the aircraft relies on ground power for its operation. A hydrogen fuel cell does not pollute, is not noisy, and may operate throughout the flight. Its operational efficiency is 40% and at idle is 50%. Its output besides electricity is water, which could save an aircraft up to 700 kg of takeoff weight.

Motorized nose wheel powered by fuel cells will enable taxiing with engine off, as demonstrated in an A320 ATRA on July 1, 2011, by DLR German Aerospace Center. "The potential saving at Frankfurt Airport from the use of electrically driven nose wheels for Airbus A320 class aircraft is about 44 tons of kerosene per day," which is

Table 4.16 Operational characteristics of fuel cells

Type	Operating temperatures	System outputs	Efficiency
Alkaline	90–200 °C 200–400 °F	10–100 kW	60–70%
Molten carbonate	650 °C 1200 °F	<1 MW (250 kW typical)	60–65% 85% (wastes used)
Phosphoric acid	150–200 °C 120–210 °F	50 kW–1 MW (250 kW typical)	46–42% (alone) 80–85% (co-generation)
Proton exchange membrane (PEM)	50–100 °C 120–210 °F	<250 kW	40–60%
Solid oxide (SOFC)	650–1000 °C 1200–1800 °F	5 kW–3 MW	60% 85% (waste used)

estimated by DLR Institute of Flight Guidance (see Quick 2011). In February 2016, EasyJet (a customer of Airbus) demonstrated the autonomous taxiing using fuel cells powering the main landing-gear wheels.

Airbus/DLR German Aerospace Center/Parker Aerospace is working on a multifunctional fuel cell (MFCC) system that could provide 100 kW of electricity to cut aircraft pollution and noise emissions.

4.7 Fuels for the Compact Fusion Reactor (CFR)

Fusion is the process that powers the Sun and other stars. Fusion is a cleaner, safer form of energy than fission. The energy release by fusion is three to four times that of fission (see https://www.lockheedmartin.com/en-us/products/compact-fusion.html).

At extreme temperatures, say 100 000 000 °C, the nuclei of light atoms, instead of repulsion, fuse to form the nucleus of heavier atoms. For example, deuterium (^{2}H) and tritium (^{3}H), which are heavy isotopes of hydrogen (^{1}H), can fuse to form helium (^{4}He), as shown in Figure 4.27 and a high-speed neutron. This process releases tremendous energy that binds the nucleus. Two of the most promising reactions in the compact fusion reactor involve the heavy isotopes of hydrogen, deuterium and tritium. The two reactions are:

1) DT reaction, with deuterium fusing with tritium to make helium and a neutron
2) DD reaction, which fuses deuterium with deuterium

Deuterium is a naturally occurring isotope of hydrogen that may be extracted from sea water. D_2O is referred to as *heavy water*. Each ton of sea water contains nearly 33 g of deuterium. This means that ~0.015% of hydrogen in sea water is deuterium, which makes it a limitless supply of fusion reactor fuel. The extraction of deuterium from sea

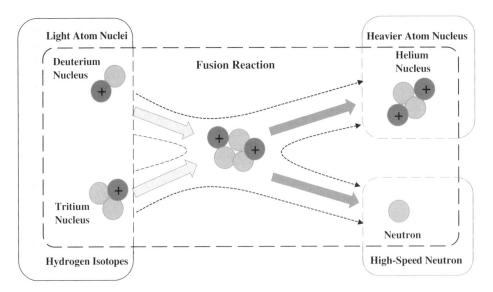

Figure 4.27 DT reaction of atomic nuclei of deuterium-tritium, in a fusion reaction, creates helium and an energetic neutron.

water is an easily accomplished and well-proven industrial process (see Chen 1974 and Arnoux 2011), which again bodes well for this fuel.

Tritium is the heavier isotope of hydrogen, which has a half-life of 12 years. This heavy isotope is not naturally occurring (in large quantities) and thus needs to be bred from lithium (see David MacKay 2009, or Francis Chen 1974). The DT fusion reactions and their energies are noted to be (from Chen 1974):

$$D + T \rightarrow {}^4He\left(3.5\,MeV\right) + n\left(14.1\,MeV\right) \tag{4.6}$$

$$n + {}^6Li \rightarrow {}^4He\left(2.1\,MeV\right) + T\left(2.7\,MeV\right) \tag{4.7}$$

where n is the neutron. The DT fusion reactions described by Eqs. (4.6) and (4.7) require the lowest *ignition temperature*, produces the highest energy, and has the lowest confinement requirement of other fusion reactions. There are three requirements for the successful fusion reaction:

1) Very high temperature (of order of 150 000 000 °C)
2) Sufficient plasma particle density for the probability of collision to increase
3) Sufficient confinement time

Intense R&D in compact fusion reactor (CFR) design in industry (e.g. Lockheed Martin) and academia (e.g. MIT) is rapidly progressing (see Talbot 2014). The abundance of nuclear fuel in CFR is clearly a positive attribute of this technology.

4.8 Summary

Sustainable aviation requires sustainable fuel. Fortunately, there are alternatives to the current fossil fuel in use in commercial aircraft. The prominent among these are biofuels, LNG, and hydrogen. Forecasting the technology maturity of AJFs, i.e. TRL, and commercial adoption in civil aviation is complicated by the myriad of factors that affect its realization and thus tentative markets; regulations and government investment and incentives influence these factors. The most promising among these are drop-in AJFs that are renewable and can help reduce GHG emissions and improve air quality. These fuels are estimated to achieve wide market adoption (TRL 8–9) by 2040. The hybrid-electric propulsion is expected to achieve technology readiness by 2025–2030 (first pursued in the regional jet sector) and be widely adopted by 2050. Liquid natural gas and liquid hydrogen are both cryogenic fuels that require a new fleet of commercial transports (due to lower VED). They are available in nature in abundance. The cost per unit energy and safety concerns for routine commercial aircraft fuel delivery is, however, major drawbacks of these fuels. Under optimistic forecasts, these are expected to achieve wide use in 50 years. Electric propulsion will focus on GA aircraft first to gain market share, rapid development and certification and advance the SOA in battery systems. It is anticipated that electric propulsion will mature and will include medium-range, single-aisle commercial aircraft by 2060–2080.

The hydrogen fuel cell is destined to replace the kerosene-fueled APU within the next 10 years. However, its role as the primary propulsion system requires new aircraft, as noted earlier, is perhaps 50–100 years into the future.

The fuels for fusion reactor are deuterium and tritium bred from lithium, and these are found in abundance in nature and their processing is widely available. Under optimistic scenarios, CFR-powered aircraft for heavy-lift, long-endurance applications may reach the runway by the turn of the century, 2100.

References

Achten, W.M.J., Verchot, L., Franken, Y.J. et al. (2008). Jatropha bio-diesel production and use. *Biomass and Bioenergy* 32 (12): 1063–1084.

Anderson, B. 2014. Alternative-Fuel Effects on Contrails and Cruise Emissions (ACCESS-II) Flight Experiments. http://science.larc.nasa.gov/large/data/ACCESS-2.

Arnoux, R., 2011. Deuterium: a precious gift from the Big Bang. ITER Newsline No. 167, 11 March. https://www.iter.org/newsline/167/631

ASTM D7566-16b (2016). Standard Specification for Aviation Turbine Fuel Containing Synthesized Hydrocarbons. West Conshohocken, PA, https://www.astm.org/: ASTM International.

Bauen, A., Howes, J., Bertuccioli, L., and Chuzdiak, C. (2009). Review of the Potential for Biofuels in Aviation," Final Report Prepared for the Committee on Climate Change. Switzerland: E4Teck.

Burkhardt, U. and Kärcher, B. (2011). Global radiative forcing from contrail cirrus. *Nature Climate Change* 1: 54–58.

Chen, F.F. (1974). Introduction to Plasma Physics. New York: Plenum Press.

Cuttita, C. 2010. Air Force completes second biofuel test. (https://www.wpafb.af.mil/News/Article-Display/Article/400012/air-force-completes-second-biofuel-test/.

Del Rosario, R. 2011. Propulsion Technologies for Future Aircraft Generations: Clean, Lean, Quiet and Green. Paper presented at the 20th ISABE Conference, Gothenburg, Sweden (September 2011).

Department of Energy (DOE) (2002). A National Vision of America's Transition to a Hydrogen Economy – to 2030 and Beyond. Washington, DC: DOE.

Deutsches Zentrum für Luft-und Raumfahrt, DLR. 2015. A journey through an exhaust plume – DLR flight tests for alternative fuels.

Edwards, J.K., Shafer, L.M., and Klein, J.K. 2012. U.S. Air Force Hydro-Processed Renewable (HRJ) Fuel Research, USAF Report Number. *AFRL-RQ-WP-TR-2013-0108.*

Energy Information Administration. 2005. *Annual Energy Outlook 2005*, www.eia.doe.gov.

Epstein, A. (2014). Aeropropulsion for commercial aviation in the twenty-first Century and research directions needed. *AIAA Journal* 52 (5): 901–911.

Expert Group on Future Transport Fuels (2015). State of the Art in Alternative Fuels Transport Systems, Final Report. The European Commission, EU.

Foss, M. (2003). LNG Safety and Security. Center for Energy Economics at the Bureau of Economic Geology, The University of Texas at Austin, https://www.beg.utex.edu/energyecon/Ing/documents/CEE/LNG Safety and Security.pdf.

Friedrich, A., Wagner, N., and Bessler, W. (2012). Entwicklungsperspektiven von Li-Schwefel- und Li-Luft- Batterien. In: Energie Speicher Symposium 2012. Stuttgart: DLR.

Girishkumar, G., McCloskey, B., Luntz, A.C. et al. (2010). Lithium-air battery: promise and challenge. *Journal of Physical Chemistry Letters* (1): 2193–2203.

SAAB. (2017). Grippen completes test flights with 100% biofuel. https://saabgroup.com/media/stories/stories-listing/2017-04/gripen-completes-test-flights-with-100-biofuel/.

Guynn, M.D. and Olson, E.D. 2002. Evaluation of an Aircraft Concept with Over-Wing, Hydrogen-Fueled Engines for Reduced Noise and Emissions. *NASA/TM-2002-211926*.

Han, J., Elgowainy, H., Cai, H., and Wang, M.Q. (2013). Lifecycle analysis of bio-based aviation fuels. *Bioresource Technology* 150: 447–456.

Han, J., Tao, L., and Wang, M. (2017). Well-to-wake analysis of ethanol-to-jet and sugar-to-jet pathways. *Journal of Biotechnology for Biofuels* 10 (1): 21.

Hileman, J., Stratton, R.W., and Donohoo, P. (2010). Energy content and alternative jet fuel viability. *Journal of Propulsion and Power* 26 (6): 1184–1196.

International Air Transport Association (2015). IATA Sustainable Aviation Fuel Roadmap, 1e. Montreal: International Air Transport Association.

James, B.D., Baum, G.N., Perez, J. and Baum, K.N. 2009. Techno economic Boundary Analysis of Biological Pathways to Hydrogen Production. *NREL/SR-560-46674*, Golden, Colorado.

Johnson, L. 2010. The Viability of High Specific Energy Lithium Air Batteries. Symposium on Research Opportunities in Electrochemical Energy Storage – Beyond Lithium-Ion, Materials Perspectives, Tennessee, October.

Joint European Commission. 2014. Well-to-Wheel Report Version 4.a, JEC Well-to-Wheels Analysis. *JRC Technical Reports*, European Commission.

Kim, H.D., Felder, J.L., Tong, M.T., and Armstrong, M. 2013. Revolutionary Aeropropulsion Concept for Sustainable Aviation: Turboelectric Distributed Propulsion. Paper presented at ISABE Conference, Paper Number 2013-1719.

Kinder, J.K. and Rahmes, T. 2010. Evaluation of Bio-Derived Synthetic Paraffinic Kerosene (Bio-SPKs). *ASTM Research Report*, May 2010. Addendum published, October 2010.

Lane, J. 2015. A new technology debuts for renewable jet fuel. *Biofuels Digest*, March 11, 2015.

Levy, Y., Erenburg, V., Sherbaum, V., Ocharenko, V. and Roizman, A. 2015. The use of ethanol as an alternative fuel: droplet formation and evaporation. Proceedings of ASME, Paper Number GT2015-42965, June.

Lipman, T. 2011. An Overview of Hydrogen Production and Storage Systems with Renewable Hydrogen Case Studies. A Clean Energy States Alliance Report, May.

MacKay, D.J.C. (2009). Sustainable Energy – Without the Hot Air. Cambridge, UK: UIT Cambridge Ltd.

Milbrandt, A. and Mann, M. (2007). Potential for Hydrogen Production from Key Renewable Resources in the United States," NREL/TP-640-41134. Golden, CO: National Renewable Energy Laboratory, www.nrel.gov/docs/fy07osti/41134.pdf.

Milbrandt, A. and Mann, M. 2009. Hydrogen Potential from Coal, Natural Gas, Nuclear and Hydro Resources. *Technical Report NREL/TP-560-42773*.

Milbrandt, A.R., Heimiller, D.M., Perry, A.D., and Field, C.B. (2014). Renewable energy potential on marginal lands in the United States. *Renewable and Sustainable Energy Reviews* 29: 473–481.

Moore, M.D. and Fredericks, W.J. 2014. Misconceptions of Electric Propulsion Aircraft and Their Emergent Aviation Markets. 52nd Aerospace Sciences Meeting, AIAA SciTech Forum, AIAA Paper Number 2014-0535.

National Academy of Science and National Academy of Engineering (2004). The Hydrogen Economy: Opportunities, Costs, Barriers, and R&D Needs. National Academies Press.

NREL Fact Sheet: *Photobiological Production of Hydrogen*, FS-560-42285, Colorado, November 2007.

Penner, J.E. (1999). Aviation and Global Atmosphere. Cambridge, UK: Cambridge University Press.

Quick, D. (2011). Airbus and DLR testing fuel cell technology to cut aircraft pollution and noise emissions. http://newatlas.com/airbus-and-dlr-testing-fuel-cell-technology/19159/.

Rao, A.G. (2016). AHEAD: Advanced Hybrid Engines for Aircraft Development. Delft University of Technology: Netherlands, see www.ahead-euproject.eu.

Sehra, A.K and Shin, J. 2003. Revolutionary Propulsion System for 21st Century Aviation. *NASA/TM-2003-212615*.

Spear, K. 2018. Virgin Atlantic flight from Orlando is the first-ever to run on synthetic fuel. *Orlando Sentinel*, October 23. https://www.orlandosentinel.com/news/.

Stoll, A.M., Bevirt, JB, Moore, M.D., Fredericks, W.J., and Borer, N.K. 2014. Drag reduction through distributed electric propulsion. Aviation Technology, Integration, and Operations (ATIO) Conference, Atlanta, Georgia, 16–20 June.

Stratton, R.W., Wong, H.M., and Hileman, J.I. 2010. Lifecycle Greenhouse Gas Emissions from Alternative Jet Fuels. Partnership for AiR Transportation Noise and Emissions Reduction, PARTNER, *Report No. PARTNER-COE-2010-001*.

Stückl, S., van Toor, J., Lobentanzer, H.2012. Voltair – the all-electric propulsion concept platform – a vision for atmospheric friendly flight. Paper presented at the 28th International Congress of the Aeronautical Sciences (ICAS).

Talbot, D. (2014). Does Lockheed Martin Really Have a Breakthrough Fusion Machine? *Technology Review* (October 20).

Tennekes, H. (2009). The Simple Science of Flight, Revised and Expanded Edition. Cambridge, MA: MIT Press.

Thielmann, A., Isenmann, R., and Wietschel, M. (2010). Technologie-Roadmap Lithium-Ionen-Batterien 2030. Karlsruhe, Germany: Frauenhofer-Institut für System- und Innovationsforschung ISI.

US Energy Information Administration. 2007. *Annual Energy Outlook 2007 with Projections to 2030.* www.eia.doe.gov.

US Energy Information Administration (EIA). 2008. The Impact of Increased Use of Hydrogen on Petroleum Consumption and Carbon Dioxide Emissions. *Report Number SR-OIAF-CNEAF/2008-04*, www.eia.doe.gov/oiaf/servicerpt/hydro/appendixc.html.

Visco, S., Nimon, E., Katz, B. et al. (2006). High energy density lithium-air batteries with no self discharge. In: Proceedings of the 42nd Power Source Conference, 201–203. The Electrochemical Society.

Wang, M. 2011. Greenhouse Gasses, Regulated Emissions, and Energy Use in Transportation (GREET) Model, Version GREET1_2011. *Center for Transportation Research, Argonne National Laboratory*, Argonne, IL.

Winchester, N., McConnachie, D., Wollersheim, C., and Waitz, I.A. 2013. Market Cost of Renewable Jet Fuel Adoption in the United States. *PARTNER Project 31 Report, PARTNER-COE-2013-001*, Cambridge, MA, March.

5

Promising Technologies in Propulsion and Power

5.1 Introduction

Aviation contributes to about 12% of the greenhouse gas emissions in the transportation sector (i.e. road transport, rail, aviation, and shipping). Transportation sector itself accounts for about 31% of the US carbon dioxide emissions based on Environmental Protection Agency (EPA) report (2016). Hence, aviation's share in anthropogenic CO_2 emissions is currently about 4%. The importance of aviation to environmental concerns is thus not tied to its current 4% value rather to it is rising trends. The long-term impact potential is extremely harmful when the estimated 16B passengers take to sky in 2050. Within the aviation sector, commercial aviation contributes 93% to carbon dioxide emissions, whereas business jets, general aviation aircraft, helicopters, and UAS contribute the remaining 7%, as shown in Figure 5.1. Thus, to create a low-carbon aviation future, we focus on promising low-carbon-footprint propulsion and power systems for commercial aviation. We expect that the spinoffs of this large-scale technology development effort in commercial aviation to impact all sectors of flight.

The road to sustainable propulsion and power systems in commercial aviation goes through advanced gas turbine (GT) engines either as fully integrated propulsion-airframe system or as hybrid electric propulsion system (HEPS). The rationale for continued investment and reliance on advanced gas turbine (GT) engines, in particular, advanced turbofan (TF) engines, is based on their steady improvements and unique qualities that are developed over the past 100 years. Briefly, some of the main GT attributes are found in their:

- Affordability, reliability, and scalability
- Steady gains in GT component efficiencies
- Steady gains in GT cycle propulsive, thermal and overall efficiencies
- Steady reductions in GT emissions, e.g. NO_x, CO, unburned hydrocarbon (UHC), particulate matter (PM) (soot and smoke)
- Steady reductions/mitigation of GT engine, including jet noise
- Emergence of drop-in alternative jet fuels or biofuels with reduced lifecycle carbon emissions
- Enjoying the highest power density (i.e. power-to-weight ratio) compared to alternative propulsion and power systems/concepts
- Maintainability and Long life that impact reliability and affordability

Future Propulsion Systems and Energy Sources in Sustainable Aviation, First Edition. Saeed Farokhi.
© 2020 John Wiley & Sons Ltd. Published 2020 by John Wiley & Sons Ltd.
Companion website: www.wiley.com/go/farokhi/power

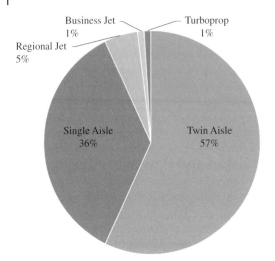

Business Jet
1%

Turboprop
1%

Regional Jet
5%

Single Aisle
36%

Twin Aisle
57%

Figure 5.1 Worldwide fuel burn by aircraft type. *Source:* Data from Yutko and Hansman 2011.

Sehra and Shin (2003) present an overview of the revolutionary propulsion systems for the twenty-first century aviation, which presents NASA's vision in 2003. Emissions and efficiency are the challenges and thus the drivers for the new technologies. Epstein (2014) presents research directions needed for high-efficiency aeropropulsion in commercial aviation in the twenty-first century, which presents industry's perspective. Lord et al. (2015) discuss engine architecture that meets the NASA N+3 goal for 30% TSFC reduction (from 737-800/CFM56-7B baseline) for entry into service (EIS) of 2035. The European vision on sustainable aviation is detailed in *Flightpath 2050* document that outlines the critical areas of research and development, e.g. alternative jet fuel, advanced propulsion and power concepts and electrification that help achieve zero-carbon aviation. These roadmaps all lead to one destination: *greener aviation*.

Efficiency being the hallmark of future development, we start with the efficiency of simple Brayton and compounded cycles, followed by the advances in aeroengine architecture.

5.2   Gas Turbine Engine

5.2.1   Brayton Cycle: Simple Gas Turbine Engine

Aircraft gas turbine engines, e.g. turbojets, operate on the basis of Brayton cycle. The ideal thermal efficiency of Brayton cycle is based on isentropic compression (both through the inlet and the compressor/fan), constant total pressure combustion and isentropic expansion through the (uncooled) turbine and the exhaust nozzle. It is expressed as (see Oates 1988, Kerrebrock 1992 or Farokhi 2014, among others):

$$\eta_{th} = 1 - \frac{T_0}{T_3} = 1 - \frac{1}{\tau_r \tau_c} = 1 - \frac{1}{\left(\pi_r \pi_c\right)^{\frac{\gamma-1}{\gamma}}} \tag{5.1}$$

At sea-level static takeoff condition, $\pi_r = 1$ and for compressor total pressure ratio $\pi_c = 55$, ideal thermal efficiency is ~68%. For Mach 2.0 cruise in lower stratosphere, with $\pi_c = 20$, the ideal thermal efficiency increases to about 77%. In these cycles, we have kept compressor discharge temperature to around 900 K. At this temperature, we can still use titanium for the compressor disk and blade material. At higher compressor exit temperatures, e.g. $T_{t3} = 1000$ or 1200 K, we have to use nickel-based alloys that are suitable for turbines for the high-pressure compressor (HPC). Alternatively, we may provide the aft stages of the HPC with cooling, e.g. fuel-based regenerative cooling.

5.2.2 Turbofan Engine

The need for advancements in GT core will be critical to higher cycle efficiencies. For turbofan engines, Koff (1991) defines a *core thermal efficiency*, a transfer efficiency and propulsive efficiency that describe the turbofan propulsion system more effectively than the simple Brayton cycle efficiency that describes a turbojet engine. The core thermal efficiency is defined as the power produced by the core as a fraction of the fuel thermal power (see Koff 1991). These efficiencies are listed here for reference:

$$\text{Core thermal efficiency} \equiv \frac{E_{\text{core}}}{E_{\text{fuel}}} = \frac{\dot{m}_f Q_R - \dot{m}_0 h_0 \left(\frac{\tau_c}{\tau_r} - 1\right)}{\dot{m}_f Q_R} = \frac{f Q_R - h_0 \left(\frac{\tau_c}{\tau_r} - 1\right)}{f Q_R} \tag{5.2}$$

$$\text{Transfer efficiency} \equiv \frac{E_{\text{jets}} - E_{\text{inlet}}}{E_{\text{core}}} = \frac{\left[\alpha \frac{V_{19}^2}{2} + (1+f)\frac{V_9^2}{2}\right] - (1+\alpha)\frac{V_0^2}{2}}{f Q_R - h_0 \left(\frac{\tau_c}{\tau_r} - 1\right)} \tag{5.3}$$

$$\text{Propulsive efficiency} \equiv \frac{F_n V_0}{E_{\text{jets}} - E_{\text{inlet}}} \tag{5.4}$$

The engine thermal efficiency based on core is currently about 55%. There are promising technologies in advanced manufacturing and flow control that promote higher thermal efficiencies in the core. One example is the aspiration flow control technology applied to the casing and the blades on the HPC, among others (see Merchant et al. 2004 and Wilfert et al. 2007).

In Figure 5.2a, we show the state-of-the-art (SOA) in turbofan cycle efficiencies and in Figure 5.2b, adapted from Epstein (2014), we project possible trajectories in aeropropulsion development. These point to potential gains in core thermal efficiency as well as propulsive and transmission efficiencies that yield overall efficiencies in 70%[+] range.

For another perspective, we may gain insight by examining the Carnot cycle, which represents the maximum thermal efficiency of heat engines operating between two limit temperatures (or reservoirs), T_0 and T_{t4}. The highest T_{t4} in conventional GT engines corresponds to the stoichiometric combustion of hydrocarbon fuels, which is about 2500 K. This level of T_{t4} at takeoff gives thermal efficiency of 88% and the cruise thermal efficiency exceeds 90%. The simple Brayton cycle approaches the Carnot efficiency with increasing cycle pressure ratio. This brings us back to the needs for

(a)

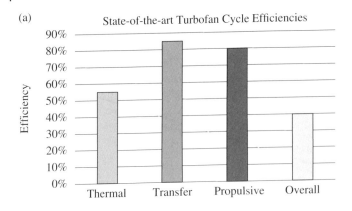

(b)

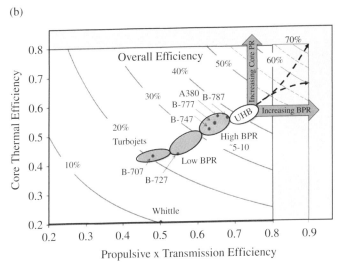

Figure 5.2 Core thermal, propulsive and overall efficiencies of commercial aircraft Engines at Mach 0.8 cruise, 35 kft ISA, uninstalled. (a) State-of-the-art Turbofan Cycle Efficiencies. *Source:* Data from Whurr 2013. (b) Projections in GT core thermal and propulsive x transmission efficiencies in turbofan engines. *Source:* Adapted from Epstein 2014 and Lord et al. 2015.

advancements in the GT high-pressure core. These topics are discussed in great detail in Kerrebrock (1992), Koff (1991) and Epstein (2014).

Increasing bypass ratio (BPR) and the core pressure ratio both shrink the core size. The overall pressure ratio (OPR) of ~80$^+$ is deemed feasible according to Maynard 2015. However, in shrinking the core size, there is a transition point (of 3.5 lbm s^{-1} corrected mass flow rate at the compressor exit) that calls for a change in engine architecture from all-axial to mixed axial-centrifugal compression system (see Lord et al. 2015).

The propulsive efficiency of airbreathing jet engines depends on the engine BPR, α, as well as the exhaust jet-to-flight speed ratio, in the core and the fan, V_9/V_0 and V_{19}/V_0, respectively. Farokhi (2014) shows that for perfectly expanded nozzles and neglecting

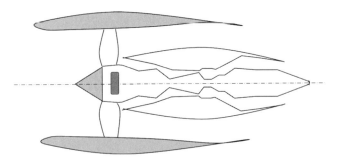

Figure 5.3 Schematic drawing of a geared ultra-high bypass ratio turbofan engine.

fuel-to-air ratio ($f \ll 1$), a separate-flow turbofan engine propulsive efficiency may be expressed as:

$$\eta_p \approx \frac{2\left[\left(\dfrac{V_9}{V_0}\right)+\alpha\left(\dfrac{V_{19}}{V_0}\right)-(1+\alpha)\right]}{\left(\dfrac{V_9}{V_0}\right)^2+\alpha\left(\dfrac{V_{19}}{V_0}\right)^2-(1+\alpha)} \tag{5.5}$$

Assuming the fan and core nozzles produce the same velocity ratio, i.e.

$$V_{19}/V_0 = V_9/V_0 \tag{5.6}$$

Equation (5.5) reduces to the familiar equation for the propulsive efficiency of a single-stream jet engine

$$\eta_p \approx \frac{2}{\dfrac{V_9}{V_0}+1} \tag{5.7}$$

Although BPR is explicitly missing from Eq. (5.7), it is, however, implicit in the nozzle velocity ratio, V_9/V_0. As the BPR increases, the jet speed ratio drops and in the limit of V_9/V_0 approaching 1, the propulsive efficiency approaches 100%. There are several modern examples in this category, such as ultra-high bypass (UHB) turbofan engine and advanced open-rotor propulsion system. In UHB architecture, the BPR exceeds 12 and in the case of open-rotor propulsion system, the BPR varies between 35 and 60. The enabling technologies of a geared TF-system in UHB and advanced propellers that employ sweep in a thin 3D woven composite propeller structure in open-rotor architecture are responsible for the successful development of these propulsion systems. The propulsive efficiency of current high-BPR turbofan engines is about 80% (see Whurr 2013). Figure 5.3 shows the schematic drawing of a geared, UHB turbofan engine.

The first generation geared turbofan (GTF) with BPR of 12 earns a ΔSFC of -15% relative to V2500 with BPR of 5 and OPR of 36. The future generation of UHB will feature BPR of 14–20 and OPR of 60$^+$. The new generation UHB is estimated to produce ΔSFC of -25% and correspondingly (fuel) ΔCO_2 of -25% relative to V2500 engine. Safran's open-rotor engine, with 30$^+$ BPR, is shown in Figure 5.4. It is expected to produce 30%

Figure 5.4 Counter-rotating open-rotor (CROR) engine architecture with bypass ratio in 30^+ range features 12 front and 10 aft rotor blades, 4 m diameter and 100-kN (22 000 lbf) thrust @ takeoff (SSL). *Source:* Courtesy of Safran Aircraft Engines.

reduced fuel burn compared to CFM56 turbofan engine. The engine has planned flight test for 2024–2025, development beginning in 2030 and EIS in 2035 (see Risen 2018). Comparing a GTF engine and the counter-rotating open rotor (CROR), we note that the ducted configuration in GTF could incorporate acoustic liners in the nacelle to mitigate noise, whereas the open-rotor engine has no (outer) duct, which makes it inherently noisier than the GTF. Consequently, the advanced rotor design, in open-rotor configurations, is aerodynamically optimized for reduced noise operation. The design parameters that prove beneficial in noise reduction of counter-rotating rotors are thin blades, aft blade clipping, blade count increase, and blade spacing increase. In addition, noise shielding by the aircraft wing and empennage is viewed as a noise mitigation design tool in modern commercial aviation. Safran CROR program envisions 15% fuel-burn reductions relative to the 2016 CFM LEAP engine by the year 2030–2035. The same CROR engine in boundary-layer ingestion (BLI) configuration is expected to produce 25% fuel-burn reduction by the year 2040–2045. Another challenge for CROR program, in addition to noise, is the blade-out contingency certification (e.g. through bird strike). Shielding of the airframe and critical systems to contain (the damage of) released blade adds weight to the aircraft that offsets the weight savings in open-rotor configurations. Envia (2010) discusses the open-rotor noise research at NASA.

5.3 Distributed Combustion Concepts in Advanced Gas Turbine Engine Core

A simple gas turbine engine has a single combustor in its core that is placed between the compressor and the turbine. The entire thermal power input by the fuel to the engine is $\dot{m}_f Q_R$, which is transferred to the fluid in one spot at burner efficiency, η_b. The resulting

(a)

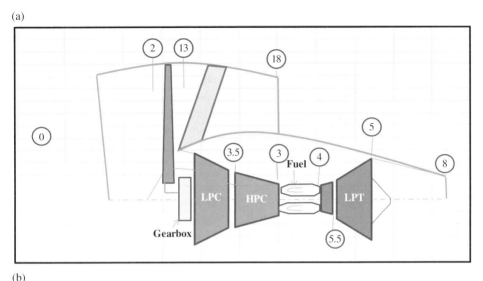

(b)

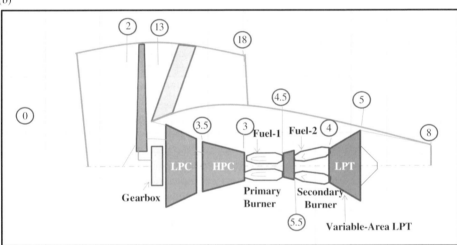

Figure 5.5 An alternative GT architecture to the conventional gas turbine design. (a) Conventional turbofan with a single combustion chamber (baseline). (b) Unconventional turbofan engine with two combustion chambers.

turbine entry temperature, TET, is limited by the turbine material and cooling technology, on the low side, and by the stoichiometric fuel-to-air ratio on the upper end. The current SOA places the TET at takeoff to around 1900–2000 K, whereas the stoichiometric flame temperature for Jet A (in air at 800 K initial temperature and 25 atm pressure) is around 2500 K (see Blazowski 1985, or Farokhi 2014). Figure 5.5 shows the two GT engine architectures. Part (a) is the conventional GT engine core and part (b) shows the GT core with two combustors, one between the HPC and high-pressure turbine (HPT) and the second one between the high-pressure spool and the low-pressure turbine (LPT), i.e. between the HPT and LPT.

Table 5.1 Common parameters in the two UHB turbofan architecture.

Bypass ratio	14	Fan pressure ratio	1.36
LPC pressure ratio	4.6	HPC pressure ratio	8.0
Core mass flow rate	$41 \, \mathrm{kg \, s^{-1}}$	Fuel Q_R (LHV)	$43\,124 \, \mathrm{kJ \, kg^{-1}}$
Takeoff thrust (SSL)	140 kN (31 500 lbf)	OPR @ Takeoff	50
The core and fan nozzles are both convergent with			$\pi_{fn} = \pi_{cn} = 0.96$

Table 5.2 Parameters unique to the two UHB engines.

One-burner UHB (baseline)		Two-burner UHB		
TET (T_{t4})	2100 K	$T_{t4.5}$	1600 K	Burner #1
		T_{t4}	1600 K	Burner #2
HPT	6–7% Cooling air fraction	HPT		Uncooled
LPT	3% Cooling air fraction	LPT		Uncooled
η_{HPT}	0.75[a]	η_{HPT}		0.94
η_{LPT}	0.85[a]	η_{LPT}		0.95
π_b	0.97	$\pi_{b1} = \pi_{b2}$		0.97

[a] The assumption of ~3% turbine efficiency loss per cooling percent follows Kerrebrock (1992).

Table 5.3 Performance gains in two-burner UHB at takeoff.

Δ (SFC)	~16% lower than baseline; which means 16% less fuel burn
Δ (Thermal Efficiency)	~6% higher than the conventional GT
Δ (CO_2 Emissions)	~16% lower due to reduced fuel burn
NO_x	Reduced due to lower TET (1600 K versus 2100 K)

To demonstrate the potential performance gains in the two-combustor system, we performed a simple cycle analysis at takeoff (standard sea level) condition for an UHB ratio turbofan engine (uninstalled). The cycle parameters are summarized in Tables 5.1 and 5.2. The gas is modeled in a zonal approach, namely cold and hot with:

$$\gamma_c = 1.4 \quad \gamma_t = 1.36 \quad c_{pc} = 1,004.5 \, \mathrm{J/kgK} \quad c_{pt} = 1,084 \, \mathrm{J/kgK}$$

The cycle analysis demonstrates the superiority of the dual-combustor GT architecture over the baseline engine, which employs single-combustor design. Compared to the conventional UHB, major reduction in fuel burn, significantly higher thermal efficiency and improved emissions are calculated for the two-burner UHB *at takeoff*. These are summarized in Table 5.3.

The thermodynamic state of the gas in the GT engine is shown in Figure 5.6. The two engines produce the same thrust (30 000 lbf). The dual-combustor engine produces a

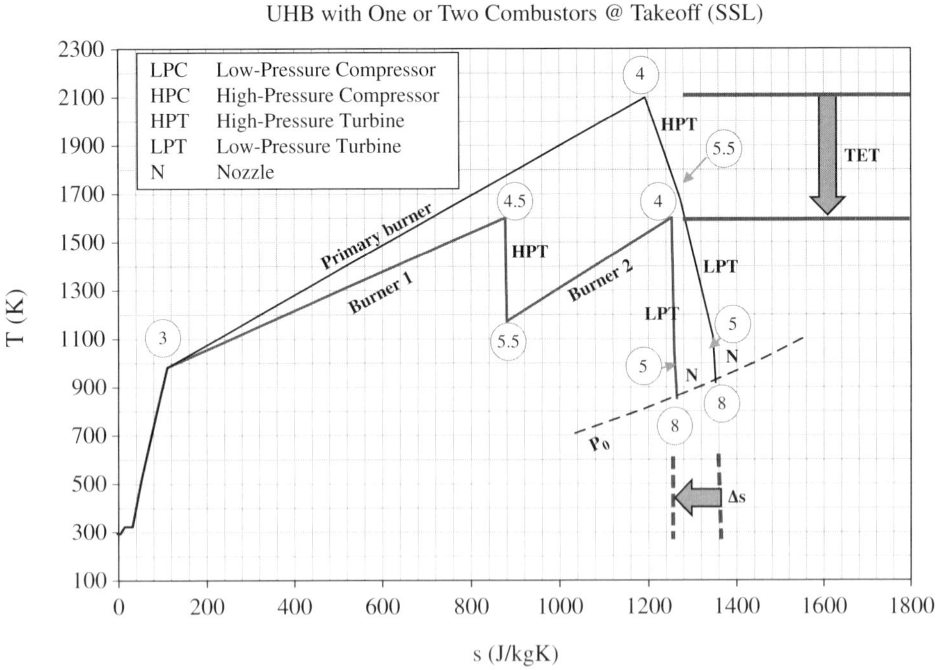

UHB with One or Two Combustors @ Takeoff (SSL)

Figure 5.6 Thermodynamics of a UHB-TF gas turbine engine cycle with a single- and dual-combustion system @ Takeoff, standard seal level ($T_0 = 288$ K and $p_0 = 101$ kPa), uninstalled.

more *Carnot-looking* cycle than the corresponding Brayton cycle. We also note that the conventional GT engine needed to operate at TET of 2100 K to produce the required takeoff thrust of 140 kN (31 500 lbf), whereas the dual combustor architecture had its TETs set at 1600 K, which corresponds to uncooled turbines using modern materials and thermal barrier coating (TBC). Higher thermal efficiency of nearly 6% and fuel-burn reduction of nearly 16% is calculated for the distributed combustion system over the conventional GT engine.

An engine simulation code by Joachim Kurzke (2017), known as GasTurb 13, is also used to validate the results of the simple baseline model. The relative error in thrust was 0.03%, the fuel-to-air ratio had a relative error of 1% and naturally the TSFC had a relative error of nearly 1%.

A final remark about this example is that the estimated performance gains in TSFC (of −16%) and thermal efficiency (of +6%) are not based on any optimization process. Consequently, higher performance gains is expected if the propulsion system is optimized. Figure 5.7 (from Clarke et al. 2012) shows projections for material and TET, or gas temperature, up to year 2030 for conventional and advanced high-temperature materials and for different cooling techniques. Second generation single crystal (nickel-based super-alloys) with TBC may operate at TET of 1500 °C (2732 °F) with no blade cooling by the year 2030. The highest TET in Figure 5.7 is projected for advanced cooling with TBC and ceramic matrix composite (CMC). The years 2030–2035 correspond to the EIS dates of NASA N+3 subsonic transport technology goals (see Del Rosario et al. 2012). The turbine cooling effectiveness parameter, ɸ, is shown for different

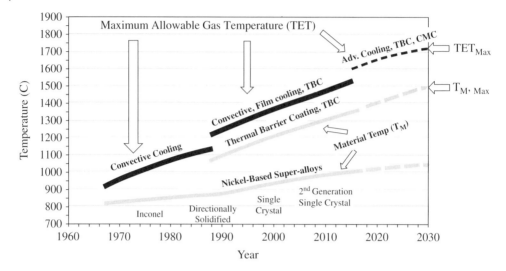

Figure 5.7 Material, gas or turbine entry temperature up to year 2030. *Source:* Adapted from Clarke et al. 2012.

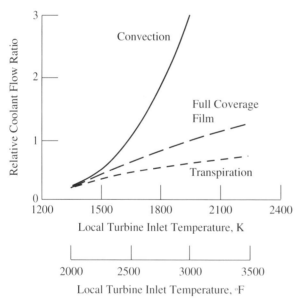

Figure 5.8 Turbine cooling methods and relative cooling flow requirement. *Source:* Adapted from Fraas 1980.

cooling methods in gas turbine engines in Figure 5.8 (adopted from Fraas 1980) shows the effect of cooling method on relative coolant flow in gas turbine blades. Moustapha et al. 2003, also address cooling effectiveness parameter, ϕ, as a function of cooling type and coolant percentage.

Additional distributed combustor configurations of gas turbine engines with interturbine burner, interstage burner, or sequential combustion are studied by Vögeler (1998), Sirignano and Liu (1999), Liu and Sirignano (2001), and Liew et al. (2003, 2005). These cycles will not be discussed further for brevity.

5.4 Multifuel (Cryogenic-Kerosene), Hybrid Propulsion Concept

The role of cryogenic fuel, such as liquid hydrogen (LH_2) and liquefied natural gas (LNG), is explored in a novel hybrid engine concept that uses two combustors (see Rao et al. 2014). The primary combustor uses cryogenic fuel whereas the second combustor, which is placed between the HPT and LPT, uses kerosene in a flameless burner. The cryogenic fuel is chosen for its environmental benefits, i.e. through low-carbon content in LNG and no-carbon content in LH_2. The kerosene/biofuel is chosen in the secondary flameless combustor for low NOx, soot, and UHC. The cooling capacity of the cryogenic fuels is further explored for bleed air-cooling, i.e. to use cryogenic fuel as the heat sink for the cooling air. The challenge of low volumetric efficiency of LH_2 and LNG is addressed by combining it with a secondary drop-in fuel (kerosene/biofuel) as well as integrating it in a volumetrically efficient blended wing-body (BWB) transport aircraft designed for long-range (14000 km or 7500 nautical miles) flight. This configuration designed for the 300-passenger aircraft provides ample volume for multifuel storage in a long-range transport. Figure 5.9 shows the cross-sectional view of the concept engine in separate- and mixed-flow turbofan engine as well as the BWB aircraft designed for the multifuel propulsion system; *AHEAD MF-BWB* aircraft (from Grewe et al. 2016 and Rao et al. 2014). The temperature-entropy (T-s) diagram for the hybrid engine for LNG and LH_2 is shown in Figure 5.10. The 300-passenger BWB aircraft designed for long-range mission is shown in Figure 5.11 (from Rao et al. 2014).

Compared to a baseline engine (GE90-94B), the hybrid multifuel engine produced superior performance results that are shown in Table 5.4 (from Rao et al. 2014). Liquid hydrogen/kerosene produced 12% reduction in fuel burn and 94% reduction in CO_2. The TSFC reduction in LNG/kerosene hybrid engine over the baseline is 4%. The reduction in CO_2 in this multifuel hybrid engine is reported as 57%. These are clearly promising results for advanced gas turbines with the potential to lead to sustainable aviation.

A thorough climate impact study of BWB aircraft with twin-hybrid, multifuel engines identified a reduction in global warming effect due to reduced CO_2, NO_x, and contrails. As expected, burning hydrogen or low-carbon fuel (e.g. LNG) results in an increase in water vapor deposition that leads to increased warming effect. These climate impact results (shown in %) are summarized in Table 5.5. The baseline aircraft in comparison is B787. For environmental impact study of BWB aircraft with two multifuel hybrid engines see Grewe et al. (2016) and Grewe and Linke (2017).

5.5 Intercooled and Recuperated Turbofan Engines

The thermodynamic advantage of intercooler is in its potential to produce higher cycle thermal efficiency, higher OPR, and lower specific fuel consumption (SFC). One concept uses a part of the fan bypass duct for air cooling of the intermediate pressure compressor (IPC) that loops through an advanced design air heat exchanger back to the HPC. The separate cooling bypass duct for the intercooler produces total pressure loss. The turnaround duct/manifold for the IPC intercooler also experiences some total

(a)

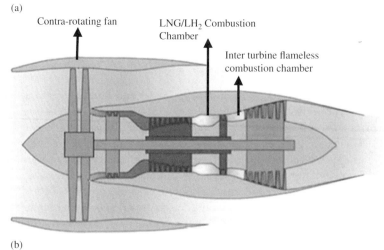

Contra-rotating fan

LNG/LH$_2$ Combustion
Chamber

Inter turbine flameless
combustion chamber

(b)

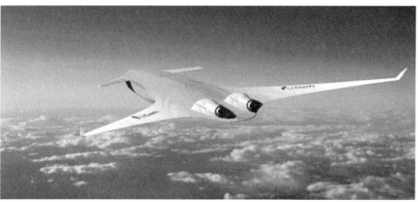

(c)

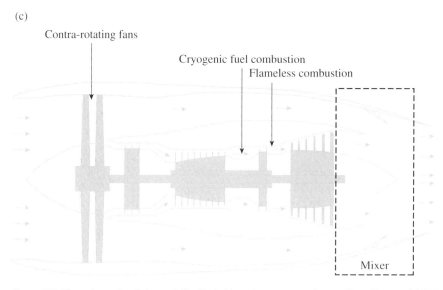

Contra-rotating fans

Cryogenic fuel combustion

Flameless combustion

Mixer

Figure 5.9 The schematic of the multifuel hybrid engine concept. *Source:* From Rao et al. 2014.
(a) Separate-flow turbofan engine with dual combustor. (b) Blended wing–body aircraft designed
for the new multifuel propulsion system (MF-BWB). (c) Mixed-flow turbofan concept with multifuel
architecture.

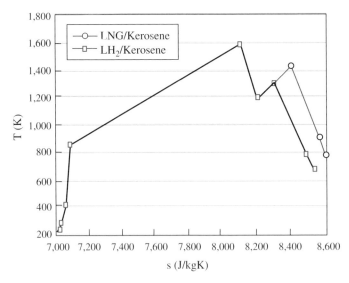

Figure 5.10 T-s diagram of the multifuel hybrid gas turbine engine that uses one cryogenic fuel (LNG and LH$_2$) and kerosene in two combustors. *Source:* Data from Rao et al. 2014.

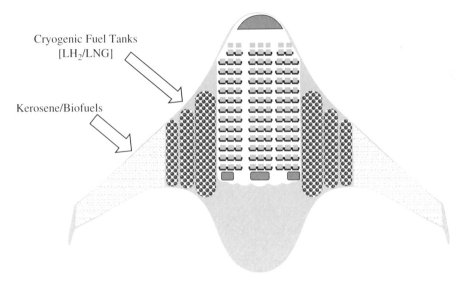

Figure 5.11 A Multifuel (cryogenic/kerosene) BWB aircraft concept. *Source:* From Rao et al. 2014.

pressure loss. The higher thermal efficiency of the intercooled cycle, in this concept causes about 4%, reduction in fuel burn according to Whurr (2013). Figure 5.12 shows thermal efficiency gains with intercoolers and recuperator. This level of performance gain seems to be insufficient to overcome the disadvantages of the concept, in its present configuration. However, the intercooling/thermal management concept will remain a promising technology. The development of high surface area, ultra-lightweight

Table 5.4 Impact of multifuel engine architecture on performance parameters.

Performance change (%)	CO_2	SFC	Specific thrust
LH$_2$/Kerosene	−94	−12	−1
LNG/Kerosene	−57	−4	+5

Table 5.5 Climate impact (%).

Climate change (%)	CO_2	NO_x	Contrails	H_2O	Total
LH$_2$	−6	−29	−15	+25	−25
LNG	−0.6	−28	−16	+12	−32

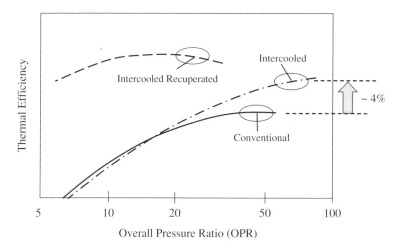

Figure 5.12 Thermal efficiency of different gas turbine cycles. *Source:* Adapted from Wilfert et al. 2007.

heat exchangers holds the key to the future use of intercooled and recuperated aeroengines. Figure 5.13 shows the schematic drawing of an intercooled turbofan engine.

One additional configuration using intercoolers and recuperator is shown in Figure 5.14 (from Xu et al. [2013]). The intercooling loop in Figure 5.14 is between the centrifugal compressor (IPC) and the HPC centrifugal compressor. The recuperation loop between the HPC and the burner provides for higher thermal efficiency through reduced fuel burn.

Another concept for the intercooler uses a cryogenic fuel as the heat sink and a recuperative that both contribute to reduced fuel burn (see Figure 5.15). The estimated 22% reduction in fuel burn is reported by Kyprianidis (2011) in a geared intercooled, recuperated turbofan engine that is indeed promising.

In UHB turbofan engines, the disparity between the fan diameter and the small core diameter becomes even more profound. As a result, the whole engine modeling is

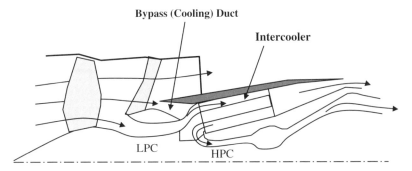

Figure 5.13 Flowpath in an intercooled turbofan engine.

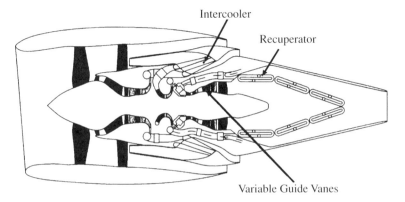

Figure 5.14 Intercooled recuperated turbofan engine concept by MTU Aero Engines. *Source:* From Xu et al. 2013.

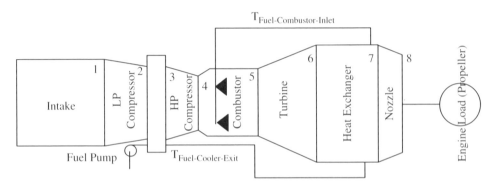

Figure 5.15 A cryo-intercooler concept with recuperative cycle. *Source:* Adapted from Whurr 2013.

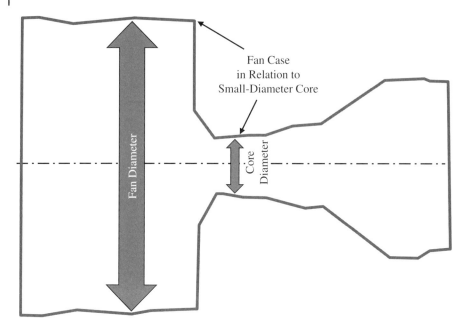

Figure 5.16 Schematic drawing of a UHB showing the relative proportion of fan and core diameters.

needed to assess small core aeromechanical response to thrust and maneuver loads. Figure 5.16 shows the schematic drawing of a candidate UHB and the disparity between the fan and core diameters.

5.6 Active Core Concepts

Active tip clearance and surge control are powerful means of improving HPC aero-thermodynamic efficiency. As the core size shrinks in high OPR gas turbine engines, the need for tip clearance control becomes more evident. These improvements coupled with advances in flow control blade design, for example through aspiration concept, and active cooling of the turbine cooling air, are the major elements of smart HPC development in aircraft engine industry today. The schematic drawings of these concepts are shown in Figure 5.17 (from Bock et al. 2008).

The impact of active core technology on turbofan SFC, is significant. For each percent improvement in component efficiency or turbine cooling with $-100\,K$ cooling air is captured in Figure 5.18 by Bock et al. 2008. The 15% reduction in cooling is achieved by $-100\,K$ cooling air. The reduction in SFC yields 6% CO_2 and 16% NO_x reduction, according to Bock et al. 2008.

5.7 Topping Cycle: Wave Rotor Combustion

Conventional throughflow combustors in gas turbine engines are ideally, inviscid, low-speed, and steady flow reactors. Under these conditions, chemical reaction takes place at constant pressure. In reality, the combustion takes place at finite Mach number with

(a)

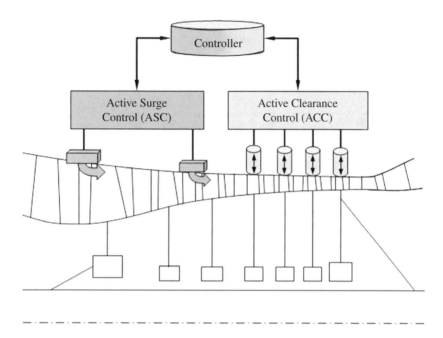

(b)

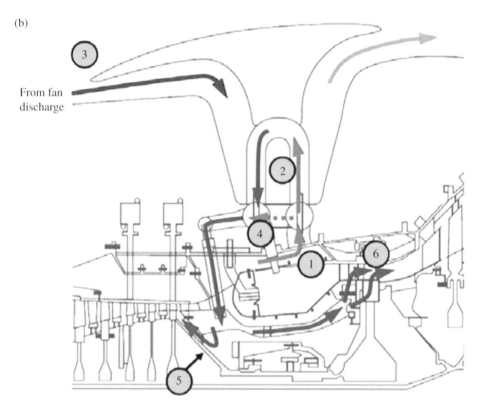

Figure 5.17 Active core concepts. *Source:* From Bock et al. 2008. (a) Closed-loop active clearance and surge control. (b) Schematic drawing of active cooling concept of the turbine cooling air.

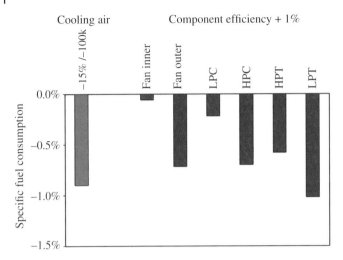

Figure 5.18 Impact of component efficiency improvements and −100 K reduced temperature cooling air on turbofan specific fuel consumption. *Source:* from Bock et al. 2008.

turbulent mixing of fuel-air, coolant-gas, and viscous drag imposed at its boundary, i.e. the combustor liner. Thus, the reacting flow process in the combustor suffers from total pressure loss, i.e. $\pi_b < 1$, that is the measure of irreversibility in the process. Kerrebrock (1992) proposed a simple model that related the total pressure loss to the mean dynamic pressure in the combustor according to:

$$1 - \pi_b \sim \varepsilon \left(\frac{\gamma}{2} M_b^2 \right) \tag{5.8}$$

where $1 < \varepsilon < 2$ and M_b represents the mean Mach number of the reacting gases in the combustor.

In gas dynamics, the study of shock tube taught us the principles of one-dimensional, unsteady wave motion. We learned that when a high-pressure (driver) gas is separated from a low-pressure (driven) gas by a diaphragm (or valve), sudden opening causes gas dynamic waves to be generated. We learned that these waves are initially infinitesimal compression and expansion Mach waves. The compression Mach waves catch up with each other, i.e. they all catch up with the head compression Mach wave and form a normal shock. The compression Mach waves thus coalesce to form the shock. In contrast, the expansion Mach waves spread and form a wave train. Hence, an incipient shock is formed that propagates in the driven gas and thus transfers energy to the gas as well as increase its pressure. Thermodynamic state across the shock is governed by Hugoniot equation:

$$e_2 - e_1 = \frac{p_1 + p_2}{2} \left(\frac{1}{\rho_1} - \frac{1}{\rho_2} \right) \tag{5.9}$$

Assuming calorically perfect gas, we may relate pressure and density ratio across a propagating shock according to:

$$\frac{p_2}{p_1} = \frac{\left(\dfrac{\gamma+1}{\gamma-1}\right)\dfrac{\rho_2}{\rho_1} - 1}{\left(\dfrac{\gamma+1}{\gamma-1}\right) - \dfrac{\rho_2}{\rho_1}} \tag{5.10}$$

The expansion wave train propagates is entropically in the opposite direction, i.e. in the high-pressure driver gas, to accelerate it and relieve its pressure. Unsteadiness, as a means of energy transfer in gas dynamics, is best captured by the energy equation written for a fluid particle in inviscid flow with no body force (see Greitzer et al. 2004):

$$\rho \frac{Dh_t}{Dt} = \frac{\partial p}{\partial t} \tag{5.11}$$

Classical textbooks by Shapiro (1953), Liepmann and Roshko (1957), and modern classics such as Anderson (2003), among other gas dynamics books, e.g. Kentfield (1993), provide the background on unsteady gas dynamics, nonlinear waves and energy transfer. Kentfield (1995) examines the feasibility of pressure gain combustors for gas turbine applications. A historical perspective is produced by Hawthorne (1994) on aircraft gas turbines in United Kingdom.

In gas turbines, we generate high-pressure gas at the compressor exit; thus, a rotating multi-passage channel located between the compressor exit and the combustor could simulate the opening and closing "diaphragm" action, as in a shock tube. Hence, compression and expansion waves are generated that promote unsteady energy transfer to the compressor discharge gas. Thus, enter the *wave rotor* as a power exchange device with a potential to enhance the performance of gas turbine propulsion and power systems. The discharge of the wave rotor enters a combustor with both ends simultaneously closed, which in essence it becomes a constant-volume combustor with pressure gain. The ideal version of this cycle is known as the Humphrey cycle. The high-pressure and temperature gas at the end of combustion suddenly faces a low-pressure passage, as the wave rotor spins. The expansion wave train causes temperature drop and thus the gas at the turbine inlet has a lower temperature and higher pressure than achieved in the constant-volume combustion section. Thus, the new cycle that incorporates a wave rotor, in this capacity, is called the *topped* cycle in relation to the baseline cycle. In another application, since the working principle behind a wave rotor is the energy transfer between longitudinally traveling waves and the gas in shrouded passages, it is possible to convert the wave-power into a net shaft power. Figure 5.19 shows the wave-rotor hardware, the test rig at NASA-Glenn Research Center (NASA-GRC) and schematic drawing wave rotor integration in gas turbine engines.

The variants of wave rotor application in gas turbines is summarized by Welch et al. (1995). These include pressure gain wave rotor with combustion on the rotor, wave engine with combustion on the rotor, pressure gain wave rotor with conventional burner, and wave engine with conventional burner. Figure 5.20 shows the schematic drawing of these variants.

(a)

Figure 5.19 The rotating drum in a wave rotor combustion system and the test rig. *Source:* Courtesy of NASA-Glenn Research Center. (a) wave rotor hardware (rotating drum); (b) wave Rotor Rig at NASA-Glenn Research Center; (c) schematic for combustion wave rotor integration into gas turbine engines. *Source:* Adapted from Akbari et al. (2007).

(b)

(c)

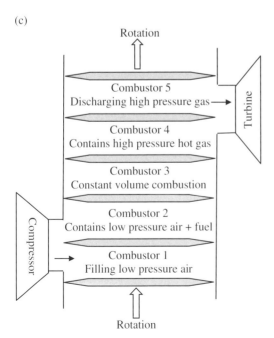

Rotation

Combustor 5
Discharging high pressure gas →

Combustor 4
Contains high pressure hot gas

Combustor 3
Constant volume combustion

Combustor 2
Contains low pressure air + fuel

Combustor 1
Filling low pressure air

Rotation

Turbine

Compressor

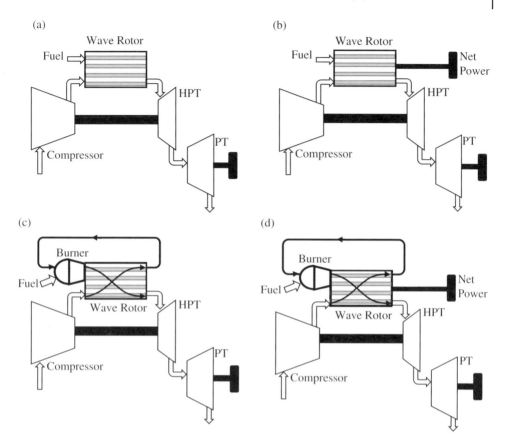

Figure 5.20 Wave rotor/gas turbine engine integration variants (HPT is high-pressure turbine, PT is power turbine): (a) pressure gain wave rotor with combustion on the rotor; (b) wave engine with combustion on the rotor; (c) pressure gain wave rotor with Conventional burner; (d) wave engine with conventional burner.

Schematic drawing of a wave-rotor topped turboshaft engine is shown and compared to a baseline turboshaft engine in Figure 5.21. The baseline engine is composed of a gas generator on the same shaft followed by the power turbine on a separate shaft. The SFC for the wave rotor topping cycle for a small turboshaft engine at 85% power, with no cooling flow extraction from the wave rotor, is shown in Figure 5.22 (data from Jones and Welch 1996). The wave rotor topped engine shows that for the same shaft power, nearly 13% reduction in fuel burn is achieved.

The pressure gain in the wave rotor is expressed as the wave rotor pressure ratio, PR_W, which is a design parameter, with the corresponding wave rotor temperature ratio, TR_W. These parameters are defined as:

$$PR_W \equiv \frac{p_{4A}}{p_{3A}} \tag{5.12}$$

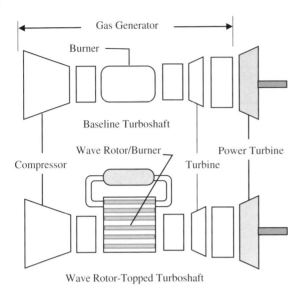

Figure 5.21 Schematics drawing of a wave-rotor topped turboshaft engine.

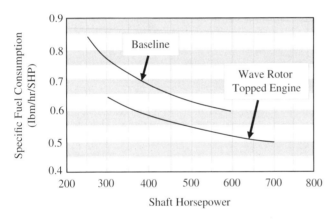

Figure 5.22 Off-design (i.e. 85% power) performance benefit of a wave rotor topped turboshaft engine compared to baseline (inlet corrected mass flow rate @ design point is 5 lbm s^{-1}). *Source:* Data from Jones and Welch 1996.

$$TR_W \equiv \frac{T_{4A}}{T_{3A}} \tag{5.13}$$

Based on the station numbers defined in Figure 5.23 for a four-port wave rotor and burner system.

The design point pressure ratio of the wave rotor is a function of the wave rotor temperature ratio, as shown in Figure 5.24.

The wave rotor-enhanced turbofan engine is also evaluated. A high-BPR turbofan engine with an OPR = 39 and TET of 3200 °R (1778 K) at 80000–100000 lbf thrust class

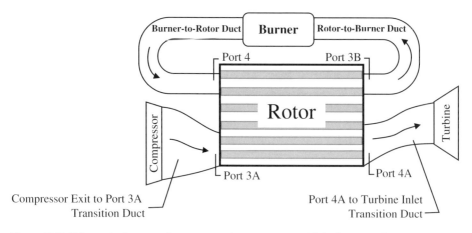

Figure 5.23 Schematic diagram of a wave rotor burner system with its four-port designations.

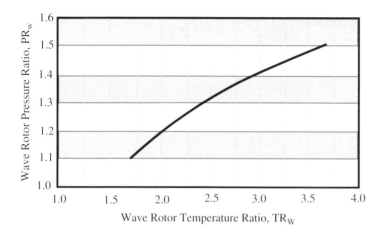

Figure 5.24 Design-point wave rotor pressure ratio variation with temperature ratio. *Source:* Adapted from Jones and Welch 1996.

is considered as *the baseline engine*. Wave rotor pressure ratios of $PR_W = 1.08$ and 1.15 are determined to be appropriate for this application. While maintaining the same fuel flow as the baseline case, wave rotor-enhanced cycle produces higher cycle pressure ratio (PR) and thrust, and thus, a lower TSFC. Figure 5.25 (from Jones and Welch 1996) show the schematic drawings of the two turbofan engines.

The added weight of the wave rotor is estimated to be 25% of the basline core engine weight. In the large turbofan engine example that was studied by Jones and Welch, the estimated wave-rotor weight was 1650 lbs. (750 kg). To establish the impact of the wave rotor-enhanced tubofan engine on aircrtaft performance, the baseline engine had a BPR of 7, net thrust at takeoff (hot day) was assumed to be 86 820 lbf, engine weight 20 430 lbf with an inlet air flow rate of 2800 lbms s^{-1} (1270 kg s^{-1}). The coolant fraction (bleed) was assumed to be 20%, which is typical for this class of engine. The fan pressure ratio (FPR) was 1.59, the low-pressure compressor (LPC) had 1.55 and the HPC had 15.8 pressure

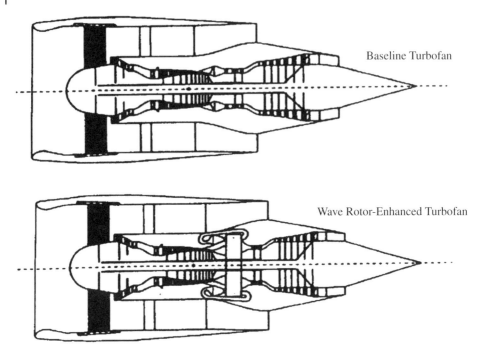

Baseline Turbofan

Wave Rotor-Enhanced Turbofan

Figure 5.25 Schematics of baseline and wave rotor-topped turbofan engines. *Source:* From Jones and Welch 1996.

ratio, which gives the OPR of 39. The aircraft cruised at 40 kft altitude at Mach 0.85 for a range of 6500 nm. The aircraft wing loading (W/S) was 130 lbf ft^{-2} and thrust-to-weight ratio (T/W) at takeoff was 0.30.

The reduction in the aircraft takeoff gross weight (TOGW) relative to the baseline for different wave rotor weights is plotted in Figure 5.26 (data from Jones and Welch 1996). With the estimated wave rotor weight of 1650 lbs., the relative TOGW of wave rotor-enhanced turbofan is reduced by ~3% for PR$_W$ of 1.08 and by ~6% for *PR$_W$* of 1.15. The net thrust for the wave rotor topped engine increased to 88 370 and 89 470 lbs. for *PR$_W$* = 1.08 and 1.15, respectively. These increased thrust levels for the same fuel flow results in TSFC reduction of ~1.7% and ~3%, respectively, for the two-wave rotor pressure ratios of 1.08 and 1.15. These modest gains in performance, namely increased thrust (and reduced fuel burn), translates into reduced corrected flow (for the same thrust). In turn, component sizes are reduced. Smaller turbine, as noted by Jones and Welch (1996) experiences about 8% lower AN^2, or centrifugal stress than the baseline engine. The weights of HPT and LPT are reduced by about 6% and 10%, respectively.

Wave rotor-enhanced turboshaft engine for the auxiliary power unit (APU) application on aircraft is also considered. The APU function is to provide pressurized bleed air for the aircraft pneumatic and environmental control system (ECS) as well as shaft power for the electric generator. Two different approaches to APU with wave rotor technology is studied. The baseline and the two wave-rotor variant configurations are shown in Figure 5.27.

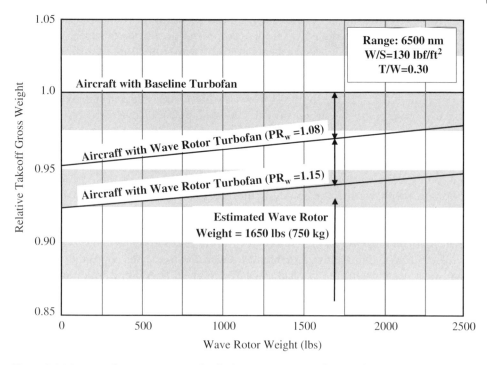

Figure 5.26 Impact of wave rotor topped turbofan engine on aircraft TOGW for wave rotor $PR_W = 1.08$ and 1.15. *Source:* Data from Jones and Welch 1996.

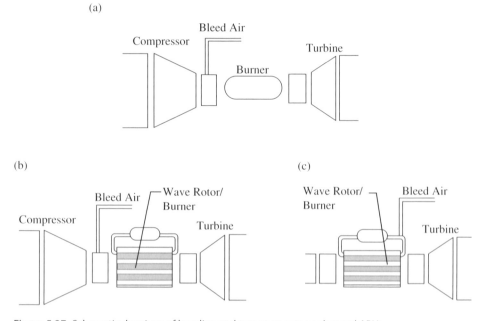

Figure 5.27 Schematic drawings of baseline and two wave rotor-enhanced APUs.

The baseline configuration is shown with bleed air taken from the compressor exit and the turbine provides the power to the compressor in the gas generator as well as providing shaft power to an electric generator. The wave rotor-topped APU generates higher pressure ratio and thus achieves improved performance over the baseline. The bleed air is taken at station 3A, i.e. before the wave rotor. In the second variant, the wave rotor eliminates the need for compressor and the bleed air is taken in station 3B. In the absence of compressor, the turbine provides the shaft power to an electric generator. To achieve the necessary wave rotor PR, the fuel burn in this APU is higher than the baseline. However, the APU system weight reduction, maintenance, and reliability is improved. Another possibility would have eliminated the turbine as well, since the wave rotor can be designed to produce shaft power. The summary of the APU cycle and performance with wave rotor are show in Tables 5.6 and 5.7 (data from Jones and Welch 1996).

For the review of the wave rotor combustion technology see Akbari et al. (2006, 2007). For the mathematical model of multi-port wave rotors based on the conservation laws for one-dimensional unsteady gas dynamics Welch (1996) may be consulted. For additional wave rotor/engine study results see Welch et al. (1995) and Jones and Welch (1996).

As a simple example of a higher efficiency cycle that takes advantage of constant-volume combustion, we examine the Humphrey cycle in the next section. Strictly speaking, pulse detonation engines (PDEs) are not constant-volume combustion devices, but their efficiencies are closely approximated by the Humphrey cycle.

Table 5.6 APU cycle parameters.

	Baseline APU	Wave rotor APU Topped cycle	Wave rotor APU No compressor
Inlet flow	$6.02\,\text{lbm s}^{-1}$ ($2.73\,\text{kg s}^{-1}$)	$6.02\,\text{lbm s}^{-1}$ ($2.73\,\text{kg s}^{-1}$)	$6.02\,\text{lbm s}^{-1}$ ($2.73\,\text{kg s}^{-1}$)
Inlet recovery	1	1	1
Inlet temperature	559.7 °R (311 K)	559.7 °R (311 K)	559.7 °R (311 K)
Compressor PR	4.0	4.0	—
Compressor efficiency	0.77	0.77	—
Compressor corrected flow	$6.25\,\text{lbm s}^{-1}$ ($2.83\,\text{kg s}^{-1}$)	$6.25\,\text{lbm s}^{-1}$ ($2.83\,\text{kg s}^{-1}$)	—
Bleed Flow (%)	28.4	28.4	28.4
Bleed flow rate	$1.71\,\text{lbm s}^{-1}$ ($0.775\,\text{kg s}^{-1}$)	$1.71\,\text{lbm s}^{-1}$ ($0.775\,\text{kg s}^{-1}$)	$1.71\,\text{lbm s}^{-1}$ ($0.775\,\text{kg s}^{-1}$)
Wave rotor TR	—	2.30	3.73
Turbine inlet temperature	2086 °R (1159 K)	2086 °R (1159 K)	2085 °R (1158 K)
Turbine efficiency	0.83	0.83	0.83

Table 5.7 APU performance results.

	Baseline APU	Wave rotor APU Topped cycle	Wave rotor APU No compressor
Wave rotor PR	—	1.25	1.24
Turbine expansion ratio	3.26	4.13	1.09
Turbine corrected flow	2.34 lbm s^{-1} (1.06 kg s^{-1})	1.83 lbm s^{-1} (0.83 kg s^{-1})	7.27 lbm s^{-1} (3.3 kg s^{-1})
Shaft power	60 HP (44.7 kW)	187 HP (139.4 kW)	60.0 HP (44.7 kW)
Fuel flow	281 lbm h^{-1} (127 kg h^{-1})	281 lbm h^{-1} (127 kg h^{-1})	449 lbm h^{-1} (204 kg h^{-1})
Bleed flow pressure	51 psia (352 kPa)	51 psia (352 kPa)	548 psia (378 kPa)

5.8 Pulse Detonation Engine (PDE)

The thermodynamics of PDE is analyzed by Heiser and Pratt (2002). The ideal cycle thermal efficiency of the PDE is estimated between 40% and 80% for hydrocarbon fuels. The specific impulse, I_S, is estimated in the range of 3000–5000 seconds. These performance parameters exceed their equivalent Brayton cycle counterparts. As a precurser to PDE studies, we start with the analysis of Humphrey cycle, which is often studied as a surrogate to PDE.

5.9 Humphrey Cycle vs. Brayton: Thermodynamics

An ideal Humphrey cycle is shown in Figure 5.28a as the pressure-volume and Figure 5.28b, as the temperature-entropy diagrams. Combustion takes place at constant-volume in a Humphrey cycle that results in pressure gain whereas it takes place at constant pressure in an ideal Brayton cycle. The static pressure rise in a constant-volume combustion can easily be seen on the pressure-volume diagram of the Humphrey cycle operating between the same temperature limits of a Brayton cycle, as in Figure 5.28a.

We utilize the definition of cycle efficiency and thermodynamic principles to get Brayton and Humphrey cycle efficiencies. The ideal cycle thermal efficiency of a constant-pressure combustion (Brayton) cycle: 1-2-5-6-1 (shown in Figure 5.28), is rewritten from Eq. (5.1), as

$$\eta_{th} = 1 - \frac{T_1}{T_2} \left[\text{Ideal Brayton cycle thermal efficiency} \right] \tag{5.14}$$

Note that ideal Brayton cycle thermal efficiency is explicitly independent of γ. The thermal efficiency of an ideal constant-volume combustion cycle, that results in pressure gain, known as the Humphrey cycle: 1-2-3-4-1 (shown in Figure 5.29) is

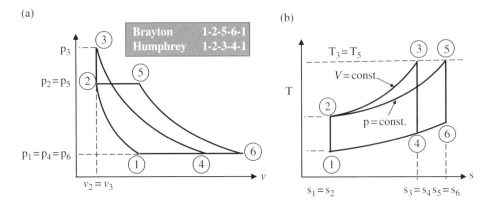

Figure 5.28 Constant-volume and constant-pressure combustion cycles in p-v and T-s diagrams operating between the same temperature limits: (a) p-v diagram of Brayton and Humphrey cycles; (b) T-s diagram of the Brayton and Humphrey cycles.

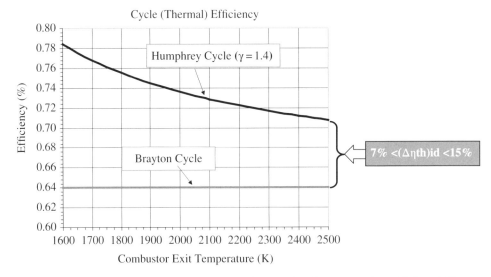

Figure 5.29 Thermal efficiency of ideal Humphrey (for $\gamma = 1.4$) and Brayton cycles, with $T_1 = 288$ K, $T_2 = 800$ K and T_3, combustor exit temperature, that varies between 1600 and 2500 K.

$$\eta_{th} = 1 - \gamma \frac{T_1}{T_2} \left[\left(\frac{T_3}{T_2} \right)^{\frac{1}{\gamma}} - 1 \right] / \left[\frac{T_3}{T_2} - 1 \right] \quad \text{[Ideal Humphrey cycle thermal efficiency]} \quad (5.15)$$

where γ is the ratio of specific heats. Thermal efficiency in the Humphrey cycle depends on T_1/T_2 and on the temperature ratio T_3/T_2 (in effect p_3/p_2). Figure 5.29 shows the ideal cycle thermal efficiency of a Brayton and a Humphrey cycle (for $\gamma = 1.4$) and $T_1 = 288$ K, $T_2 = 800$ K, and T_3 varies between 1600 and 2500 K. Thermal efficiency in the Humphrey cycle is higher than Brayton by as much as 15% (see Figure 5.29).

The ratio of entropy rise across the combustors in the two cycles, written based on the Gibbs equation, may be used to compare the efficiency of energy transfer to the fluid in the two combustors according to

$$\frac{\Delta s_{2-3}}{\Delta s_{2-5}} = \frac{\int_2^3 c_v \frac{dT}{T} - Rln\frac{v_3}{v_2}}{\int_2^5 c_p \frac{dT}{T} - Rln\frac{p_5}{p_2}} \approx \frac{\bar{c}_v}{\bar{c}_p} \approx \frac{1}{\bar{\gamma}}$$

(5.16)

For the mean ratio of specific heats in the combustor, $\bar{\gamma} = 1.33$, we get the ratio of entropy rise across the constant-volume combustor to the constant-pressure combustor as 0.75. This means that 25% lower entropy is generated in a pressure-gain combustor as compared to a constant-pressure burner operating between the same temperature limits. The lower entropy generation is then related to higher thermal efficiency of the Humphrey as compared to the Brayton cycle operating at the same temperature limits.

5.9.1 Idealized Laboratory PDE: Thrust Tube

An idealized PDE is a tube, aka *thrust tube* or detonation tube, with one end closed and one end open. The tube is filled with a combustible mixture of fuel/air or fuel/oxidizer. One or more powerful electric discharge serve as the ignitor near the closed end of the tube that initiates the chemical reaction in the mixture. Depending on the thrust tube design, either a deflagration or a detonation combustion wave is formed. Deflagration corresponds to the low-speed (subsonic) combustion, as in a pulsejet or conventional GT engines, whereas detonation is a supersonic propagating wave in the combustible mixture, as in PDE. The detonation (shock) wave propagates in the tube toward the open end. Detonation wave transfers energy to the gas and raises its temperature, pressure, and density. In addition, the chemical energy due to combustion is released and the stagnation temperature of the gas increases. In this sense, detonation shock wave is nonadiabatic and is supported by the energy release in the combustion process. The Mach number of the normal detonation shock, known as the Chapman-Jouguet (CJ) wave, is the ratio of the wave speed relative to the unburned gas to the speed of sound in the unburned gas. This is derived, among others, by Shapiro (1953), Strehlow (1984), and Pratt et al. (1991), as

$$M_{CJ}^2 = (\gamma + 1)\frac{\tilde{q}}{\psi} + 1 + \sqrt{\left[(\gamma + 1)\frac{\tilde{q}}{\psi} + 1\right]^2 - 1}$$

(5.17)

where $\tilde{q}$ is the nondimensional heat release due to combustion of fuel, defined as

$$\tilde{q} \equiv \frac{fQ_R}{c_p T_0}$$

(5.18)

with f as the fuel-to-air or fuel-oxidizer ratio and Q_R is the heat of reaction of the fuel. Finally, ψ, in Eq. (5.17) is defined as

$$\psi \equiv \frac{T_3}{T_0}$$

(5.19)

where T_3 is the combustor inlet (static) temperature and T_0 is the flight static temperature. The detonation wave Mach number is in the 5–10 range and the combustion gas downstream of the shock attains sonic speed relative to the shock. This thermal choking of the combustion gas in detonation wave theory is known as Chapman-Jouguet rule. The PDE cycle thermal efficiency is related to Chapman-Jouguet Mach number and the nondimensional heat release as

$$
\eta_{th} = 1 - \frac{\left[\dfrac{1}{M_{CJ}^2} \left(\dfrac{1+\gamma M_{CJ}^2}{\gamma+1} \right)^{\frac{\gamma+1}{\gamma}} - 1 \right]}{\tilde{q}}
\tag{5.20}
$$

The value of $\tilde{q}$ for hydrocarbon fuels ($Q_R \simeq 41\,800\,\mathrm{kJ\,kg^{-1}}$) in stoichiometric combustion with air at SSL conditions is about 10. The value of $\tilde{q}$ for hydrogen (based on the lower heating value of hydrogen, $Q_R \simeq 120\,000\,\mathrm{kJ\,kg^{-1}}$) in stoichiometric combustion in SSL air is about 12. The range of ψ that is considered in performance comparisons between the PDE, Humphrey, and Brayton cycles is between 1 and 5 (see Heiser and Pratt 2002).

The detonation wave causes mass motion of the gas to trail the shock. Thus, a trail of chemically reacting combustion gases follow the detonation wave toward the open end. The detonation wave is reflected from the open end as an expansion wave propagating backward into the tube. This follows the principle of "unlike" reflections from fluid surfaces in gas dynamics, where a shock reflects as expansion wave and an expansion (Mach) wave reflects as compression Mach wave. The expelled combustion gases from the tube then enter an exhaust nozzle for further expansion and thrust production. The reflected shock that propagates as expansion wave back in the tube reaches the closed end and the cycle repeats after a fresh mixture is drawn in due to reduced pressure and detonated. Therefore, the basic cycle of *fill-ignite-exhaust* characterizes the fundamental processes in PDE. The schematic of an idealized PDE (thrust tube) is shown in Figure 5.30. The complete (five-step) PDE cycle is shown in Figure 5.31 (McCallum 2000) for reference.

The pulse frequency in PDE is a cycle design parameter that is inversely proportional to the filling time of combustible mixture in the tube. This is typically in the range of 50–300 Hz, i.e. number of detonations per second. Since an impulse is produced per cycle of PDE operation, then thrust is directly proportional to the pulse frequency. The unsteady nature of thrust generation in PDE has consequences in noise and structural fatigue as design-critical issues. In addition, the intake and exhaust system design in a

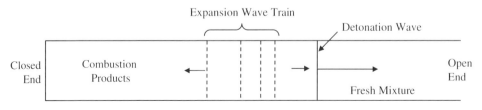

Figure 5.30 The wave system in an idealized PDE thrust tube, in the laboratory frame of reference.

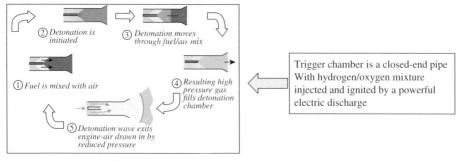

② *Detonation is initiated*
③ *Detonation moves through fuel/air mix*

① *Fuel is mixed with air*
④ *Resulting high pressure gas fills detonation chamber*

Trigger chamber is a closed-end pipe With hydrogen/oxygen mixture injected and ignited by a powerful electric discharge

⑤ *Detonation wave exits engine-air drawn in by reduced pressure*

PDE Wave Cycle

Figure 5.31 The pulse detonation engine with a trigger chamber. *Source:* From McCallum 2000.

PDE is subject to the same challenges in unsteady gas dynamics and efficiency of the compression and expansion processes that lead to an efficient propulsion system. The underlying efficiency of pressure-gain combustion is at the center of this technology's appeal. Some of the promising applications of this technology can be realized in pulse detonation ramjets (PDR), afterburning turbofan engine with PDE, pulse detonation rocket engines (PDREs) and the hybrid turbine PDE.

5.9.2 Pulse Detonation Ramjets

Conventional ramjets have an air intake system, a (steady) throughflow (deflagration-based) combustor, and an expansion nozzle. The cycle that describes the state of the gas in the engine is the Brayton cycle. Since this system lacks mechanical compression, cycle pressure ratio at takeoff is one and hence its thermal efficiency (and specific impulse) is zero; therefore a conventional ramjet produces no thrust at takeoff. The ideal combustion in a Brayton cycle is a low-speed (deflagration type), constant-pressure process. Its constant-volume counterpart is the Humphrey cycle, as discussed earlier. The constant-volume Humphrey cycle produces static pressure rise and is thus thermally more efficient than its Brayton counterpart. In fact, it is capable of producing static thrust (see Nalim 2002a, b). Figure 5.32 shows the (ideal) specific impulse of a pressure–gain combustion ramjet using hydrocarbon fuel combustion at

Figure 5.32 Specific impulse, I_{sp}, in seconds, of a pressure-gain combustor ramjet (at stoichiometric hydrocarbon fuel-air mixture) compared to a conventional ramjet. *Source:* Data from Nalim 2002a, b.

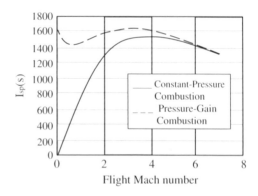

Figure 5.33 The first PDE-powered aircraft; the Long E-Z, flew on January 31, 2008. *Source:* Courtesy of USAF.

stoichiometric conditions, as compared to a conventional ramjet operating with the same fuel and at stoichiometric conditions from takeoff to Mach 7 (from Nalim 2002a, b). In this ideal case, the peak performance occurs at about Mach 3 with 1600^+ seconds of specific impulse. At Mach 2, an ideal combustion-gain ramjet produces about 25% higher specific impulse than its conventional combustion ramjet. Povinelli (2002), accounting for real gas effects, finds the specific impulse of PDE compared to a ramjet using stoichiometric propane-air combustion is superior up to flight Mach number of 2.3. Real gas effects account for dissociation and recombination of products of combustion in chemical equilibrium.

An experimental PDE was flight tested on a modified Scaled Composite Long E-Z aircraft, as shown in Figure 5.33, on January 31, 2008. This historic flight used a series of open (thrust) tube PDEs, as seen in the aft section of Long E-Z in Figure 5.33 (see Barr 2008).

5.9.3 Turbofan Engine with PDE

The integration of PD combustion in the fan (bypass) duct of a turbofan engine was studied by Mawid et al. (2003). When compared to an afterburning turbofan engine, the pulse detonation turbofan showed superior performance for the operating frequency of 100 Hz and higher. At 200 Hz frequency, the turbofan performance was doubled using the pulse detonation in the fan duct. However, these results are based on idealized assumptions about filling, ignition, discharge of the PDE in the bypass duct, and its interaction with the fan and the exhaust system. In particular, the challenges of integrating a pulse detonation thrust tube in the bypass duct of a turbofan engine are:

- Potential fan stall due to oscillating (unsteady) backpressure
- Liquid fuel injection, evaporation, mixing, and ignition in the fan bypass duct

- Pre-detonator requirements needing further investigation
- System integration

The promise of PDE in compounded cycles with air-breathing propulsion goes beyond the above challenges. Higher efficiency of PDE due to pressure-gain combustion, the versatility of multifuel use from hydrocarbon fuels to hydrogen, and flight Mach range from takeoff to Mach 4 all make PDE a promising technology for the future of high-speed flight.

5.9.4 Pulse Detonation Rocket Engine (PDRE)

The fuel of choice in PDE is hydrogen that is mixed with air (in a ramjet) or oxygen (in a rocket) to form the combustible mixture of gases in the thrust/detonation tube. Multiple detonation chambers may replace the conventional rocket combustion chamber to create a PDRE. An arrangement of the propellant feed system, multiple detonation combustor integration into a rocket is schematically shown in Figure 5.34 (from Bratkovich et al. 1997).

The advantages of the PDRE are in higher pressure ratio that is created in a PDE pressure gain system, higher-efficiency combustion in PDE, and anticipated weight reduction in the propellant pumping system due to lower pressure ratio requirement in PDRE, among other improvements. Bratkovich et al. (1997) suggest that the detonation compression ratio in the combustion chamber with H_2/O_2 detonation ranges between 7 and 15; consequently, the reduction in the weight of the turbo-pump system will be significant. The chamber temperature in PDRE is also higher by about 10–20% due to

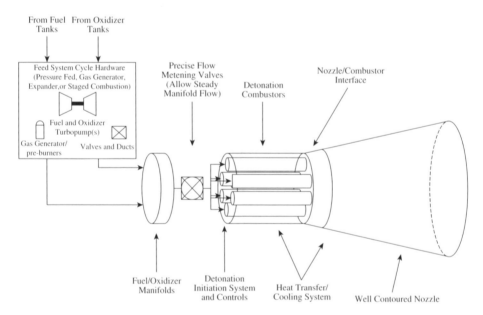

Figure 5.34 Schematic drawing of the components in a pulse detonation rocket engines (PDRE). *Source:* From Bratkovich et al. 1997.

pressure gain combustion. Higher temperatures then create more dissociation and thus lower molecular weight (see Povinelli 2001). The specific impulse gain of the PDRE is estimated at 5–10%.

5.9.5 Vehicle-Level Performance Evaluation of PDE

The best three application vehicles for PDE were determined to be (Kaemming 2003):

1) Supersonic tactical aircraft
2) Supersonic strike missile
3) Hypersonic single-stage-to-orbit (SSTO) vehicle

For the first category, i.e. supersonic tactical aircraft, a high-altitude Mach 3.5 reconnaissance strike aircraft was envisioned. The aircraft had 700-nm mission radius. The turbine-based propulsion system for this vehicle was turbo-ramjet (TRJ). Kaemming's study identified 11–21% TOGW benefit to the PDE. Lower SFC and reduced vehicle drag were the main sources of PDE superiority in this application. The combination of reduced TOGW and the fuel usage resulted in 4% lower lifecycle cost (LCC) for the PDE vehicle. Figure 5.35 (from Kaemming 2003) compares the vehicle cross-sectional area distribution for the PDE and TRJ-powered supersonic tactical aircraft. The Sears-Haack body represents the optimal aerodynamic shape in supersonic flight, for a given volume, and is shown in Figure 5.35 for comparative purposes. The area distribution of PDE provides for a better match with the optimal shape, thus lower wave drag. Vos and Farokhi (2015) discuss optimal aerodynamic shapes, including Sears-Haack body, in their transonic aerodynamics book.

The lower fuel burn in PDE as compared to TRJ is shown in Figure 5.36 at Mach 3.5 as percent net propulsion force (NPF), from Kaemming (2003).

The use of air-breathing PDE and the rocket-mode PDE allow for takeoff to suborbital hypersonic flight (Royal Aeronautical Society 2018). The dual-mode PDE-engine is under development by PD Aerospace in Japan. The stated goal is for the sub-scale model to operate in 2019, with piloted test flight in 2021 and the commercial space tourism operation in 2023. Japan's All Nippon Airlines (ANA) is the investor and the first customer for the low-cost flight to space. Figure 5.37 shows the two-pilot, six-passenger spacecraft designed by (Terumasa Koike) PD AeroSpace for ANA.

Shapiro (1953) presents the basic treatment of non-adiabatic shock waves that is fundamental to detonation theory. The scientific principles behind PDE are presented in Zel'dovich and Raizer (2002), as well as Heiser and Pratt (2002). Roy et al. (2004) described pulsed detonation propulsion technology, its challenges, and future perspectives.

5.10 Boundary-Layer Ingestion (BLI) and Distributed Propulsion (DP) Concept

The BLI and distributed propulsion (DP) prove most effective when they are combined. The BLI concept mainly affects aircraft (viscous) drag (hence fuel burn) as well as noise in the landing–takeoff (LTO) cycle. Distributed propulsion facilitates BLI in

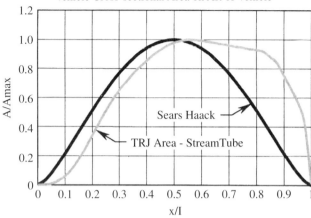

Turbo-Ramjet

• Longer Engine and Larger X-area Increase
Vehicle Cross-sectional Area on Aft of Vehicle

Sears Haack

TRJ Area - StreamTube

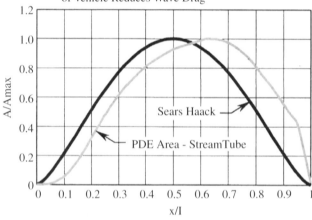

PDE

• Shorter Engine Permits Forward Placement
• Engine Trailing Edge Aligned
• "Smoother" Area Distribution on Aft Portion
of Vehicle Reduces Wave Drag

Sears Haack

PDE Area - StreamTube

Figure 5.35 Comparison of aircraft area distribution. *Source:* From Kaemming 2003.

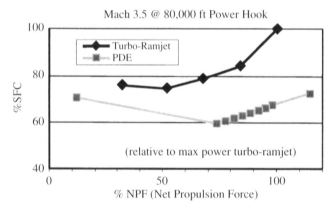

Mach 3.5 @ 80,000 ft Power Hook

Turbo-Ramjet
PDE

(relative to max power turbo-ramjet)

Figure 5.36 Thrust-specific fuel consumption for PDE and TRJ along baseline mission. Normalized with respect to SFC of TRJ @ 100% thrust. *Source:* From Kaemming 2003.

Figure 5.37 The PDE-based spaceplane developed for space tourism. *Source:* Designed by Terumasa Koike; PD AeroSpace, Japan.

addition to enhancing reliability through redundancy, among other benefits. We address the BLI concept first, followed by the synergy that the distributed propulsion offers in modern aviation.

5.10.1 Aircraft Drag Reduction Through BLI

Aircraft BLI is the ultimate form of the airframe and propulsion integration that reduces vehicle drag and improves propulsive efficiency. The earliest investigation dates back to 1947, by Smith-Roberts. Smith (1993) presents the propulsion benefits in wake ingestion. The airframe drag due to boundary layer formation ideally can be eliminated if the boundary layer is embedded in the captured streamtube that enters the jet engine. An ideal scenario is shown in Figure 5.38. In part (a), there is momentum deficit in the wake, which in essence is balanced by the momentum excess in the jet. These two sources often are physically separated from each other through a podded installation. In part (b), all the momentum deficit due to wake and the excess due to jet have ideally resulted in uniform flow that is equal to the flight speed. This is the *wake-filling* concept.

Drela (2009) presents a control-volume approach to aircraft with propulsion system and flow control that is based on power balance instead of momentum approach with thrust/drag formulation. Drela's approach is best-suited for the analysis of BLI, since the distinction between the drag and thrust becomes *fuzzy* at best in a highly integrated (coupled) BLI system. Recently, Hall et al. (2017) detailed the BLI propulsion benefits for transport aircraft. Understandably, the impact of BLI propulsion on reduced fuel burn is directly proportional to the extent of BLI into the engines as percent of the airframe boundary layer affected through BLI-airframe integration.

However, there is a fundamental challenge in the design of BLI propulsion system, which is distortion-tolerant fan and the associated high-cycle fatigue. In propulsion, we learned that Inlet distortion reduces the fan *stall margin* and if the distortion level is severe, then the fan enters into an unstable flow regime, known as stall or surge. Fan efficiency is also degraded by distorted inlet flow. For this purpose, inlet boundary-layer management, primarily through suction, is a practical/promising approach to distortion level control at the engine face. A conceptual aircraft design, credited to NASA/MIT/Aurora Flight Sciences, which is called Double Bubble, D8 series, is based on the

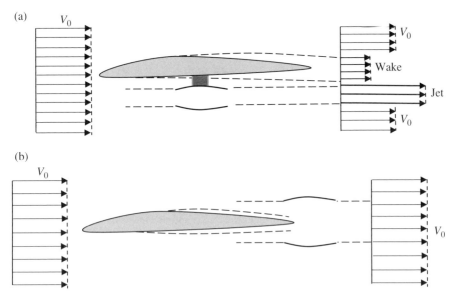

Figure 5.38 Comparison between an ideal jet-wake structure in a conventional podded propulsion installation and a candidate BLI propulsion system. (a) Aircraft with conventional pylon-mounted propulsion system. (b) Aircraft with boundary-layer ingestion propulsion system (ideal jet-wake profile).

Figure 5.39 Double bubble, D8, advanced civil transport. *Source:* Courtesy of NASA/MIT/Aurora Flight Sciences.

BLI concept, as shown in Figure 5.39. This aircraft features unswept wings for cruise at Mach 0.72 and double-bubble fuselage for carryover lift. The BLI engines digest 40% fuselage boundary layer, which is 17% of total airframe (see Hall 2016). The fuel burn in D8 series is reduced by 8.5%, according to NASA publication (2018), 8.7% according to Blumenthal et al. (2018) and 9% according to Hall et al. (2017).

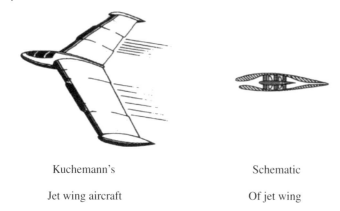

Kuchemann's   Schematic

Jet wing aircraft   Of jet wing

Figure 5.40 Küchemann's jet-wing concept with distributed propulsion inside the wing with thick trailing edge. *Source:* From Attinello's 1957 paper.

(a)   (b)

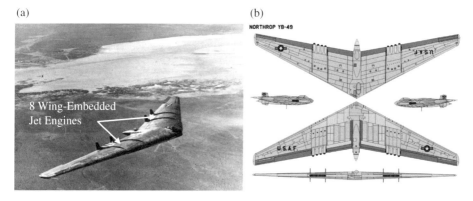

Figure 5.41 Northrop YB-49 uses eight wing-embedded jet engines. *Source:* Courtesy of USAF.

Historically, the idea of aircraft drag reduction by integrated jet-wing design was originally proposed by Küchemann in 1938. The concept of integrating jet engines that are embedded in aircraft wings where the jet engines' nozzles are placed at the wing's thick trailing edge is shown in Figure 5.40 (from Attinello 1957). Küchemann's jet-wing concept was posthumously published in 1978 in his *aerodynamic design of aircraft* book. Ko et al. (2003) and Dippold et al. (2004) provide much more aero-propulsive detail on the concept of jet-wing. Another application of this concept is found in Northrop YB-49, where completely embedded jet engines are integrated in a flying wing as shown in Figure 5.41.

5.10.2 Aircraft Noise Reduction: Advanced Concepts

Turbulent boundary layer on an aircraft wing/body and in the wake emit broadband noise due to turbulent pressure fluctuations. The boundary layer and wake noise contribute to the overall effective perceived noise (EPNdB) level emitted from aircraft. The EPNdB in takeoff, landing, and sideline impacts communities near airports and is thus regulated by Federal Aviation Administration (FAA) and International Civil Aviation Organization (ICAO), as presented in Chapter 3. BLI and wake-filling propulsion concept then directly impacts noise emitted from the aircraft wing/body and its wake. Therefore, it is entirely justifiable to integrate BLI and wake-filling propulsion in the design of *Silent Aircraft*. Indeed, *Silent Aircraft* Initiative is a joint project (launched in 2003) between Cambridge University (UK) and MIT (US) (see Dowling et al. 2006–2007). The team designed a 215-passenger jet aircraft of BWB design for high lift-to-drag ratio and with BLI propulsion system for efficiency and low noise. The *Silent Aircraft eXperimental* design, called *SAX-40*, is expected to produce 30% reduction in EPNdB in LTO cycles and to operate with 25% reduction in fuel burn. These monumental performance gains are made possible due to advanced aerodynamic design and BLI distributed propulsion. SAX-40 is a BWB aircraft with aft-fuselage mounted BLI distributed propulsion system (see Popular Mechanics 2016). Hall and Crichton (2005, 2007) detailed engine design and installation for the *Silent Aircraft*, and is recommended for further reading. The role of flow control on both aircraft drag reduction and noise mitigation is shown in Figure 5.42 (from Anders et al. 2004).

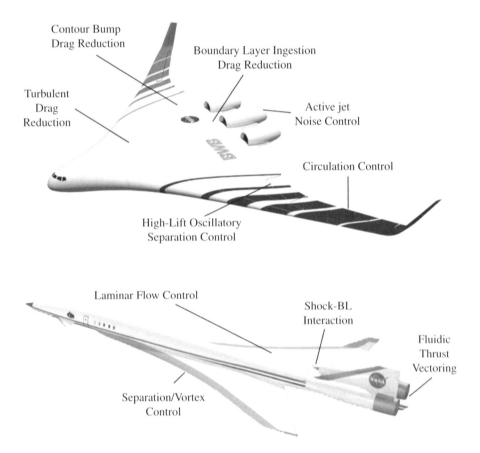

Figure 5.42 The impact of flow control on aircraft drag reduction and noise mitigation. *Source:* From Anders et al. 2004. Courtesy of NASA.

BWB aircraft, e.g. SAX-40, offers a lifting body, thus high L/D and exposes a large surface area for the aft BLI-DP propulsion system to reduce noise and aerodynamic drag. The wake-filling concept takes the nozzle from the conventional round configuration to high-aspect ratio rectangular nozzles to reduce drag and noise. The aerodynamic design of *transition ducts*, from circular-to-rectangular in the exhaust system, and rectangular-to-circular for the intake system, plays a critical role in BLI-airframe integration. Some design guidelines based on computational simulations as well as experimental research are developed at NASA. Figure 5.43 shows a round-to-square transition duct as well as design guidelines for the constant-area transition duct (from Abbott et al. 1987).

The effect of transition duct length, exit aspect ratio and 2-D offset on stall margin for the neutral and accelerating transition ducts is shown in Figure 5.44 (from Farokhi et al. 1989). The inlet blockage parameter, δ/R_i, is assumed to be 10%. Inlet Reynolds number based on the inlet radius, R_i, is 200 000.

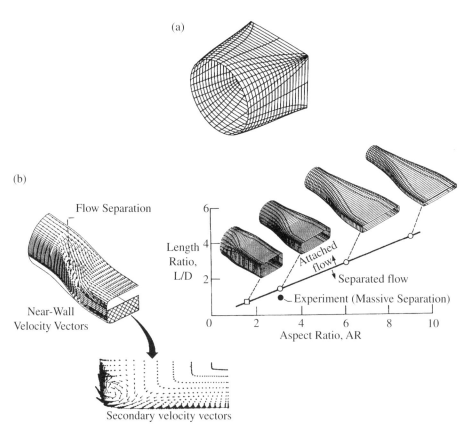

Figure 5.43 Constant-area transition duct geometry, flowfield, and design guidelines. *Source:* From Abbott et al. 1987. (a) Round-to-square transition duct with constant cross sectional area. (b) Flowfield and design guidelines.

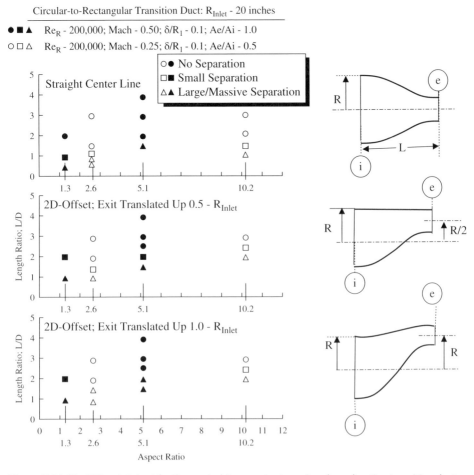

Figure 5.44 "Stall Margin" chart for the neutral (i.e. constant-area) and accelerating transition ducts (i.e. nozzles) with straight and S-centerlines with 2-D offset. *Source:* From Farokhi et al. 1989.

5.10.3   Multidisciplinary Design Optimization (MDO) of a BWB Aircraft with BLI

Liebeck of Boeing (2004) detailed the design of BWB aircraft and called it "a potential breakthrough in subsonic transport efficiency." Leifsson et al. (2005) applied multidisciplinary design optimization (MDO) framework to a BWB aircraft with distributed propulsion and compared its performance with the same BWB aircraft using conventional pylon-mounted UHB turbofan engines. The optimized BWB aircraft is shown in Figure 5.45 from Leifsson et al. (2005).

There are four advantages of distributed propulsion, in addition to reduced noise (see Leifsson et al. 2005):

1) The improved load redistribution along the wing that offers potential alleviation of the gust load /flutter problems

(a)

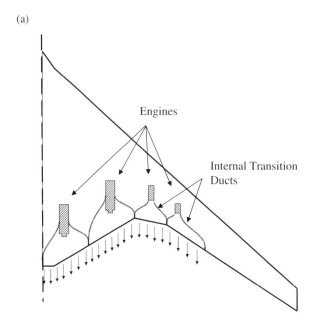

(b)

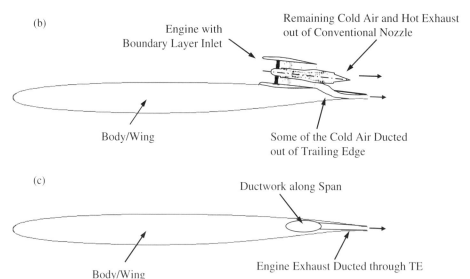

(c)

Figure 5.45 Distributed propulsion with BLI in a BWB aircraft: (a) BWB aircraft with eight distributed BLI propulsion system; (b) Streamwise cut through a section with an engine; (c) Streamwise cut through a section between engines. *Source:* From Leifsson et al. 2005.

2) Improved safety in engine-out condition due to engine redundancy in distributed propulsion system versus the conventional configurations that use two to four large turbofan engines
3) Smaller, easily interchangeable engines make the distributed propulsion aircraft more affordable than conventional configurations with a few large engines
4) Tailless configuration is feasible

Figure 5.46 BWB subsonic transport aircraft with conventional Aft-mounted propulsion system. *Source:* Courtesy of Boeing.

The BWB aircraft designed with distributed propulsion uses internal ducting to divert some of the cold fan flow inside the wing and through the trailing edge where high aspect ratio rectangular nozzles are employed. The remaining exhaust flow is directed through conventional nozzles, as shown in Figure 5.45. The trailing edge jets could be used as high-lift devices similar to jet-wing and jet-flap concepts and eliminate the need for conventional trailing-edge flaps. The BWB aircraft with conventional propulsion is shown in Figure 5.46 (courtesy of Boeing). The placement of the engines on the upper surface provides for low noise operation due to shielding.

5.11 Distributed Propulsion Concept in Early Aviation

The concept of distributed propulsion, like BLI, is not new in aviation (see Epstein 2007). Based on the aircraft TOGW, multiple engines may be required to power the aircraft in takeoff. The demand on the number of engines also depends on the available thrust and type of suitable engines. Some of the classical examples of aircraft with multiple propulsors are found in the world of bombers, such as Convair B-36 with 10 engines, Boeing B-52 with 8 engines, and the North American XB-70 with 6 engines. These aircraft are shown in Figure 5.47a–c for historical reference only, as there are other examples of multiengine aircraft, e.g. B-47 Stratojet, among others.

In modern times, distributed propulsion is more than just the number of engines, thus it is not a mere function of the TOGW in an aircraft. We shall discuss the modern view of DP in the next section.

(a)

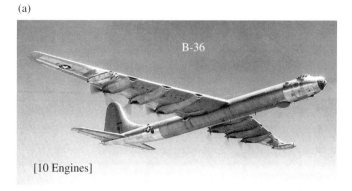

[10 Engines]

(b)   (c)

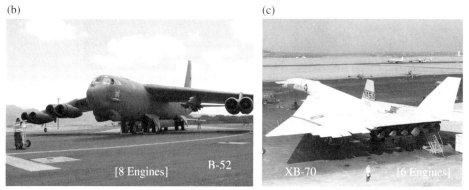

[8 Engines]   B-52   XB-70   [6 Engines]

Figure 5.47 Examples of the US bomber aircraft that used 6–10 engines: (a) Convair B-36 bomber used 10 (mixed) engines, 4 outboard turbojet (TJ) engines and 6 inboard piston prop engines (max speed 435 mph); (b) Boeing B-52 bomber used 8 turbofan engines; Mach 0.86; (c) North American XB-70, Mach 3 strategic bomber, used 6 engines *Source:* Courtesy of USAF.

5.12 Distributed Propulsion in Modern Aviation

Large-diameter aircraft gas turbine engines are more efficient than small engines, but they are noisy and have integration problems with the aircraft. In underwing installation, they encounter limitations with ground clearance and at takeoff rotation. They are also prone to foreign object damage due to their close proximity to ground. In over the wing installation, we encounter cabin noise problem, among other stability and control issues (with center of gravity shift in the vertical direction). In addition, Federal Aviation Regulation (FAR) Part 25 imposes climb gradient requirements on aircraft with one or more engine-out situations. An aircraft with few engines, e.g. 2–4, is clearly at a disadvantage in fraction of thrust lost compared to an aircraft with many engines, e.g. 10–40. The additional safety afforded by the number of engines is best studied by the probability analysis of one or more engine failure(s) in a multiengine aircraft.

Probability of engine failure is based on statistical measures, which is expressed as the frequency of engine failure per flight hour. The reliability of modern gas turbine engines places this probability at:

$$\text{Probability of engine failure} \equiv \mathbf{P}_{\text{ef}} = 5 \times 10^{-5} \text{ per flight hour} \qquad (5.21)$$

The binomial coefficient $\begin{pmatrix} N \\ m \end{pmatrix}$ applied to the problem of engine failure in a multiengine aircraft describes the number of m engines that could be inoperative in an aircraft with N engines. The binomial coefficient is defined as:

$$\begin{pmatrix} N \\ m \end{pmatrix} \equiv \frac{N!}{m!(N-m)!} \tag{5.22}$$

For example, in an aircraft with four engines, $\begin{pmatrix} 4 \\ 2 \end{pmatrix} = \frac{4!}{2!2!} = 6$ describes the number of possible combinations of two engines that may be inoperative, namely six possible combinations. This is graphically shown in Figure 5.48.

The probability of two engines inoperative is the square of the probability of one engine inoperative. In general, the probability of m engines inoperative is the mth power of the probability of a single engine failure, i.e.

$$\text{Probability of } m \text{ engine failure} = \mathbf{P_{ef}}^m \tag{5.23}$$

Therefore, the probability of m engine(s) failure on an aircraft with N engines is expressed as:

$$\begin{pmatrix} N \\ m \end{pmatrix} P_{ef}^m = \frac{N!}{m!(N-m)!} P_{ef}^m \tag{5.24}$$

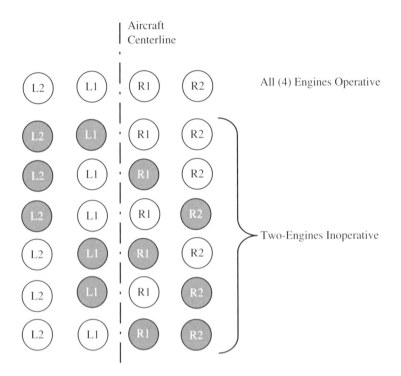

Figure 5.48 Six possibilities for two-engines inoperative on a four-engine aircraft.

Table 5.8 Probability of engine(s) failure on aircraft with multiple engines.

Aircraft with	2 Engines	4 Engines	10 Engines	20 Engines	40 Engines
Probability of engine failure/ flight hour	5×10^{-5}	5×10^{-5}	5×10^{-5}	5×10^{-5}	5×10^{-5}
Probability of 1+ engine failing	10^{-4}	2×10^{-4}	5×10^{-4}	10^{-3}	2×10^{-3}
Probability of 2+ engine failing	2.5×10^{-9}	1.5×10^{-8}	1.1×10^{-7}	4.7×10^{-7}	1.9×10^{-6}
Probability of more than half engines failing	2.5×10^{-9}	5×10^{-13}	3.3×10^{-24}	8.2×10^{-43}	6.3×10^{-80}

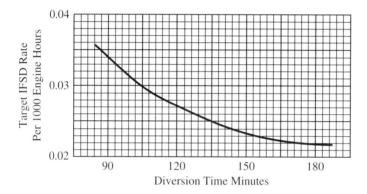

Figure 5.49 Target engine in-flight shutdown rate per 1000 engine hours. *Source:* From Civil Aviation Authority, *CAP 513 ETOPS document*, 2002.

Table 5.8 shows the likelihood of one/multiple engine(s) failure on an aircraft with 2, 4, 10, 20, and 40 engines. The extreme unlikelihood of more than half the engines failing on an aircraft with distributed propulsion is quantified on the last row of Table 5.8.

A final note on the probability of engine failure is that the in-flight shutdown rate (IFSD) is even lower than the noted P_{ef}. Today's SOA in engine technology is noted by Epstein (2014) to be better than 0.002/1000 hours. This remarkable level of reliability is achieved in commercial aircraft propulsion. It owes its success mainly to decades of intense investment in R&D by the engine manufacturers, the US government (NASA), as well as the European Union and other countries.

IFSD is the key parameter for the risk assessment and certification of Extended Range Twin Operations (ETOPS). Figure 5.49 shows the target IFSD rate with diversion time (from Civil Aviation Authority, CAP 513 ETOPS document, 2002).

In distributed propulsion using many propulsors, the probability of a single engine failure, or IFSD, increases with the number of engines. For example, the probability of a single engine failure on a 20-propulsor DP aircraft is 10 times more than a single engine failure on a twin. The assumption here is that the technology level employed in the small and the big engines result in the same probability of engine failure, i.e.

$$\left(P_{ef}\right)_{\text{small-engine}} \cong \left(P_{ef}\right)_{\text{large-engine}} \tag{5.25}$$

Although this assumption is debatable, we assume that small gas turbine engines can eventually achieve the same level of reliability as the current large turbofan engines in commercial aviation by investment in this sector mainly by the engine manufacturers, NASA, and the European Union (through Advisory Council for Aeronautics Research in Europe, ACARE). The higher rates of the IFSD on a multi-propulsor DP aircraft follows the same argument as probability of engine failure. On a 20-propulsor DP aircraft, the IFSD of a single propulsor is 10 times more than one engine IFSD on a twin. The value of a 20-propulsor DP aircraft is, however, in the loss of thrust of a single engine on this aircraft that leaves 95% of total thrust still available. The same argument is supported by the probability of more than half the engines failing, shown in the last row of Table 5.8. However, since distributed propulsion often includes electric ducted fans (as in hybrid-electric or turboelectric propulsion), the failure probability analysis must include the failures of the associated high-power electric grid, the cryo-cooling system (with superconducting wiring, motors, and generators) and the power management system/subsystems (see Armstrong 2015).

Additional benefits of distributed propulsion is inherent in propulsion-airframe integration (PAI) that yields structural efficiency and thus optimized propulsion system weight. The use of BLI is closely tied to distributed propulsion and thus reduces the fuel burn due to drag reduction. The low-PR distributed fans are less noisy than their higher-PR counterparts, and thus community noise issue is mitigated. The asymmetric thrust is reduced (or even eliminated through thrust redirection/reallocation) with increasing number of engines, and thus the vertical tail area is reduced (or even eliminated, i.e. tailless configuration) in distributed propulsion architecture (see Armstrong 2015). Since the electric fans' exhaust jets are cold, thrust vectoring devices can be integrated in the propulsive architecture using lightweight material, as noted by Kim et al. (2013).

Finally, there is another advantage that is realized using distributed turboelectric fans. Since the total air mass flow rate through the fans is proportional to the number and size of the propulsors, there is an increase in the BPR; in essence, the new design produces a *multiplier effect* in BPR. This new parameter is known as the *effective bypass ratio*, eBPR, of the propulsion system, which is defined as:

$$\text{Effective bypass ratio}\left(eBPR\right) \equiv \alpha_{\text{eff}} \equiv \frac{\sum_{i=1}^{n}\left(\dot{m}_{\text{fan}}\right)_i}{\dot{m}_{\text{core}}} \tag{5.26}$$

where n fans are driven by a single core. The propulsive efficiency improves with higher BPR, as elementary considerations demonstrate this principle in aircraft propulsion books, e.g. see Farokhi (2014), among others.

5.12.1 Optimal Number of Propulsors in Distributed Propulsion

The goal of distributed propulsion is improved efficiency:

- Improved fuel burn, i.e. lower TSFC at cruise
- Lower emissions in CO_2, NO_x, and other pollutants
- Lower noise in LTO cycle
- Higher mission reliability and safety

- Reduction/elimination of control surfaces through differential thrust for yaw control, through vector thrust/powered lift for roll and pitch control
- Reduction in structural loads
- Improvements in maintainability

The thrust in air-breathing engines scales with air-mass-flow rate; therefore, it scales as length-squared, or r^2. Engine weight scales with the volume; hence, it scales as length-cubed, or r^3. Therefore, thrust-to-weight F/W ratio scales as

$$\frac{F}{W} \sim \frac{1}{r} \sim \frac{1}{\sqrt{F}} \tag{5.27}$$

This is known as the cube-square scaling law, which is not uniformly valid (see Chan 2008). Engine accessories do not follow this simple rule, for example. Also, casing weight to contain the fan blade separation in larger engines shifts the balance toward higher weight with size. Initially, thrust-to-weight ratio increases with the number of engines. However, there are viscous effects that dominate as the size of the propulsor shrinks, or equivalently the number of engines increases. This Reynolds number effect is caused by the appearance of extended laminar boundary layers on blades and end-walls, followed by the unstable transitional state of the boundary layer. The engine components that are subject to adverse pressure gradient (e.g. inlets and compressors) thus experience a lower static pressure rise capability and higher losses. With these competing effect, (i.e. F/W and Reynolds number impact on efficiency), the number of propulsors on a DP aircraft depends on the range and the mission of the aircraft. There is no one-size-fit-all rule in DP. The optimal number of propulsors has to be determined in a multidisciplinary optimization approach that includes on- and off-design engine simulation (to calculate the mission fuel burn), structures and materials module, auxiliary systems, and accessories module, among others.

For more accurate engine sizing and simulations, see York et al. (2017), or *Numerical Propulsion System Simulation* (NPSS) developed by Southwest Research Institute (2016).

5.12.2 Optimal Propulsor Types in Distributed Propulsion

Implicit in *greener aviation* is reduced emissions that points in the direction of electrification of the aircraft propulsion and power systems. In the hierarchy of electric propulsion, we start with a hybrid propulsion and power system, which is the aviation version of the hybrid engines in automobiles. The schematic diagram of this type of propulsion is shown in Figure 5.50 (see Del Rosario 2014 and Felder 2014). The propulsor is a low-PR ducted fan. There are two sources of energy to the propulsor, the jet fuel for the gas turbine engine and the battery or electric energy storage system, such as capacitor. The transmission lines are the electrical wirings that may use the ambient temperature conventional wiring or they may use cryogenic wiring for high-temperature superconducting transmission lines (see Del Rosario 2014 and Felder 2014).

Since in hybrid propulsion, the exhaust energy is used to generate electricity, the hybrid-electric propulsion system produces lower atmospheric heat release than the conventional turbofan engines. Also implicit in *greener aviation* is reduced noise levels during LTO cycles. The lower-pressure ratio fans are less noisy than their higher-PR

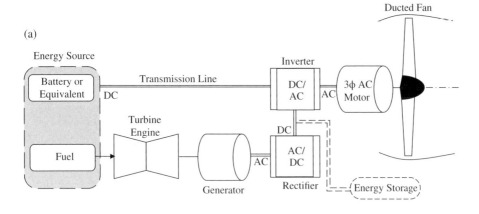

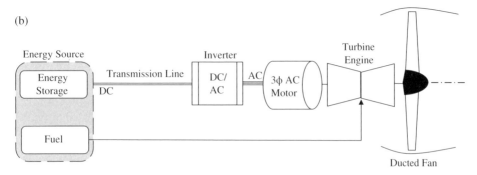

Figure 5.50 Schematic drawing of two hybrid-electric propulsion system architectures: (a) series hybrid-electric architecture with optional energy storage; (b) parallel hybrid-electric (used in Boeing SUGAR Volt – hFan; see Bradley and Droney 2011).

counterparts. In addition, since distributed propulsion is often envisioned as aft-mounted BLI on a BWB aircraft, shielding is a powerful strategy that is used to achieve *quiet aircraft* goals in hybrid-electric (or turbo-electric) DP aircraft.

In hybrid-electric propulsion, the propulsor (ducted fan) may be operated by the gas turbine, using jet fuel, or by electric motor that uses batteries or electric storage units, or combined GT and electric motor as in parallel hybrid propulsion. Unique to this architecture is the need for variable-core nozzle in a parallel hybrid-electric propulsion system to accommodate the combined GT-electric motor operation and the transition between GT and the electric motor (see Bradley and Droney 2011). An electric motor is connected to the LP spool via a gearbox. From taxi and takeoff to climb, cruise, approach, and landing, the combined energy use and range is optimized by an electric power management hub.

"Propulsive fuselage" hybrid-electric concept proposed by Bauhaus Luftfahrt in Germany for the Airbus A330 type twin-aisle aircraft is a part of Clean Sky 2 research program funded by the European Union that runs from 2014–2024. Propulsive fuselage belongs to the broader *tail-cone thruster propulsion* concept (see Dyson 2017). The goal is to mature promising technologies for an advanced airliner for EIS in 2030. The propulsive fuselage concept uses a third gas turbine engine that is embedded in the modified tail cone and powers an UHB fan in BLI configuration. This concept is shown in Figure 5.51. Although in this configuration, the geared UHB is driven by the LPT, the electric-motor-driven version is clearly the goal, which will be considered. Since the tail propulsor is in the aft fuselage-BLI position, it will reduce fuselage drag and engine noise. The preliminary studies showed an improvement in efficiency of 10%. This is a promising concept where BLI, geared-UHB/electric fan is used to reduce fuel burn and noise in an advanced hybrid-electric airliner. The tail-cone thruster concept was developed by NASA and is shown in Figure 5.52 (see Welstead et al. 2017 and Welstead and Felder 2016) for EIS of 2035. The single-aisle turboelectric aircraft with aft-boundary-layer propulsion, known as STARC-C ABL, uses the conventional tube-and-wing aircraft configuration with twin underwing turbofan engines that generate electric power for the aft-fuselage mounted electric fan. The two underwing engines and the aft electric fan each provide one-third of the cruise thrust. The FPR for the two underwing engines was chosen at 1.3 and the electric BLI fan had a PR of 1.25. The STARC-C ABL is shown to provide 7–12% reduction in block fuel (where 7% corresponds to the economic mission block fuel reduction for the 900 nm mission and the 12% is the design mission fuel-burn reduction for the 3500 nm mission) as compared to conventional configurations (Welstead and Felder 2016). The electric transmission lines are not superconducting and the total electrical generation/transmission efficiency is assumed to be 90%. These are promising results for the year 2035 EIS that assumed technology readiness level (TRL) of 6 for the key technologies by the year 2025. Since the aircraft in STARC-C ABL is the conventional tube-and-wing, the fuel burn reduction (of 7–12%) is almost entirely due to turboelectric propulsion with aft BLI fan. The fully integrated propulsion-airframe aircraft design with BLI, such as N3-X contributes an additional 50% to block fuel reduction (see Jansen et al. 2016). In this context, 7–12% achieved in STARC-C ABL is promising.

In advanced propulsion concepts, gas turbines are used as turbo-generators with the sole function of providing power to electric generators. Electric fans are driven by electric motors that receive their power from the electric generator. In essence, the power producing unit (turbo-generator) is decoupled from the thrust producing units (electric fans), as noted by Kim et al. (2008). The new architecture allows for both units to operate at peak efficiency. The GT-power unit is a turboshaft core that drives an electric generator. The turboshaft core(s) may be placed in the aft fuselage or at the wingtips. Figure 5.53 shows the schematic drawing of a turboelectric propulsion system with an optional energy storage system (see Del Rosario 2014; Felder 2014 and Armstrong 2015). The electric components between the turbine and the fan compose the new elements of the *propulsive drivetrain* (shown as dashed line) in turboelectric propulsion system. Power density and efficiency of the motor, generator, and power conversion units (rectifier, inverter), energy density of the batteries and other energy storage units, and the cooling system requirements and efficiency are all Key Performance Parameters (KPPs) in the propulsive drivetrain. The failure probability of each element and system reliability are critical to the viability of turboelectric propulsion system. Fault-tolerant design and system redundancy are the main requirements of the hybrid and turboelectric propulsion system.

(a)

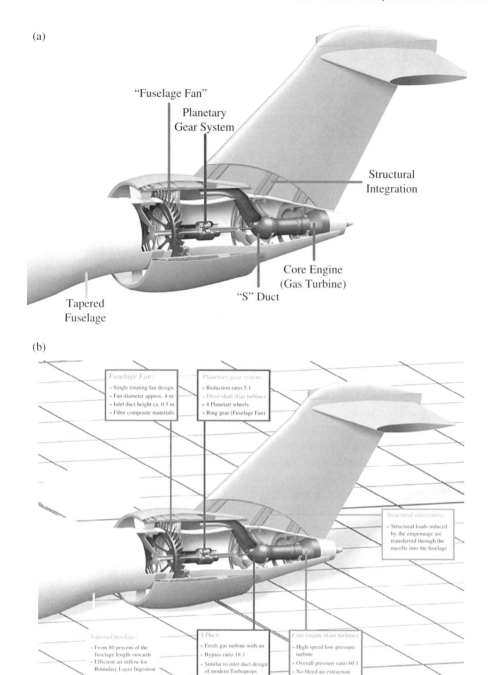

"Fuselage Fan"

Planetary
Gear System

Structural
Integration

Core Engine
(Gas Turbine)

"S" Duct

Tapered
Fuselage

(b)

Fuselage Fan:
» Single rotating fan design
» Fan diameter approx. 4 m
» Inlet duct height ca. 0.5 m
» Fibre composite materials

Planetary gear system:
» Reduction ratio 5:1
» Drive-shaft (Gas turbine)
» 4 Planetart wheels
» Ring gear (Fuselage Fan)

Structural integration:
» Structural loads induced
by the empennage are
transferred through the
nacelle into the fuselage

Tapered fuselage:
» From 80 percent of the
fuselage length onwards
» Efficient air inflow for
Boundary Layer Ingestion

S-Duct:
» Feeds gas turbine with air
» Bypass ratio 18:1
» Similar to inlet duct design
of modern Turboprops

Core engine (Gas turbine):
» High-speed low-pressure
turbine
» Overall pressure ratio 60:1
» No bleed air extraction

Figure 5.51 "Propulsive fuselage" hybrid-electric concept by Bauhaus Luftfahrt: (a) "propulsive fuselage" components; (b) details of the propulsive fuselage, engine-airframe integration elements.

(a)

(b)

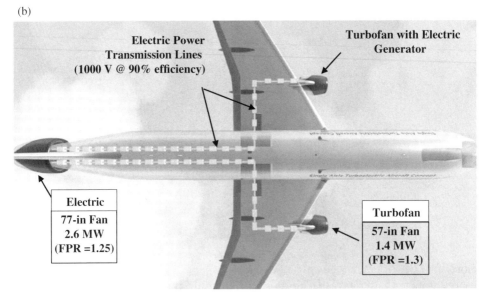

Figure 5.52 Single-aisle turboelectric aircraft with aft-boundary-layer ingestion propulsion system: (a) tail-cone thruster concept in STARC-C ABL aircraft; (b) turboelectric propulsion concept used in STARC-C ABL aircraft with design mission of 3500 nm. *Source:* From Welstead et al. 2017, www1.grc. nasa.gov.

There are other types of hybrid or turboelectric propulsion architecture, namely *partial turboelectric propulsion system* shown in Figure 5.54. In this architecture, the turbine engine powers the electric generator as well as a fan. The variable-frequency AC power output of the GT-generator is converted into DC in a rectifier. The DC power is transmitted to an inverter, which converts DC to three-phase AC power. The electric

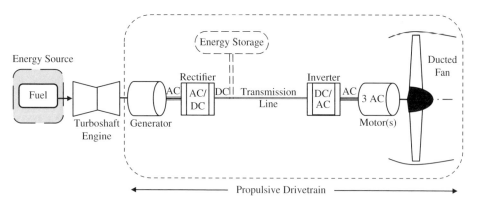

Figure 5.53 Schematic drawing of a turboelectric propulsion system driving a single ducted fan (with optional energy storage).

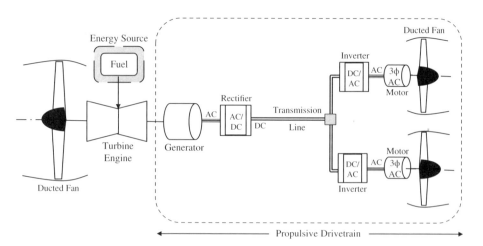

Figure 5.54 Schematic drawing of partial turboelectric propulsion system driving distributed electric fans.

motor that drives the ducted fan uses a three-phase AC power input. The optional DC energy storage (batteries, capacitors) is an element of the power management system that is inherent in all hybrid-electric propulsion architecture.

Technology acceleration exists across all areas of hybrid-electric propulsion. The KPPs are the specific energy, specific power, and efficiency. These are noted here from various NASA programs, e.g. Advanced Air Transportation Technology (AATT) Project:

Battery	currently at 250 Wh kg^{-1} [in battery lab cell tests, currently at 400 Wh kg^{-1}], it is anticipated to go to 500 Wh kg^{-1} by 2025]
Generator	reach 8 hp lb^{-1} at 96% efficiency by 2025
Electric Motor	currently tested at 4–6 hp lb^{-1} specific power with 95–97% efficiency anticipated to go to 8 hp lb^{-1} by 2025

Controllers	currently tested at $10–20\,hp\,lb^{-1}$ with extreme high precision rpm capability
Inverter (DC/AC converter)	$10\,hp\,lb^{-1}$ at 98% efficiency by 2025 is not a superconducting device, but yet cryogenic (as it reduces the losses in the inverter, according to Kim et al. 2013). The output is three-phase AC power
Rectifier (AC/DC Converter)	Each phase of power requires two rectifiers, one for + voltage and one for – voltage, therefore a three-phase AC power uses six rectifiers.

All-electric propulsion is solely powered by batteries or equivalent energy sources, as schematically shown in Figure 5.55. The carbon/environmental footprint of electric propulsion is mainly in the type of charging systems used in charging the batteries. For example, using the coal-fired (fossil-fuel) based power grid or possibly the renewable sources such as solar or wind to charge the batteries impact the carbon footprint. In addition, environmental impact of mining for the minerals/metals used in batteries, e.g. cadmium or lithium, needs to be accounted in the environmental lifecycle assessment of electric propulsion.

The network of transmission lines, motors, and generators that use high-temperature superconductivity (HTS) are encased in a single *cryostat* for high-efficiency, high-voltage power transmission in hybrid and turboelectric propulsion system (see Kim et al. 2013). Here, we briefly describe the SOA in superconducting technology as it plays an integral role in the future of sustainable aviation.

Superconductivity (SC) is a property that certain material (i.e. conductors) exhibit at very low (critical) temperatures near absolute zero, i.e. $T \leq 4.2\,K$ where they lose resistance to the flow of free electrons. Physicist Heike Kamerlingh Onnes discovered SC in a laboratory setting in 1911, but it was not until 1957 that three physicists, John Bardeen,

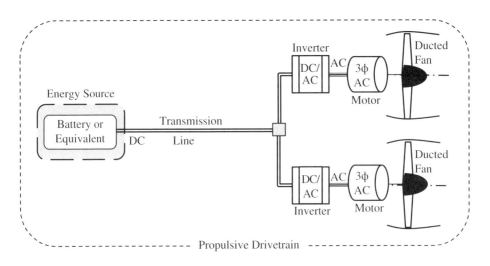

Figure 5.55 Schematic drawing of an all-electric propulsion system (here shown with two ducted fans).

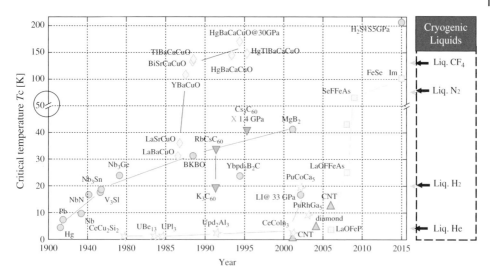

Figure 5.56 The timeline of superconducting materials and some cryogenic liquids with their boiling point. *Source:* From Ray 2015.

Leon Cooper, and John Robert Schrieffer published the first microscopic theory of SC (known as BCS theory), later receiving the Nobel Prize in 1972 for their theory.

The next leap came in 1986 with the discovery of HTS, where certain materials exhibit zero electrical DC resistance. The critical temperature where superconducting state is reached was raised to 92 K, which makes it available for practical applications (see Wu et al. 1987) using cryogenic fluids, as refrigerant. The high-temperature SC is now produced at 133–138 K and rising. The state of zero resistance to electric current makes the voltage drop nearly zero and thus a *supermagnet* is born, where the current is sustained without any significant voltage drop (Berlincourt 1987). All these discoveries and new technologies point to high-efficiency electric power generation and transmission, which make the hybrid-electric or turboelectric propulsion feasible. The rapid progress in SC from 1900 to 2015 is shown in Figure 5.56 (from Ray 2015). The suitable cryogenic fluid that may be used as the refrigerant is shown on the right-hand side in Figure 5.56. We note that the lowest temperature in cryogenics (i.e. their boiling point) is achieved in liquid helium, followed by liquid hydrogen, liquid nitrogen, and liquid CF_4. Furthermore, note the break in scale after 50 K on the critical temperature (vertical) axis. Rapid progress and investment in lightweight HTS technology makes hybrid and electric propulsion a viable alternative to the current SOA in aircraft propulsion and power systems.

A conventional separate-flow turbofan engine is schematically compared to a high-temperature superconducting electric motor and ducted fan in Figure 5.57. The fan diameter is designated as D_{fan}. The design FPR is reduced for low-noise and high-propulsive efficiency. The lower FPR results in reduced fan jet speed, V_{19}, and recalling the eighth-power law for the subsonic jet noise intensity according to Lighthill's theory, we expect that the fan jet noise to be exponentially reduced.

According to Lighthill's acoustic analogy (1952, 1954), the (subsonic) jet-noise intensity, I, is proportional to $(V_j/a_0)^8$. Therefore, the change in sound pressure level, ΔSPL in dB, relative to a reference jet speed is expressed as the log of their ratio according to:

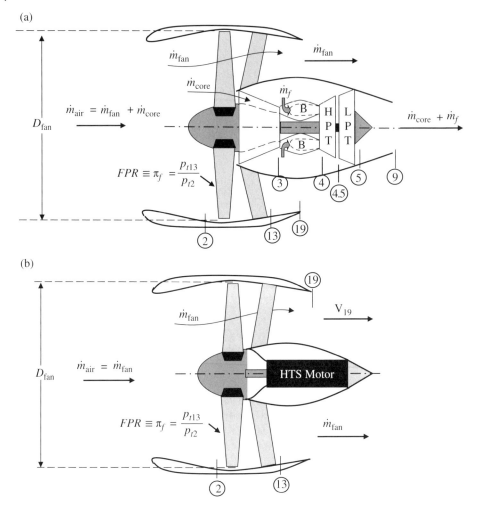

Figure 5.57 Schematic drawing of a GT-based-TF engine and an electric fan driven by HTS motor: (a) conventional GT-based turbofan engine and its mass flow distribution in the core and the fan; (b) high-temperature superconducting (HTS) electric motor and ducted fan.

$$\Delta SPL\left(dB\right) = 10\log_{10}\left(\frac{I}{I_{ref}}\right) = 80\log_{10}\left(\frac{V_j}{V_{j,ref}}\right) \tag{5.28}$$

The fan jet speed, V_{19}, is a function of the FPR, and the fan and nozzle efficiencies. For a ballpark estimate of the fan jet noise sensitivity to FPR, we ignore the fan and nozzle efficiencies, i.e. assume isentropic compression and expansion. Under takeoff condition and perfectly expanded nozzle, we derive a simple expression for V_{19}/a_0 as a function of FPR, as well as an expression for the change in fan jet noise level, according to:

$$\pi_f = \frac{p_{t19}}{p_0} \quad \text{Design choice} \tag{5.29a}$$

$$\tau_f = \pi_f^{\frac{\gamma-1}{\gamma}} \quad \text{Fan total temp ratio} \tag{5.29b}$$

$$M_{19} = \sqrt{\frac{2}{\gamma-1}\left[\pi_f^{\frac{\gamma-1}{\gamma}} - 1\right]} \quad \text{Fan jet Mach number} \tag{5.29c}$$

$$\frac{V_{19}}{a_0} = M_{19}\sqrt{\left(\frac{2}{\gamma-1}\right)(\tau_f - 1)} \quad \text{Fan jet speed ratio} \tag{5.29d}$$

$$\Delta SPL(dB) = 80\log_{10}\left(\frac{V_{19}/a_0}{V_{19,ref}/a_0}\right) \quad \text{Change in fan jet noise in dB} \tag{5.29e}$$

We have taken the reference FPR as 1.2 (as the lower end of the FPR used for ducted fans in distributed propulsion) and calculated the effect of FPR on fan jet Mach number and noise reduction with respect to FPR of 1.2. The ratio of specific heats, γ, is taken as 1.4. The result is graphed in Figure 5.58. For example, note that for the design FPR of 1.3, which is proposed for turboelectric distributed propulsion (TeDP) concept aircraft as compared to the modern high-BPR TF engines of FPR = 1.7–2.0, the fan in DP produces ~13–18 dB reduction in fan jet noise, respectively.

The HTS motor driving an electric fan is shown in Figure 5.59 (courtesy of Rolls-Royce). The status of HTS technology for all-electric aircraft applications in presented by Masson and Luongo (2007). This technology has reached operational level for ships, as the review by Kalsi et al. from American Superconductor (2006) shows. The power range of interest for naval propulsion is in 40–50 MW class, whereas for aircraft, the desired power range is 1–5 MW. For example, the aft BLI fan in STARC ABL has a design shaft power of 2.6 MW. The TRL for HTS propulsors in ships is at level 9 (i.e. operational level); however, the weight-volume critical aircraft application of the HTS motors places this technology at the anticipated level of 4–6 by the year 2025 (see Kim et al. 2013) in MW-class applications.

The electrical component specific power, specific weight, and their corresponding efficiencies for NASA STARC ABL are summarized in Table 5.9 (Welstead et al. 2017). TRL for these electrical components in 2025 is assumed to be 6 and the project EIS is assumed to be 2035.

Table 5.10 compares NASA's goals for electric propulsion and those expressed by Department of Energy (DOE) for the electric car. For reference, electric cars account for 1% of global auto sale in 2018. The power scale indicates MW for the electric aircraft as compared to kW scale for the electric car. The specific power, i.e. power per unit weight, is 10-fold for the aircraft (i.e. 16 kW kg^{-1} for aircraft vs 1.6 kW kg^{-1} for the electric car). The efficiency for both is targeted at 98% – i.e. the ratio of power-out to power-in is 98%. This comparison highlights the area of specific power (sometimes referred to as power density) as the critical focus area in EP.

The KPPs goals for power electronics in electric aircraft and electric car is shown in Table 5.11. The specific power goal on aircraft, in power electronics, is 19 kW kg^{-1} versus 14.1 kW kg^{-1} in electric cars. Both systems operate at 99% efficiency. The aircraft industry is setting higher goals than NASA; for example, Boeing is developing cryogenic converters with the goal of 26 kW kg^{-1} and 99.3% efficiency. A broad program overview

(a)

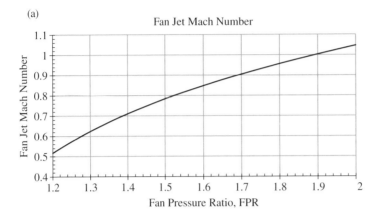

(b)

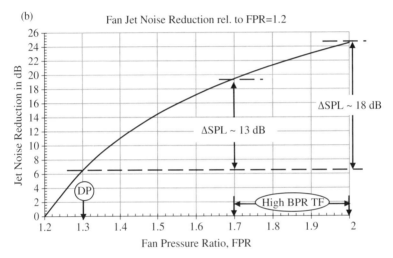

Figure 5.58 The change in ideal jet Mach number and fan jet noise (expressed as SPL [dB]): (a) the effect of FPR on jet Mach number (for ideal expansion in nozzle); (b) the effect of FPR on fan jet noise.

(a)   (b)

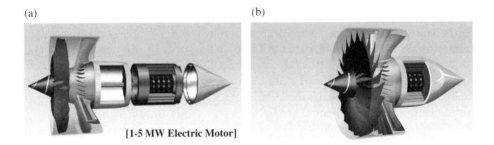

Figure 5.59 HTS motor driving an electric fan: (a) electric fan-electric motor (exploded view); (b) Assembled electric fan/motor system. *Source:* Courtesy of Rolls-Royce.

Table 5.9 NASA STARC ABL electrical component assumptions (non-superconducting).

Component	Specific power/weight	Efficiency (%)	Power(kW)
Generator (7000 rpm)	$13\,\text{kW}\,\text{kg}^{-1}$	96	1400
Rectifier (AC/DC power conversion/conditioning)	$19\,\text{kW}\,\text{kg}^{-1}$	99	1386
Cable	$170\,\text{A}/(\text{kg}\,\text{m}^{-1})$	99.6	1380
Circuit protection	$200\,\text{kW}\,\text{kg}^{-1}$	99.5	1373
Inverter (DC/AC power conversion/conditioning)	$19\,\text{kW}\,\text{kg}^{-1}$	99	2719
Electric motor (2500 rpm)	$13\,\text{kW}\,\text{kg}^{-1}$	96	2610
Thermal system	$0.68\,\text{kW}\,\text{kg}^{-1}$		

Table 5.10 Key performance parameters in electric aircraft and electric car.

Goals for electric propulsion	NASA (electric aircraft)	DOE (electric car)
Power	MW-scale	kW-scale
Power/Weight, $\text{kW}\,\text{kg}^{-1}$	16	1.6
Efficiency, %	98	98

Table 5.11 Key performance parameters for power electronics in electric aircraft and electric car.

Goals for power electronics	NASA (electric aircraft)	DOE (electric car)
Power/Weight, $\text{kW}\,\text{kg}^{-1}$	19	14.1
Efficiency, %	99%	99%

by Jankovsky et al. (2016) presents KPP, with short and long-term goals, in hybrid and turboelectric aircraft program at NASA. For more details on component weights and efficiencies in turboelectric aircraft propulsion system, see Brown (2011).

Figure 5.60 shows the internal assembly of a fully superconducting electric motor/generator from NASA (see Felder et al. 2011).

Zunum Aero 12 seater hybrid-electric aircraft, shown in Figure 5.61, targets the commuter, short-haul regional (700 mi), market for 2022 delivery. The aircraft uses lithium-ion batteries in the wings for supplemental power, especially at takeoff. In addition to propulsion batteries, the cruise power is provided by a gas turbine engine that provides shaft power to a generator (see Knapp and Said 2018 and https://zunum.aero). Zunum is developing a 1-MW powertrain, with the initial 500-kW system to fly on a testbed aircraft in 2019. A new turboshaft engine from Safran Helicopter, Ardiden 3 engine with 1700–2000 shaft horsepower will power the electrical generator of Zunum by the early 2020s.

Figure 5.60 Fully superconducting electric motor/generator. *Source:* From Felder et al. 2011.

Figure 5.61 Zunum Aero hybrid-electric regional aircraft (12 seater). *Source:* Courtesy of Zunum Aero, https://zunum.aero.

5.13 Interim Summary on Electric Propulsion (EP)

Let us recap the interest in electric propulsion in the three categories of *benefits, challenges,* and the *technology paths* that could lead to real game changers offered by the EP as disruptive technology in aviation (see Moore 2016). The benefits and challenges are self-explanatory and are noted here for reference.

Benefits of Electric Propulsion
1) Scale independence, i.e. scale effects are absent in electric propulsion
2) Efficiency in 95–97% range, which is >20 points higher than advanced turbofan cycles
3) 6x the motor power to weight
4) High-efficiency in part-load, i.e. in 30–100% power setting

5) Low cooling drag
6) No power lapse with altitude or hot day
7) Extremely compact
8) High reliability
9) Safety through redundancy
10) Lower community noise levels (>15 dB reduction in effective perceived noise level [EPNL])
11) Reduction of engine-out sizing penalty due to number of propulsors
12) Effective PAI, for higher $C_{L,\max}$, (C_L/C_D)
13) Zero in-flight carbon emission, since no hydrocarbon energy source used for propulsion
14) Aerodynamic benefits of wingtip-mounted propellers and turbines reduce induced drag, improves lift, and improves the wing (root bending moment) structural loads in cruise (see Patterson and Bartlett 1987 or Miranda 1986).

Challenges of Electric Propulsion
1) Battery-specific energy (kWh/kg), shown in Table 5.12 (Jet fuel packs 55 times more energy, per unit mass, than lithium-ion batteries.)
2) Battery system cooling
3) Battery charging systems, compatibility of quick-charge or battery swap with commercial aviation
4) High-voltage (kV) transmission/distribution
5) Thermal management of low quality heat
6) Power/fault management
7) Safety/reliability; robust power electronics
8) Integrated controls
9) Uncertainty of cost
10) Certification/Safety

Technology Path to Electric Propulsion: Early Adopters
1) All-electric secondary power system on aircraft (e.g. instead of hydraulic and pneumatic secondary power)
2) Fuel-cell APU
3) Small aircraft EP research: Early adopters
4) Large battery mass fraction aircraft [battery specific energy, 400 kWh kg^{-1}]
5) Small range extenders by providing 50% cruise power
6) Incentivize low-carbon aviation

Figure 5.62 shows the pathway to electric propulsion through Airbus E-Fan-X (source: Airbus) flight demonstrator. An earlier small electric propulsion aircraft (4-seater) is discussed by Yerman (2015).

Table 5.12 Battery vs jet fuel.

Energy density	MJ kg^{-1}
Zinc-air battery	1.6
Lithium-ion battery	0.8
Jet fuel	44

Figure 5.62 Airbus E-Fan-X flight demonstrator. *Source:* Courtesy of Airbus.

5.14 Synergetic Air-Breathing Rocket Engine; *SABRE*

An ultra-lightweight precooler heat exchanger uses a closed-cycle helium loop to cool the air from 1000 to $-150\,°C$ in a fraction of a second (actually, in 10 ms). This innovative counterflow precooler/heat exchanger technology is at the heart of an innovative air-breathing rocket engine that is capable of horizontal takeoff, climb, and acceleration to Mach 5.5^+ using subcooled air in its rocket engines and then transitioning to pure rocket mode above 26 km altitude. The air intake system uses a translating cone that completely closes the inlet in the pure rocket mode. Due to its versatility, this combined cycle engine is dubbed Synergetic Air Breathing Rocket Engine (SABRE) and is developed by Reaction Engines Ltd. in the United Kingdom, www.reactionengines.co.uk.

The breakthrough in *SABRE* is in thermal management of air in the airbreathing phase of the combined cycle that starts at takeoff and ends in Mach 5.5^+ and at 26 km altitude. A closed-loop cryogenic helium cycle precools the inlet air to $-150\,°C$ in an ultra-lightweight heat exchanger. Liquid hydrogen is used as the fuel in the common-core rocket thrust chamber, which is preheated prior to injection in the rocket combustion chamber by cooling down the helium in the helium cycle. The pure rocket mode uses onboard liquid oxygen in LOX/LH_2 rocket thrust chamber and propels the vehicle to Mach 25 orbital speeds, to a 300-km circular orbit. Figure 5.63 shows the components in *SABRE* (courtesy of Reaction Engines, Ltd., UK, www.reactionengines.co.uk). Figure 5.64 shows the closed loop heat exchanger systems in *SABRE* that include helium for the air intake system and hydrogen for the helium loop heat exchanger and injection into the rocket thrust chamber. Figure 5.65 shows the aerospace plane *SKYLON* that is designed around *SABRE* for reusable SSTO capability. A comparison of propulsion concepts for SSTO reusable launchers is presented by Varvill and Bond (2003).

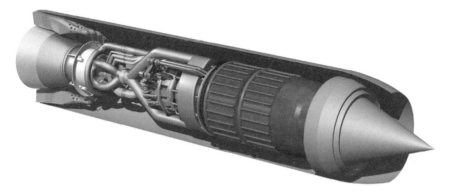

Figure 5.63 Critical components in *SABRE* engine. *Source:* Courtesy of Reaction Engines Ltd., www.reactionengines.co.uk.

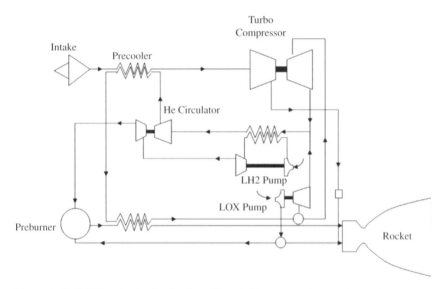

Figure 5.64 *SABRE* cycle showing its closed loop helium pre-cooler and the engine system.

Figure 5.65 The aerospace plane "SKYLON." *Source:* Courtesy of Reaction Engines, Ltd., UK, www.reactionengines.co.uk).

5.15   Compact Fusion Reactor: *The Path to Clean, Unlimited Energy*

The grand challenge of twenty-first-century engineering is the design, construction, and operation of compact fusion reactor (CFR) for aircraft propulsion and power. The technological barriers are in heating a light gas (usually deuterium and tritium) to 100 million °C, containing the hot plasma with adequate density in a strong magnetic field and shielding against neutron radiation. There are at least four practical engineering challenges in CFR:

1) Superconducting magnets must operate close to absolute zero whereas the plasma temperature must reach many millions of Kelvin.
2) There is a "blanket" component that lines the reactor vessel with the function of capturing the neutrons and making them collide with lithium atoms to make tritium to fuel the reactor. The blanket is estimated to have a 80–150 cm thickness that weighs 300–1000 tons. Refueling blankets are also an issue.
3) Plasma density is critical to maintaining steady fusion reaction.
4) Radiation and lifetime performance are concerns. Neutron radiation shielding, using cryogenics, is estimated to require about 1-m thick cryogenic shielding with 15 T magnets. The estimated weight is about 200 metric tons.

The military application of CFR will create a platform that can stay in the air (essentially) forever, or equivalently, it will have unlimited range. The civilian application of CFR will eliminate carbon emissions altogether and afford sustainability in aviation. On the horizon: Nuclear fusion reactor under research at LM Skunkworks in Palmdale, prototype under development, with applications to ships and planes. CFR of Lockheed Martin is designed to produce 100 MW of power and will need <20 kg of D-T fuel/year (www.lockheedmartin.com). To put these numbers in perspective, the maximum take-off gross weight (MTOGW) of Airbus A380 is 617 tons.

In 10 years, LM predicts that its 100 MW CFR can be built to fit in the back of a truck ($7' \times 10'$). This clearly represents significant engineering and material science challenge.

For early development work and the principles of nuclear (fission) propulsion, see Thornton 1963. Figure 5.66 shows the schematic drawing of a turbojet with its combustor replaced with a nuclear reactor. Figure 5.67 shows the superconducting magnets in LM's CFR research rig.

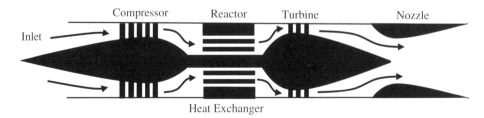

Figure 5.66   Schematic drawing of a turbojet engine with a nuclear reactor replacing its combustor.

Figure 5.67 Superconducting magnets in LM fusion research rig to confine plasma (High beta fusion reactor). *Source:* Courtesy of Lockheed Martin, www.lockheedmartin.com.

5.16 Aircraft Configurations Using Advanced Propulsion Systems

For this section, we have only sampled four configurations. These are: (i) turbo-electric distributed propulsion concept aircraft from Rolls-Royce; (ii) N-3X turboelectric aircraft from NASA; (iii) E-Fan X from Airbus; and (iv) X-57 from NASA. Throughout the book, we have looked at other modern aircraft with advanced propulsion systems.

Figure 5.68 distributed turboelectric propulsion is shown in an innovative design, by Airbus. The technology uses:

- Decoupled power production and propulsion functions
- Coupled propulsion and aircraft aero functions
- Optional alternative source for energy storage

Figures 5.69 and 5.70 show the GT-based superconducting electric generator in the aft fuselage position providing electric power through a SC-network of transmission lines.

An example of turboelectric BLI-DP is NASA N3-X program. Due to extensive use of superconducting technology, N3-X uses LH_2 for direct cooling of the HTS network, SC-motor, SC-generator. After its cooling cycle, hydrogen is then compressed and used as fuel in the turbo-generators. The fuel split is thus 10% hydrogen versus 90% jet fuel. The project goals are:

- ~63% energy use reduction
- ~90% NOx reduction
- 32–64 EPNdB cum noise reduction

Figure 5.68 Distributed turboelectric propulsion concept aircraft. *Source:* Airbus.

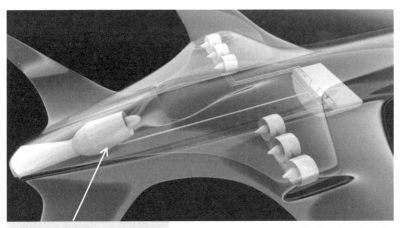

GT-Based SC Electric Generator   Turbo-electric DP

Figure 5.69 Turboelectric distributed propulsion concept. *Source:* Airbus 2017.

compared to baseline Boeing 777–200 aircraft. Some of the design-driver characteristics of N3-X that present challenges are:

- Directional stability and control thru differential thrust (for yaw trimming and active stability augmentation)
- Driving factors: safety and reliability
- Minimum weight penalty and volume requirements
- Acceptable system complexity
- Electrical system failure lead to thrust losses

Figure 5.71 shows the N3-X aircraft (courtesy of NASA). Tables 5.13 and 5.14 summarize the design point of N3-X as well as its mission specifications. The data used in the tables and the figures related to N3-X are taken from Armstrong (2015), Kim et al. (2013) and Felder (2014).

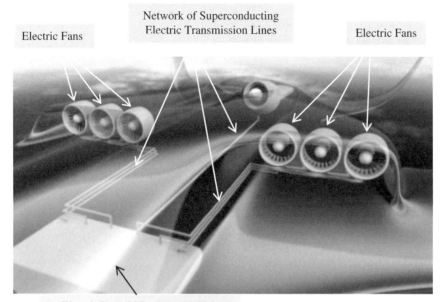

Figure 5.70 Superconducting network and electric power management. *Source:* Airbus 2017.

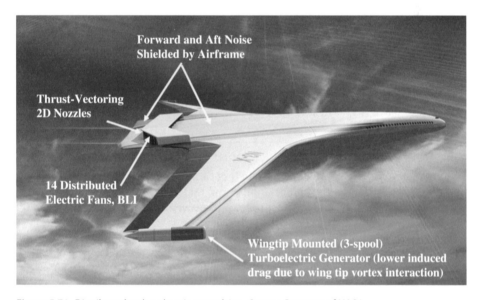

Figure 5.71 Distributed turbo-electric propulsion. *Source:* Courtesy of NASA.

Table 5.13 N3-X design point.

Aerodynamic design point (ADP)	M	Altitude (ft)
Top-of-Climb (TOC)	0.84	34 000

Table 5.14 N3-X mission parameters.

Parameters	N3-X/MgB$_2$/LH$_2$ (liquid hydrogen cooled)
Range (nm)	7500
Payload (lbm)	118 100
Empty weight (lbm)	420 000
Block fuel weight (lbm)	76 171
Fan pressure ratio (FPR)	1.3
Net thrust (lbf) – installed	85 846 (RTO) and 33 405 (TOC)
TSFC (lbm/h/lbf) – installed	0.2174 (RTO) and 0.3125 (TOC)
TSEC (BTU/s/lbf) – installed	1.1937 (RTO) and 1.727 (TOC)
Effective BPR, eBPR	36.1 (RTO) and 30.1 (TOC)
Air Mass Flow rate (lbm s^{-1})	7823 (RTO) and 3696 (TOC)
Overall pressure ratio (OPR) (turboshaft engine)	57.3 (RTO) and 84.3 (TOC)

Table 5.15 N3-X SC technology and design parameters and definitions.

MgB$_2$	is an intermediate-temperature superconductor
MgB$_2$	Magnesium Di-boride is used as the SC material (it has a critical temp of 39 K and its working temp is 28 K, thus LH$_2$ is the cryogenic liquid for this application)
RTO	Rolling takeoff (is @ sea level, Mach 0.24, ISA + 24 °R)
TOC	Top of climb (is @ 34 000 ft alt. and Mach 0.84, ISA)
TME	Total mission energy

Although the performance of superconducting turboelectric N3-X is produced here with MgB$_2$ superconductor, the choice of other SC-materials and the cooling system is still a subject of research and could produce improvements in system weight and performance.

Some of the key parameters in N3-X SC technology and design parameters are listed in Table 5.15.

Table 5.16 shows the performance of N3-X relative to the baseline (B777-200LR) aircraft. Table 5.17 lists the power parameters in the electrical systems as well as their efficiencies.

Siemens, Airbus, and Rolls-Royce are developing a single-aisle commercial aircraft powered by serial HEPS. The near-term flight demonstrator is called E-Fan X and is slated to fly in 2020 (see Figure 5.72, courtesy of Airbus). The 2-MW electric motor and

Table 5.16 N3-X performance relative to baseline[a] aircraft.

Superconducting material/cooling method	GTOW – lbm	Mission fuel consumption – lbm	Mission energy consumption – BTU	Mission energy reduction compared to 777-200LR
777-200LR Class Vehicle	768 000	279 800	5.2×10^9	—
MgB_2/LH_2	496 174	76 171	1.47×10^9	72%

[a] Baseline B777-200LR (with GE90-110B).

Table 5.17 Electrical system efficiencies for N3-X/MgB_2/LH_2.

Electrical components	Power parameter	Efficiency (baseline design value) (%)
Motor	4064 hp at 4400 rpm	99.97[a]
Generator	28 505 hp at 8000 rpm	99.98[a]
Inverter	4064 hp	99.93[a]

[a] Best-value assumption based on TRL 4–6 by the year 2025.

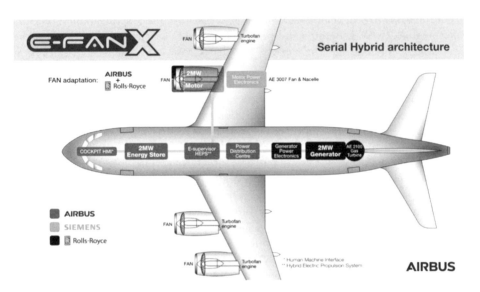

Figure 5.72 Modified BAe 146 aircraft used as hybrid electric flight demonstrator. *Source:* Airbus 2017.

Figure 5.73 NASA X-57, an all-electric distributed propulsion flight demonstrator. *Source:* Courtesy of NASA.

its power electronics control unit, including inverter DC/DC converter and the power distribution system is developed by Siemens. Rolls-Royce is responsible for the turboshaft engine, 2 MW generator and power electronics. Airbus is responsible for the overall system integration. The testbed aircraft is BAe 146, which is a short-haul, regional airliner.

NASA X-57 aircraft, called *Maxwell*, is an all-electric distributed propulsion flight demonstrator with the goal of achieving 80% reduction in energy consumption. The 12 wing leading-edge mounted high-lift electric motors power 12 folding propellers and there are two larger wingtip electric cruise motors, as depicted in Figure 5.73 (courtesy of NASA). At the top of climb, the high-lift motors deactivate and their five folding propellers fold into the nacelle to mitigate drag rise during cruise. The X-57 is modified from the baseline Italian Tecnam P2006T aircraft. This program started in 2014 by the SCEPTOR (Scalable Convergent Electric Propulsion Technology and Operations Research) project at NASA. The final version of X-57 (i.e. Modification 4 in the program) will feature a high-aspect ratio wing with integral high-lift motors. Electric motors on Maxwell use rechargeable lithium-ion batteries that weigh 860 pounds and offer 69.1 kWh of energy, of which 47 kWh is usable. The cruise motors are 60 kW and the 12 high-lift motors are 10.5 kW machines. The electric motors are air-cooled. For more details on X-57, see Gibbs (2018), or Warwick and Norris (2018).

Electric propulsion offers new aircraft design capabilities that are not possible in either conventional gas turbine-powered or reciprocating aircraft (see Moore and Fredericks 2014; Gohardani 2013; and Moore et al. 2014). The comparison between Cirrus SR22 and Leading-Edge Asynchronous Propellers Technology (LEAPTech)

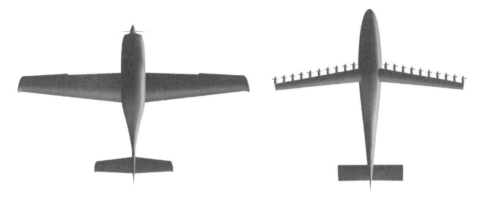

Figure 5.74 Cirrus SR22 and LEAPTech; a comparison.

Table 5.18 Comparison of the Cirrus SR22 with the LEAPTech aircraft.

	Cirrus SR22	LEAPTech
Seating capacity	4	4
Gross weight, lbf	3400	3000
Wing area, ft^2	145	55.1
Wingspan, ft	38.3	31.0
Aspect ratio	10.1	17.4
Wing loading, lbf ft^{-2}	23.5	54.4
Cruise speed, mph	211	200
Cruise C_L (12 000 ft)	0.30	0.77

aircraft is shown in Figure 5.74 and Table 5.18. LEAPTech aircraft has a 72% higher aspect ratio, wing loading that is 131%, and lift coefficient at cruise, which is 157% higher than SR22. The wetted area of the LEAPTech wing is 62% less than the SR22 aircraft wing.

5.17 Summary

Promising technologies in the propulsion system architecture are under active R&D effort at the industry, NASA, European Union's ACARE, and the participating universities. To reach the goals of sustainable aviation, we need to redesign the airframe for higher L/D and lower drag. The tube and wing is aerodynamically inefficient. BWB or hybrid wing–body (HWB) aircraft are aerodynamically more efficient and lend themselves to BLI integration and distributed propulsion. Also, the noise emission in future aircraft is subject to severe restrictions that require an all-out strategy to achieve Quiet Aircraft (similar to MIT program). Noise mitigation needs to be addressed by engine design, shielding strategy and turbulent boundary layer/separation flow management.

In addition to advanced distributed propulsion BWB aircraft design, we need FAA NextGen advances in air traffic management (ATM) and operations to achieve sustainable aviation goals.

The promising propulsion technologies include advanced core UHB-turbofan engine, multi-combustor, multifuel propulsion systems (using kerosene and cryogenic fuel), wave rotor topping cycle, intercooled and recuperated turbofan engines, PDE, i.e. tapping into the unsteadiness as a means of harvesting energy. The strategy toward electric aircraft took us through hybrid-electric architecture. We discussed the number of propulsors and related it to the probability of engine failure. Electric propulsion through batteries and fuel cells were also discussed e.g., X-57. A promising air-breathing rocket engine, called *SABRE*, was presented as the technology for SSTO travel. Finally, due to intense R&D and the promise of fusion, we presented the CFR technology that is pursued at Lockheed Martin and MIT, among others as the promising technology for nuclear propulsion flight.

References

Abbott, J.M., Anderson, B.H. and Rice, E.J., 1987. Inlets, Ducts and Nozzles. Aero-Propulsion '87. *NASA CP-3049*.

ACARE (n.d.) Flightpath 2050 Goals. http://www.acare4europe.org/sria/flightpath-2050-goals.

Airbus 2017. Airbus, Rolls-Royce, and Siemens team up for electric future partnership launches E-Fan X hybrid-electric flight demonstrator. Press release 28 November, 2017. https://www.airbus.com/newsroom/press-releases/en/2017/11.

Akbari, P., Nalim, M.R., and Müller, N. (2006). A review of wave rotor technology and recent developments. Journal of Engineering for Gas Turbines and Power 128 (4): 715–735.

Akbari, P., Szpynda, E., and Nalim, M.R. 2007. Recent Developments in Wave Rotor Combustion Technology and Future Perspectives: A Progress Review. AIAA Paper Number 2007-5055, presented at the 43rd AIAA/ASME/SAE/ASEE Joint Propulsion Conference and Exhibit, Cincinnati, Ohio (6–11 July 2007).

Anders, S.G., Sellers, W.L. and Washburn, A.E. 2004. Active Flow Control Activities at NASA Langley. AIAA Paper Number, 2004-2623. Presented at the 2nd AIAA Flow Control Conference (June 28–July 1, 2004).

Anderson, J.D. Jr. (2003). Modern Compressible Flow: With Historical Perspective, 3e. New York: McGraw-Hill.

Armstrong, M. 2015. Superconducting Turboelectric Distributed Aircraft Propulsion. Presentation at the Cryogenic Engineering Conference (July 2015).

Attinello, J.S. 1957. The Jet Wing. IAS Preprint No. 703, IAS 25th Annual Meeting (Jan. 28–31, 1957).

Bardeen, J., Cooper, L.N., and Schrieffer, J.R. (1957). Theory of superconductivity. Physical Review 108 (5): 1175–1205.

Barr, L. (2008). Pulsed Detonation Engine Flies into History. http://www.af.mil/News/Artcle-Display/Article/123534/pulsed-detonation-engine-flies-into-history.

Berlincourt, T.G. (1987). Emergence of Nb-Ti as supermagnet material. Cryogenics 27 (6): 283–289.

Blazowski, W.S. (1985). Fundamentals of combustion. In: Aerothermodynamics of Aircraft Engine Components, AIAA Education Series, 1–43. Washington, DC: AIAA, Inc.

Blumenthal, B.T., Elmiligui, A.A., Geiselhart, K.A. et al. (2018). Computational investigation of a boundary-layer ingestion propulsion system. Journal of Aircraft (3): 55, 1141–1153.

Bock, S., Horn, W., and Sieber, J. 2008. Active core – A key technology for more environmentally friendly aero engines being investigated under the NEWAC Program. Paper presented at the 26th International Congress of the Aeronautical Sciences, ICAS 2008, Anchorage, Alaska, ICAS-2008-686.

Bradley, M.K. and Droney, C.K.2011. Subsonic Ultra Green Aircraft Research: Phase I Final Report. *NASA/CR-2011-216847.*

Bratkovich, T.E., Aarnio, M.J., Williams, J.T. and Bussing, T.R.A. 1997. An Introduction to Pulse Detonation Rocket Engines (PDREs). *AIAA Paper Number 97-2742.*

Brown, G.V. 2011. Weights and Efficiencies of Electric Components of a Turboelectric Aircraft Propulsion System. *AIAA paper No. 2011-225.*

Chan, N.Y.S. 2008. Scaling considerations for small aircraft engines. Masters thesis. Department of Aeronautics and Astronautics, MIT, Cambridge, Massachusetts.

Clarke, D.R., Oechsner, M., and Padture, N.P. (October 2012). Thermal-barrier coatings for more efficient turbine engines. Materials Research Society Bulletin 37 (10): 891–899.

Civil Aviation Authority. 2002. *CAP 513 ETOPS.* www.caa.co.uk.

Del Rosario, R. 2014. A future with Hybrid Electric Propulsion Systems: A NASA perspective. Turbine Engine Technology Symposium. Strategic Visions Workshop, Dayton, OH, September.

Del Rosario, R., Follen, G., Wahls, R., and Madavan, N. 2012. Subsonic fixed-wing project overview and technical challenges for energy efficient environmentally compatible subsonic transport aircraft. 50th AIAA Aerospace Sciences Meeting, Nashville, TN (9–12 January 2012).

Dippold, V., Hosder, S., and Schetz, J.A. 2004. Analysis of Jet-Wing Distributed Propulsion from Thick Wing Trailing Edges. *AIAA Paper Number 2004–1205.*

Dowling, A., Greitzer, E., Hynes, T. et al. (2006–2007). The Silent Aircraft. MIT Aero-Astro Annual Report/magazine.

Drela, M. (2009). Power balance in aerodynamic flows. AIAA Journal 47 (7): 1761–1771.

Dyson, R. 2017. NASA hybrid electric aircraft propulsion. Presentation at the NIEA Biomimicry Summit (October 2017).

Envia, E. 2010. NASA open-rotor noise research. Paper presented at the 14th CEAS-ASC Workshop, Warsaw, Poland (October 2010).

Environmental Protection Agency, EPA 2016 Report, (1990–2014). Inventory of US Greenhouse Gas Emissions and Sinks.

Epstein, A.H., 2007. Distributed Propulsion: New Opportunities for an Old Concept. *Final Technical Report submitted by MIT Gas Turbine Laboratory to DARPA.*

Epstein, A.H. (2014). Aeropropulsion for commercial aviation in the twenty-first century and research directions needed. AIAA Journal 52 (5): 901–911.

European Commission (2011) Flightpath 2050, https://ec.europa.eu/transport/sites/ transport/files/modes/air/doc/flightpath2050.pdf.

Farokhi, S. (2014). Aircraft Propulsion, 2e. Chichester (UK): Wiley.

Farokhi, S., Sheu, W.L., and WU, C. (1989). On the design of optimum-length transition ducts with offset: a computational study. In: Computers and Experiments in Fluid Flow (eds. G.M. Carlomagno and C.A. Brebbia), 215–228. Berlin: Springer Verlag.

Felder, J.L. 2014. NASA N3-X with turboelectric distributed propulsion. Presentation at the Institution of Mechanical Engineers, London, UK (November 17–18, 2014).

Felder, J.L., Brown, G.V., Kim, H.D., and Chu, J. 2011. Turboelectric distributed propulsion in a hybrid wing body aircraft. Paper presented at the International Society of Air Breathing Engines (ISABE) Conference. *ISABE Paper No. 2011-1340.*

Fraas, A.P. 1980. Summary of Research Development Effort on Air and Water Cooling of Gas Turbine Blades. Oakridge National Laboratory Technical Memorandum, ORNL/TM-6254, March.

Gibbs, Y. 2018. NASA Armstrong Fact Sheet: NASA X-57 Maxwell. August 27, https://www.nasa.gov/centers/armstrong/news/FactSheets/FS-109.html.

Gohardani, A.S. (2013). A synergistic glance at the prospects of distributed propulsion technology and the electric aircraft concept for future unmanned air vehicles and commercial/military aviation. Progress in Aerospace Sciences 57: 25–70.

Greitzer, E.M., Tan, C.S., and Graf, M.B. (2004). Internal Flow: Concepts and Applications. Cambridge (UK): Cambridge University Press.

Grewe, V. and Linke, F. (2017). Eco-efficiency in aviation. Meteorologische Zeitschrift 26 (6): 689–696.

Grewe, V., Bock, L., Burkhardt, U. et al. (2016). Assessing the climate impact of the AHEAD multifuel blended wing body. Meteorologische Zeitschrift 26 (6): 711–725.

Hall, D.K. 2016. Boundary layer ingestion propulsion – benefits, challenges and opportunities. Presented at the 5th UTIAS International Workshop on Aviation and Climate Change, Toronto, Canada (May 2016).

Hall, C.A., and Crichton, D. 2005. Engine and installation configurations for a silent aircraft, ISABE-2005–1164. Presented at International Symposium on Air Breathing Engines, Munich, Germany.

Hall, C.A. and Crichton, D. (2007). Engine design studies for a silent aircraft. Journal of Turbomachinery 129 (7): 479–487.

Hall, D.K., Huang, A.C., Uranga, A. et al. (2017). Boundary layer ingestion propulsion benefit for transport aircraft. Journal of Propulsion and Power 33 (5): 1118–1129.

Hardin, L.W., Tillman, G., Sharma, O.P., Berton, J., and Arend, D.J. 2012. Aircraft System Study of Boundary Layer Ingesting Propulsion. AIAA 2012-3993, Joint Propulsion Conference (30 July – 01 August, 2012)

Hawthorne, W.R. (1994). Reflections on United Kingdom aircraft gas turbine history. Journal of Engineering for Gas Turbines and Power 116 (3): 495–510.

Heiser, W.H. and Pratt, D.T. (2002). Thermodynamic cycle analysis of pulse detonation engines. Journal of Propulsion and Power 18 (1): 68–76.

Jankovsky, A., Bowman, C., and Jansen, R.H., 2016. Building blocks for transport-class hybrid and turboelectric vehicles. Paper presented at the Electric & Hybrid Aerospace Technology Symposium, Cologne, Germany (November 2016).

Jansen, R.H., Bowman, C., and Jankovsky, A. 2016. Sizing power components of an electrically driven tailcone thruster and a range extender. Presented at the 16th AIAA Aviation Technology, Integration, and Operations Conference (June 2016).

Jones, S.M. and Welch, G.E., 1996. Performance Benefits for Wave Rotor-Topped Gas Turbine Engines. *NASA TM 107193.*

Kaemming, T. (2003). Integrated vehicle comparison of turbo-ramjet engine and pulsed detonation engine," Transactions of ASME. Journal of Engineering for Gas Turbines and Power 125 (1): 257–262.

Kalsi, S.S., Gamble, B.B., Snitcher, G., and Ige, S.O. 2006. The status of hts ship propulsion motor developments. *Proceedings of the IEEE PES Meeting*, Montreal, Canada (June 2006).

Kentfield, J.A.C. (1993). Nonsteady, One-Dimensional, Internal Compressible Flows. Oxford (UK): Oxford University Press.

Kentfield, J.A.C. (1995). On the feasibility of gas-turbine pressure-gain combustors. International Journal of Turbo and Jet Engines 12 (1): 29–36.

Kerrebrock, J.L. (1992). Aircraft Engines and Gas Turbines, 2e. Cambridge, MA: MIT Press.

Kim, H.D., Brown, G.V., and Felder, J.L. 2008. Distributed turboelectric propulsion for hybrid wing body aircraft. Paper presented at the 2008 International Power Lift Conference, London, England (July 2008)

Kim, H.D., Felder, J.L., Tong, M.T., and Armstrong, M. 2013. Revolutionary aeropropulsion concept for sustainable aviation: turboelectric distributed propulsion. ISABE Conference, Paper Number 2013-1719.

Knapp, M., and Said, W. 2018. Zunum Aero's hybrid-electric airplane aims to rejuvenate regional travel. *IEEE Spectrum*, 26 April

Ko, A., Schetz, J.A., and Mason, W.H. 2003. Assessment of the potential advantages of distributed-propulsion for aircraft. Paper presented at the XVIth International Symposium on Air Breathing Engines (ISABE). No. 2003-1094 (August 31–September 5, 2003).

Koff, B.L. 1991. Spanning the globe with jet propulsion. Presented at the 21st Annual Meeting and Exhibit, AIAA paper 1991–2987 (May 1991).

Küchemann, D. (1978). The Aerodynamic Design of Aircraft, 229. New York: Pergamon Press.

Kurzke, J. 2017. GasTurb 13. www.gasturb.de.

Kyprianidis, K.G. (2011). Future aero engine designs: an evolving vision. In: Advances in Gas Turbine Technology (ed. E. Benini). InTech. ISBN: 978-953-307-611-9, Available from: http://www.intechopen.com/books/advances-in-gas-turbine-technology/future-aero-engine-designs-anevolving-vision.

Leifsson, L.T., Ko, A., Mason, W.H., Schetz, J.A., Haftka, R.T. and Grossman, B., 2005. Multidisciplinary Design Optimization for a Blended Wing Body Transport Aircraft with Distributed Propulsion. MAD Center for Advanced Vehicles Report No. 2005-05-01, Virginia Polytechnic Institute & State University.

Liebeck, R.H. (2004). Design of the blended wing body subsonic transport. Journal of Aircraft 41 (1): 10–25.

Liepmann, H.W. and Roshko, A. (1957). Elements of Gas Dynamics. New York: Wiley.

Liew, K.H., Urip, E., Yang, S.L., and Siow, Y.K.2003. A complete parametric cycle analysis of a turbofan with interstage turbine burner. AIAA Paper Number AIAA-2003-0685. Presented at 41st AIAA Sciences Meeting and Exhibit, Reno, Nevada (January 2003).

Liew, K.H., Urip, E., Yang, S.L., Mattingly, J.D., and Marek, C.J. 2005. Performance Cycle Analysis of a Two-Spool, Separate-Exhaust Turbofan with Interstage Turbine Burner. *NASA/TM-2005-213660.*

Lighthill, M.J. (1952). On sound generated aerodynamically, I: general theory. Proceedings of the Royal Society of London, Series A 211: 564–587.

Lighthill, M.J. (1954). On sound generated aerodynamically. II, turbulence as a source of sound. Proceedings of the Royal Society of London, Series A 222: 1–32.

Liu, F. and Sirignano, W.A. (2001). Turbojet and turbofan engine performance increases through turbine burners. Journal of Propulsion and Power 17 (3): 695–705.

Lord, W.K., Suciu, G.L., Hasel, K.L., and Chandler, J.M. 2015. Engine architecture for high efficiency at small core size. AIAA Paper Number 2015- 0071. Presented at the SciTech Forum, Kissimmee, FL (5-9 January, 2015).

Bauhaus Luftfahrt (n.d.). Propulsive fuselage concept. https://www.bauhaus-luftfahrt.net/en/topthema/propulsive-fuselage.

Masson, P.J. and Luongo, C. 2007. HTS machines for applications in all-electric aircraft. Proceedings of IEEE PES Meeting, Tampa, FL.

Mawid, M.A., Park, T.W., Sekar, B., and Arana, C. (2003). Application of pulse detonation combustion to turbofan engines. Journal of Engineering for Gas Turbines and Power 25: 270–283.

Maynard, G. Air Travel – Greener by Design. 2015 2014–2015 Annual Report; Royal Aeronautical Society, London (UK).

McCallum, P.W. (2000). Aero-Space Base Research & Technology. https://flight.nasa.gov/events/home&home/glenn/aerosp1/sld001.htm.

Merchant, A., Kerrebrock, J.L., Adamczyk, J.J., and Braunscheidel, E. 2004.Experimental Investigation of a High Pressure Ratio Aspirated Fan Stage. *ASME Paper No. GT2004-53679.*

Miranda, L.R. 1986. Aerodynamic Effects of Wingtip Mounted Propellers and Turbines. *AIAA paper No. 86-1802.*

Moore, M.D. 2016. Distributed electric propulsion (DEP) aircraft. Presented at the 5th Symposium on Collaboration in Aircraft Design, Naples, Italy (October 2016). https://aero.larc.nasa.gov/files/2012/11/Distributed-Electric-Propulsion-Aircraft.pdf.

Moore, M.D. and Fredericks, W.J. 2014. Misconceptions of electric propulsion aircraft and their emergent aviation markets. Presented at the 52nd Aerospace Sciences Meeting, AIAA SciTech Forum, AIAA Paper Number 2014-0535.

M. D. Moore, W. J. Fredericks, N. K. Borer, A. M. Stoll and J. Bevirt. 2014. Drag Reduction Through Distributed Electric Propulsion. Aviation Technology, Integration, and Operations Conference, Atlanta.

Moustapha, H., Zelesky, M.F., Baines, N.C., and Japikse, D. (2003). Axial and Radial Turbines. Concept ETI, Inc.

Nalim, M.R. (2002a). Thermodynamic Limits of Work and Pressure Gain in Combustion and Evaporation Processes. Journal of Propulsion and Power 18 (6): 1176–1182.

Nalim, M. R. 2002b. Wave rotor detonation engine. US Patent No. 6, 460,342, filed 2002.

NASA Glenn Research Center. 2018. Boundary Layer Ingestion Propulsion. https://www1.grc.nasa.gov/aeronautics/bli.

Numerical Propulsion System Simulation (NPSS). 2016. Southwest Research Institute, San Antonio, Texas. www.swri.org.

Oates, G.C. (1988). Aerothermodynamics of Gas Turbine and Rocket Propulsion. Washington, DC: AIAA.

Patterson, J.C., and Bartlett, G.R.1987. Evaluation of Installed Performance of A Wingtip Mounted Pusher Turboprop on a Semi-Span Wing. *NASA TP-2739.*

Popular Mechanics. 2016. 10 ways to fix air travel. http://www.popularmechanics.com/flight/a2424/4237593. Accessed 05 April 2018.

Povinelli, L.A. 2001. Impact of Dissociation and Sensible Heat Release on Pulse Detonation and Gas Turbine Engine Performance. International Symposium for Airbreathing Engines. ISABE, Paper Number 1212, September.

Povinelli, L.A. 2002. Pulse Detonation Engines for High Speed Flight. *NASA/TM-2002-211908.*

Pratt, D.T., Humphrey, J.W., and Glenn, D.E. (1991). Morphology of standing oblique detonation waves. Journal of Propulsion and Power 7 (5): 837–845.

Royal Aeronautic Society (2018). Pulse detonation spaceplane. Aerospace 45 (7): 4–5.

Rao, A.G., Yin, F., and van Buijtenen, J.P. (2014). A hybrid engine concept for multifuel blended wing body. Aircraft Engineering and Aerospace Technology: An International Journal 86 (6): 483–493.

Ray, P.J. 2015. Structural investigation of La(2-x)Sr(x)CuO(4+y) – Following staging as a function of temperature. Master's thesis. Niels Bohr Institute, Faculty of Science, University of Copenhagen. Copenhagen, Denmark, November, Figure 2.4.

Risen, T. (2018). Q & A: open-rotor innovator. Aerospace America: 12–13. https:// aerospaceamerica.aiaa.org/departments/open-rotor-innovator/.

Rolt, A.M., and Baker, N.J. 2009. Intercooled turbofan engine design and technology research in the EU Framework 6 NEWAC Programme. Paper 2009-1278 presented at the International Congress of the Aeronautical Sciences, ICAS.

Rolt, A.M., and Kyprianidis, K.G. 2010. Assessment of new aero-engine core concepts and technologies in the EU Framework 6 NEWAC Programme. Paper presented at the 27th International Congress of the Aeronautical Sciences, ICAS, Paper 2010-4.6.3 (September 2010).

Roy, G.D., Frolov, S.M., Borisov, A.A., and Netzer, D.W. (2004). Pulse detonation propulsion: challenges, current status and future perspectives. Progress in Energy and Combustion Science 30: 545–672.

Sehra, A.K., and Shin, J. 2003. Revolutionary Propulsion Systems for 21st Century Aviation. *NASA/TM-2003-212615.*

Shapiro, A. (1953). The Dynamics and Thermodynamics of Compressible Fluid Flow. New York, Vol. 1 and Vol. 2: Ronald Press.

Sirignano, W.A. and Liu, F. (1999). Performance increases for gas turbine engines through combustion inside the turbine. Journal of Propulsion and Power 15 (1): 111–118.

Smith, L.H. Jr. (1993). Wake ingestion propulsion benefit. Journal of Propulsion and Power 9 (1).

Smith, A.M.O. and Roberts, H.E. (1947). The jet airplane utilizing boundary layer air for propulsion. Journal of Aeronautical Sciences 14 (2): 97–109.

Strehlow, R.A. (1984). Combustion Fundamentals, 302–307. New York: McGraw-Hill.

Thornton, G. 1963. Introduction to nuclear propulsion- introduction and background lecture 1 Feb. 26-28. *NASA Technical Reports Server.*

Varvill, R. and Bond, A. (2003). A comparison of propulsion concepts for SSTO reusable launchers. Journal of the British Interplanetary Society 56: 108–117.

Vögeler, K. 1998. The Potential of Sequential Combustion for High Bypass Jet Engines. ASME Paper Number 98-GT-311. *Proceedings of the 1998 International Gas Turbine & Aeroengines Congress and Exhibitions*, Stockholm, Sweden, (June 1998).

Vos, R. and Farokhi, S. (2015). Introduction to Transonic Aerodynamics. Berlin: Springer-Verlag.

Warwick, G., and Norris, G. 2018.NASA Shares Hard Lessons as All-Electric X-57 Moves Forward, Aviation Week & Space Technology, June 5.

Welch, G.E. 1996. Macroscopic Balance Model for Wave Rotors. NASA TM-107114 and AIAA Paper Number 96-0243. Paper presented at the 34th Sciences Meeting and Exhibit, Reno, Nevada (January 15-18, 1996).

Welch, G.E., Jones, S.M., and Paxson, D.E. 1995. Wave Rotor-Enhanced Gas Turbine Engines. *NASA TM- 106998.*

Welstead, J.R., and Felder, J.L.2016. Conceptual Design of a Single-Aisle Turboelectric Commercial Transport with Fuselage Boundary Layer Ingestion. AIAA Paper No. 2016-1027 Presented at the AIAA SciTech Conference (January 2016).

Welstead, J., Felder, J., Guynn, M., et al. (2017). Overview of the NASA STARC-C ABL (Rev. B): Advanced Concept. https://ntrs.nasa.gov/archive/nasa/casi.ntrs.nasa.gov/20170005612.pdf.

Whurr, J. (2013). Future Civil Aeroengine Architectures and Technologies. http://www.etc10.eu/mat/Whurr.pdf.

Wilfert, G., Sieber, J., Rolt, A. and Baker, N.2007. New Environmental Friendly Aero Engine Core Concepts. Paper 2007-1120 presented at the International Society for Airbreathing Engines, ISABE.

Wu, M.K., Ashburn, J.R., Torng, C.J. et al. (1987). Superconductivity at 93K in a new mixed-phase Y-Ba-Cu-O compound system at ambient pressure. Physical Review Letters 58 (9): 908–910.

Xu, L., Kyprianidis, K.G., and Grönstedt, T.U.J. (2013). Optimization study of an intercooled recuperated aero-engine. Journal of Propulsion and Power 29 (2): 424–432.

Yerman, J. Airbus Charges Forward with All-Electric Aircraft. *Industry,* May 6, 2015. https://apex.aero/airbus-electric-aircraft.

York, M.A., Hoburg, W.W., and Drela, M. (2017). Turbofan engine sizing and tradeoff analysis via signomial programming. Journal of Aircraft 55 (3): 1–16.

Yutko, B. and Hansman, J. (2011). Approaches to Representing Aircraft Fuel Efficiency Performance for the Purpose of a Commercial Aircraft Certification Standard. Cambridge, MA: MIT International Center for Air Transportation.

Zel'dovich, Y.B. and Raizer, Y.P. (2002). Physics of Shock Waves and High-Temperature Hydrodynamic Phenomena (eds. W.D. Hayes and R.F. Probstein). New York: Dover Publications.

Zunum Aero's (n.d.) website: https://zunum.aero

6

Pathways to Sustainable Aviation

6.1 Introduction

The outlook for sustainable aviation is bright. Every single sector within aviation industry, from airframe and propulsion companies to avionics (guidance, navigation, and control) and air traffic management (ATM) and operations, are all actively engaged and investing in promising research in greener aviation. Add to these efforts government agencies in the United States and Europe such as National Aeronautics and Space Administration (NASA) and Advisory Council for Aeronautics Research in Europe (ACARE), among others, as well as academia that conduct pioneering research in environmentally friendly transportation. The specific technologies in this sector, from alternative jet fuels to superconducting electric motors/generators and cryo-cooling thermal management system in hybrid and all electric propulsion aircraft, have reached different technology readiness levels (TRLs) and thus are candidates for new challenges toward certification. In this chapter, we briefly address the specific pathways to sustainable aviation in key technologies, including the challenges in regulations and certifications, public confidence, and societal acceptance.

6.2 Pathways to Certification

Greener aviation, as we discussed in Chapters 3–5, is more than just *electric propulsion*. However, with the rise of drones in the twenty-first century (with 940 000 registered unmanned aerial vehicle (UAV) operators in the United States in 2017), a transformational commercial aviation market is born that is called *urban air mobility* (UAM). Markets in this new sector include airport shuttle, air taxi, ride sharing, and air ambulance. This new sector is best-suited for electric propulsion application due to its shorter range and high utilization factor. Figure 6.1 shows the electric vertical takeoff and landing (eVTOL) aircraft concepts from Uber and Airbus for the UAM market sector. A ride-sharing concept from Uber, known as *Uber Elevate* with five electric motors powering four lift fans and one propulsor, is shown in Figure 6.1a. This project is fast reaching full-scale flight test maturity, i.e. TRL 9, with 2020 planned flight demonstration with the anticipated certification and revenue production service in 2023. Airbus Skyways is an autonomous parcel delivery UAV. Airbus eVTOL aircraft called Project *Vahana* is shown in Figure 6.1b. This autonomous aircraft uses eight rotors in a

Future Propulsion Systems and Energy Sources in Sustainable Aviation, First Edition. Saeed Farokhi.
© 2020 John Wiley & Sons Ltd. Published 2020 by John Wiley & Sons Ltd.
Companion website: www.wiley.com/go/farokhi/power

(a)

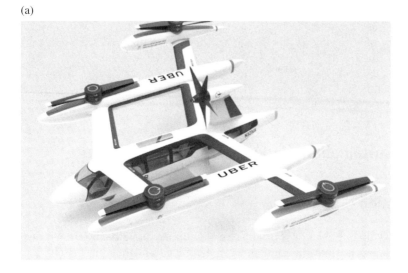

(b)

Figure 6.1 Two examples of eVTOL aircraft chosen for the urban air mobility market. (a) *Uber Elevate* UAM Project. *Source:* Courtesy of Uber. (b) Airbus *Vahana* Project. *Source:* Courtesy of Airbus.

tilt (fore and aft) wing design. The first flight test of the self-piloted *Vahana* was conducted in January 2018. The company projects marketing *Vahana* in 2020. The German startup company *E-Volo* and a startup funded by Google cofounder Larry Page, called *Zee.Aero* are among many other companies that work in UAM and eVTOL markets.

The challenges in green aviation due to airworthiness, certification (Federal Aviation Administration [FAA]), safety, and *drone traffic management* first will be explored/resolved in the UAM sector. UAM as an early adopter of green aviation technology is expected to generate benchmarks and pathways to broader sustainable aviation. For example, battery life that governs vehicle range, and operational aspects such as battery charging time or battery swap, and related ground support operations/procedures will first be addressed and streamlined in the commercial UAM sector. The high utilization factor that is envisioned in UAM commercial aviation with estimated $2.6 billion/year revenue will serve as the green technology accelerator in the twenty-first century. Societal acceptance of UAM and autonomous flight and confidence in its safety record will be essential to the future of sustainable aviation. The dual-use, commercial-military, nature of the advanced technologies that are embedded in UAM, electric propulsion, and green aviation will also serve as an important accelerator to the twenty-first century innovation revolution in aviation.

6.3 Energy Pathways in Sustainable Aviation

Promising energy sources, as in renewable jet fuels and batteries, will continue to be the cornerstone of sustainable aviation. Figure 6.2 shows the overview of many pathways to produce alternative jet fuels with similar properties to conventional jet fuel (see Brown, FAA, 2012). Feedstock availability, commercial production, and global distribution are three challenges facing the renewable jet fuel industry.

The challenge with alternative, drop-in jet fuels is in quantity production and global distribution. The cost is expected to be competitive with time, especially with

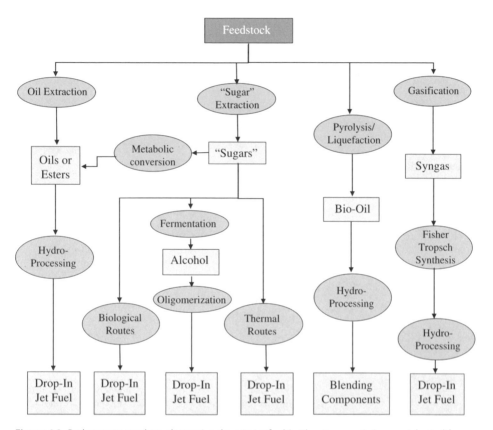

Figure 6.2 Pathways to produce alternative drop-in jet fuel (without oxygen). *Source:* Adapted from Brown 2012.

governmental incentive programs. Cryogenic fuels, in particular hydrogen, is very attractive in fuel cell applications to replace the gas turbine (GT)-based auxiliary power units (APU). Hydrogen fuel cells offer an emissions-free replacement for the small GT APU. A hydrogen fuel cell–based electric propulsion system is the natural outgrowth of APU research and development.

The certification challenge in electric propulsion aircraft is best faced by small planes in UAM and general aviation (GA) category and then the experience be matured to longer-range commercial aircraft, as noted by Moore and Fredericks 2014. Figure 6.3 shows the progression of early adopters from GA market to regional jets and larger aircraft. The metrics of aerodynamic efficiency (lift-to-drag ratio), ride quality (wing loading), and emission (overall mile per gallon efficiency per passenger) are noted by Moore and Fredericks (2014) in Figure 6.3 to determine technology market infusion.

To gain early certification, operational confidence, and in-flight experience with electric propulsion, hybrid-electric propulsion (HEP) offers the first logical step for regional and medium-range commercial aviation. Toward this goal, an advanced flight demonstrator, E-Fan X, is shown in Figure 6.4 from the team of Airbus, Rolls-Royce, and Siemens (see Airbus 2017). This quad regional aircraft with a single turbofan engine replaced with a 2-MW electric motor is scheduled for flight tests in 2020.

Boundary-layer ingestion (BLI) and distributed propulsion (DP) are a powerful means of aircraft drag mitigation and thus block fuel reduction. Aft-mounted DP aircraft defines a new design space with propulsion-airframe integration as its cornerstone. The first step in BLI is applied to the fuselage, on both sides of the Atlantic. The US effort is pursued at NASA and is called Single-aisle Turboelectric Aircraft with Aft Boundary Layer Propulsion, or STARC-ABL. The European counterpart is the Bauhaus Luftfahrt

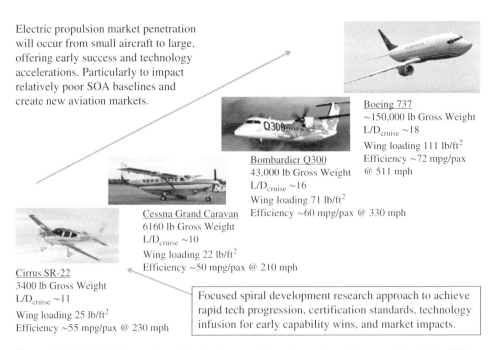

Electric propulsion market penetration will occur from small aircraft to large, offering early success and technology accelerations. Particularly to impact relatively poor SOA baselines and create new aviation markets.

Boeing 737
~150,000 lb Gross Weight
L/D_{cruise} ~18
Wing loading 111 lb/ft^2
Efficiency ~72 mpg/pax
@ 511 mph

Bombardier Q300
43,000 lb Gross Weight
L/D_{cruise} ~16
Wing loading 71 lb/ft^2
Efficiency ~60 mpg/pax @ 330 mph

Cessna Grand Caravan
6160 lb Gross Weight
L/D_{cruise} ~10
Wing loading 22 lb/ft^2
Efficiency ~50 mpg/pax @ 210 mph

Cirrus SR-22
3400 lb Gross Weight
L/D_{cruise} ~11
Wing loading 25 lb/ft^2
Efficiency ~55 mpg/pax @ 230 mph

Focused spiral development research approach to achieve rapid tech progression, certification standards, technology infusion for early capability wins, and market impacts.

Figure 6.3 Market infusion pathway by electric propulsion. *Source:* From Moore and Fredericks 2014.

Gas Turbine Air Intake to Power a
2 MW Electric Generator

2 MW Electric Motor

Figure 6.4 E-Fan X flight demonstrator is a modified BAe 146 aircraft showing one engine replaced with an electric motor driving a ducted fan. *Source:* Courtesy of Airbus.

Propulsive Fuselage concept (see www.bauhaus-luftfahrt.net/en/topthema/propulsive-fuselage; Welstead et al. n.d.). Figure 6.5 shows both BLI propulsive fuselage concepts. Moore et al. (2014), Bock et al. (2008), Bradley and Droney (2011), Armstrong 2015, Dyson (2017), and Kim et al. (2013) address HEP concepts. Epstein (2014) presents future research directions in GT aeroengines.

Higher aerodynamic efficiency in future subsonic transport will take on blended wing–body (BWB) or hybrid wing–body (HWB) configurations. At supersonic cruise speeds, it is the sonic boom alleviation that governs the current and future development. Aerion Supersonic (www.aerionsupersonic.com), Boom Supersonic (https://boomsupersonic.com), and Spike Aerospace (www.spikeaerospace.com) are actively pursuing low-boom technology. Business jet class aircraft, such as X-59, will be an early adopter of the sustainable aviation technology in supersonic flight over land. Other competing interests in high-speed commercial aviation are being pursued at Boom Supersonic, which is developing a 55-seat airliner for Mach 2.2 cruise and Spike Aerospace with cruise Mach number 1.6. These aircraft use HEP with the GT engine providing power to electric generators. Lockheed Martin is investing in hybrid airships (www.lockheedmartin.com).

6.4 Future of GT Engines

The future of GT engines in commercial aviation continues to be bright. The industry is moving in the direction of higher bypass ratios in geared, variable pitch turbofan engines with advanced core. Open rotor propulsor will present new opportunities in green aviation. In addition, new engine architecture that incorporates intercooler, recuperator, and active controls promises to save even more fuel. The long-range wide-body

(a)

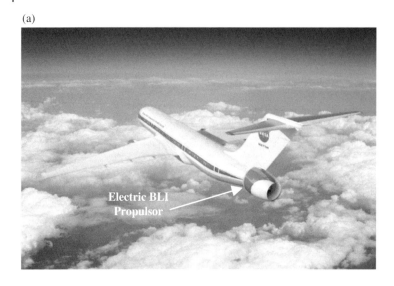

(b)

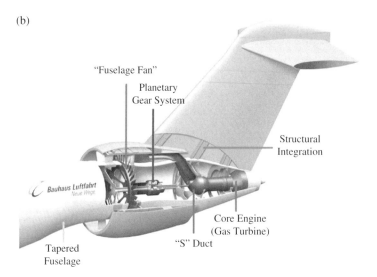

Figure 6.5 Pathway to sustainable aviation in aircraft drag reduction and systems integration: (a) NASA STARC-ABL with tail-cone thruster; (b) Bauhaus Propulsive Fuselage.

commercial aviation sector will continue to be powered by advance GT engines. Low-emission combustors, low community noise engines, and low fuel consumption are the main drivers in advanced GTs that have achieved extremely high levels of reliability and safety. The HEP that includes GT-based electric generator will continue to be a viable solution in distributed propulsion, which will be the new architecture in green aviation. Figure 6.6 based on the projections from NASA shows the pathways to sustainable aviation and the time frame to achieve it.

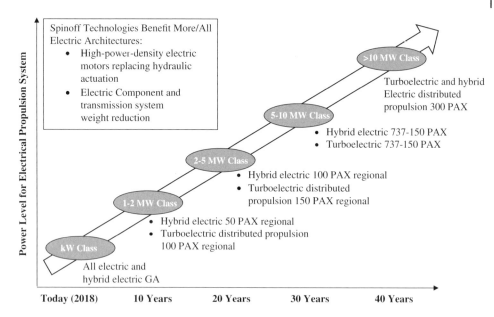

Figure 6.6 Projected time frame to achieve TRL 6 in hybrid electric propulsion. *Source:* Based on projections from Del Rosario 2014.

6.5 Summary

The future of aviation is tied to carbon-neutral growth (CNG). Renewable drop-in jet fuels and batteries are the new energy sources for sustainable aviation. The new sector in UAM will help with the certification, regulations, and operations. The electric and HEP utilizing superconducting motors/generators and network represent the propulsion revolution in the next 40–50 years. New aircraft configurations with high aerodynamic efficiency will focus on BWB and HWB aircraft designs as well as MIT's Double Bubble lifting fuselage concept. In this framework, BLI with distributed propulsion (DP) are key design characteristics that reduce aircraft drag/block fuel use. Ultra-high bypass (UHB) turbofan engines with advanced core using biofuels will serve the commercial long-range market. Advanced ATM and operations, e.g. FAA's *NextGen* program, will help minimize block fuel use while facilitating efficient air transportation.

Noise pollution will be mitigated at the source, namely at the fan, jet, core, and at the airframe level. System integration with shielding, BLI, and embedded concepts will prove essential in the future of low-noise, quiet aeronautics. The fluidic controls will minimize/eliminate control surfaces with the associated block fuel reduction and mitigation of viscous/vortex noise. The sonic boom overpressure accompanying supersonic flight will reduce to a *thump* level noise that allows supersonic cruise overland.

Beyond renewable jet fuels and batteries, the ultimate energy source in aircraft propulsion is nuclear fusion. The flight-weight compact fusion reactor (CFR) to be used in aircraft propulsion will become a reality in the twenty-first century.

References

Airbus.2017. Airbus, Rolls-Royce, and Siemens team up for electric future Partnership launches E-Fan X hybrid-electric flight demonstrator. Press release, 28 November, https://www.airbus.com/newsroom/press-releases/en/2017/11.

Armstrong, M. 2015. Superconducting Turboelectric Distributed Aircraft Propulsion. Presentation at the Cryogenic Engineering Conference (July 2015).

Bock, S., Horn, W. and Sieber, J. 2008. Active Core – A Key Technology for More Environmentally Friendly Aero Engines Being Investigated under the NEWAC Program. Paper presented at the 26th International Congress of the Aeronautical Sciences, ICAS 2008, Anchorage, Alaska, ICAS-2008-686.

Bradley, M.K. and Droney, C.K. 2011. Subsonic Ultra Green Aircraft Research: Phase I Final Report. *NASA/CR-2011-216847*.

Brown, N. 2012. Alternative Jet Fuels- Feedstock Development, Processing and Conversion Research & Regional Supply and Refining Infrastructure. FAA Presentation for Center of Excellence for Alternative Jet Fuels and Environment, November 15.

Del Rosario, R. 2014. A Future with Hybrid Electric Propulsion Systems: A NASA Perspective. Turbine Engine Technology Symposium, Strategic Visions Workshop, Dayton, OH, September.

Dyson, R. 2017. NASA Hybrid Electric Aircraft Propulsion. Presentation at the NIEA Biomimicry Summit (October 2017).

Epstein, A.H. (2014). Aeropropulsion for commercial aviation in the twenty-first century and research directions needed. *AIAA Journal* 52 (5): 901–911.

Kim, H.D., Felder, J.L., Tong, M.T., and Armstrong, M. 2013. Revolutionary Aeropropulsion Concept for Sustainable Aviation: Turboelectric Distributed Propulsion. ISABE Conference, Paper Number 2013-1719.

Moore, M.D. and Fredericks, W.J. 2014. Misconceptions of Electric Propulsion Aircraft and their Emergent Aviation Markets. 52nd Aerospace Sciences Meeting, AIAA SciTech Forum, *AIAA Paper Number 2014-0535*.

M. D. Moore, W. J. Fredericks, N. K. Borer, A. M. Stoll and J. Bevirt 2014. Drag Reduction Through Distributed Electric Propulsion. In Aviation Technology, Integration, and Operations Conference, Atlanta.

Welstead, J., Felder, J., Guynn, M., Haller, B., Tong, M., Jones, S., Ordaz, I., Quinlan, J., and Mason, B., (n.d.). Overview of the NASA STARC-C ABL (Rev. B): Advanced Concept. https://ntrs.nasa.gov/archive/nasa/casi.ntrs.nasa.gov/20170005612.pdf.

Index

Future Propulsion Systems and Energy Sources in Sustainable Aviation, First Edition. Saeed Farokhi.
© 2020 John Wiley & Sons Ltd. Published 2020 by John Wiley & Sons Ltd.
Companion website: www.wiley.com/go/farokhi/power